Photosensitisers in Biomedicine

Photosensitisers in Biomedicine

Mark Wainwright

School of Pharmacy and Chemistry,
Liverpool John Moores University, UK

WILEY-BLACKWELL

A John Wiley & Sons, Ltd., Publication

This edition first published 2009
© 2009 John Wiley & Sons Ltd.

Wiley-Blackwell is an imprint of John Wiley & Sons, formed by the merger of Wiley's global Scientific, Technical and Medical business with Blackwell Publishing.

Registered office
John Wiley & Sons Ltd, The Atrium, Southern Gate, Chichester, West Sussex, PO19 8SQ, UK

Other Editorial Offices
9600 Garsington Road, Oxford, OX4 2DQ, UK
111 River Street, Hoboken, NJ 07030-5774, USA

For details of our global editorial offices, for customer services and for information about how to apply for permission to reuse the copyright material in this book please see our website at www.wiley.com/wiley-blackwell

Library of Congress Cataloging-in-Publication Data

Wainwright, Mark.
Photosensitisers in biomedicine / authored by Mark Wainwright.
      p.   ;   cm.
  Includes bibliographical references and index.
  ISBN 978-0-470-51060-5 (cloth)
  1. Photosensitizing compounds.   2. Photochemotherapy.   I. Title.
  [DNLM: 1. Photosensitizing Agents—chemistry.   2. Photosensitizing Agents—therapeutic use.
  3. Photochemotherapy.   QV 240 W142p 2009]
  RM666.P73W35 2009
  615.8′31—dc22

                                                                                          2009004189

A catalogue record for this book is available from the British Library.

ISBN 978-0-470-51060-5 (H/B)

Set in 10.5/12.5pt Times by Integra Software Services Pvt. Ltd, Pondicherry, India.
Printed in Great Britain by CPI Antony Rowe, Chippenham, Wiltshire.

First [Impression]   2009

To

Bernadette, Michael, Lauren and Kate.
There can be no richer man than
a husband and father.

Contents

Author's introduction

Introduction

I am often puzzled that authors wait until the end of a piece of work to thank those involved or those who inspired or supported the work. This may only be tradition, but it does, occasionally, seem like an afterthought. I shall therefore thank and send my best wishes to all those who have supported and inspired my work at the outset of this piece. Research is often a long drawn-out process requiring considerable help, understanding and forbearance from family and colleagues, often with little tangible reward. At least you can find my gratitude here without having to wade through the introduction!

Photodynamic therapy (PDT) has been a clinical reality, with many thankful recipients, for around 25 years at the time of writing. Cancer is a terrible – and terrifying – disease, and its therapy is often quite as frightening in terms of side effects. In many cases, the application of a photosensitising drug and subsequent illumination of the tumour site has been sufficient to remove the malignant tissue without scarring or disfigurement – either to the body or to the psyche – and this is a major and continuing strength of the approach.

The reliance of cancer PDT on porphyrins and porphyrin-based photosensitisers is a constant theme in my writing on the subject area. This is not a personal bias, and I hope that the many 'photodynamicists' who have quite happily never gone beyond working with haematoporphyrin derivative, in whichever form, do not see my views as such. Rather, I am interested in the exploration of new avenues regarding photosensitiser development, as this is, in my own view again, the best way to make progress in drug discovery. Indeed, many who concentrate on porphyrin photosensitisers are not involved in drug discovery or development, but in optimising the use of clinically accepted compounds. In my own research since 1990, I have consciously stayed away from porphyrins as there have always been many others involved already.

The antimicrobial application of photosensitisers, i.e. to infection control, is usually seen as a development of cancer PDT. Again, in writing on this subject I am usually at pains to point out that Oskar Raab's photodestruction of unicellular organisms pre-dates the first anticancer reports, but it is an inescapable fact that anticancer PDT was first to be clinically accepted. Photoantimicrobial agents are only now, in the early twenty-first century, taking their first proper clinical steps, although it is to be hoped that photodynamic antimicrobial chemotherapy (PACT) achieves a rather quicker acceptance from clinicians than did PDT itself.

Really the evolution of photosensitisers for use in PDT and PACT should mirror the conventional drug development pattern seen with beta-lactams, fluoroquinolones, etc.

However, one of the reasons for the continuing reliance on porphyrins lies in the lack of funding for drug discovery: Given the small amount available, it makes sense to optimise proven methods and preparations. While this may be logical, it does not improve the chances for novelty, nor is it likely to provide great strides in photosensitiser application. Perhaps we should take heart from Howard Florey's struggles in funding penicillin development in the early 1940s!

There is an additional drawback in photosensitiser development, of course: the drug–dye relationship. Neither the pharmacological industry nor the clinicians are enthusiastic about coloured therapeutics, and yet the use of biological stains- either for pathology or for *in vivo* use during surgery – is quite acceptable. Consider the following two statements.

> Biological stains work by their selective uptake by target cells and the resulting difference in colour between these and non-target cells.

> Photosensitisers work by their selective uptake by target cells and the resulting difference in colour between these and non-target cells.

Admittedly this is a simplistic view, but the difference between these two approaches is that we use light to view the former and light to alter the latter.

Of course there is a danger of over-simplification here. A considerable difference exists between staining and fixing cells on a microscope slide and, for example, triggering apoptosis in tumour cells in an older patient's prostate or bacterial cell wall destruction in an MRSA-colonised wound. Here, as a systemic approach, there are considerations at least in pharmacology, pharmacokinetics and light delivery. However, in line with several PDT protocols, PACT is not proposed, in its current form at least, as a systemic approach. The local application of a photosensitiser followed by local, superficial illumination means that the aforementioned considerations do not apply.

The evolution of clinical photosensitisers from dyestuffs rather than natural precursors is another problem besetting the unwary fund-seeking researcher. It is pointless to deny the link, but the fact remains that although some standard dyes are also photosensitisers (methylene blue for one), most photosensitisers intended for clinical use would be of no use in textile dyeing. In addition, the idea that the synthesis of photosensitisers employs 'bucket chemistry' techniques is also misguided – given the paucity of chemists among the various photodynamic research groups worldwide, a considerable amount of novel, elegant synthetic work has been published. Target molecules unattainable via new methodologies obviously require the old, dye chemistry routes, but even here product purity requirements result in differences in separation and isolation. All things being equal, the only significant difference between getting a photoantibacterial candidate and a conventional antibiotic to clinic trial should be the light activation requirement for the former.

Photosensitisers must offer advantages over conventional therapy, in whichever field. Thus, within anticancer therapy, the benefits lie mainly in the lack of side effects and excellent post-treatment cosmesis. For the antimicrobial application, the efficacy of cationic photosensitisers against conventional drug-resistant organisms is the obvious gain, particularly in view of the small number of new alternatives available from the pharmaceutical industry. In both the anticancer and the antimicrobial application, of course, the photosensitisers involved operate via multiple rather than singular sites within the target cell. This must be considered a strength, particularly from the point of view of potential resistance development. With both applications, however, the problem of systemic light delivery remains – we cannot yet say that *any* malignant or microbial disease treatable by conventional chemotherapy can also be successfully treated with photosensitisers and light. The range of treatable tumours is increasing year on year and photoantimicrobials are in clinical trials or have been very recently licensed, but this does not represent front-line therapy. There are more rapid methods of progress, for example where no efficient or affordable therapeutics exist. Consequently, researchers involved in PACT may be less affected than more conventional workers by the ongoing fiscal crises of health providers and/or the continuing rise of the bacterial superbug!

In writing this book, I am attempting the impossible: to cover all aspects of photo-sensitisers in some depth. Plainly, there will be omissions after over a century of research and more than 25 years of clinical PDT. However, the main intention is to encourage further efforts in the field – particularly in chromophore diversification – and so, hope-fully, to aid in moving more of the research into the clinical arena. The book is intentionally different in area and approach to Raymond Bonnett's *Chemical Aspects of Photodynamic Therapy* of 2000, covering the full range of photosensitisers and all applications, rather than concentrating on porphyrins and phthalocyanines. In a world where the cost of launching a new therapeutic is often prohibitive, it is incumbent on those involved in drug discovery to produce the optimal derivative for use. In my opinion, this means investigating as many candidates as possible before making decisions. There are classes of photosensitiser that are mainly ignored by the majority involved in photosensitiser drug development. I hope this book will act as a reminder that there are many possible answers to our current questions in the antimicrobial and anticancer fields.

Mark Wainwright
Lea, Preston
October 2008

PART I

Introduction

1

Light

1.1 Electromagnetic radiation in everyday life

Light is essential to life as we know it. Without light there would be no edible vegetation on the planet, and thus no food chain. Nevertheless, we take light for granted. We flick a switch when we enter a dark room, and once we have banished the darkness we forget about it. Very few non-scientists appreciate the situation of visible light in the electromagnetic spectrum (Figure 1.1), and yet such an appreciation is helpful in explaining many everyday phenomena, from rainbows and the growth of garden flowers to the fading of curtains or furniture covers or the spines of books on bookcases! The relationship between wavelength and energy usually remains the property of those acquainted with photophysics and/or Max Planck, but such knowledge would undoubtedly aid in getting the message across about the dangers of overexposure to ultraviolet (UV) light, especially because most people do not make the link between sunshine and sunbeds! Conversely, the equation $E = mc^2$ can be recalled by many, not from secondary-level physics but rather from the fame of Einstein and the atom bomb. Perhaps Planck, Roentgen and co. should have produced a famous military application for their discoveries! To most people, even with the burgeoning presence of solar panels on roofs, light is light and not energy.

So what happens when electromagnetic radiation hits matter? This depends on both the energy of the radiation and the nature of the matter. Where there is some harmony between the radiation and the matter, absorption of energy is possible. Where there is no harmony, reflection or scattering may occur. Consider the following examples: having an X-ray taken, wearing a blue pinstripe suit and cooking a microwave dinner.

We all know what an X-ray film looks like: a black background with a very pale shadow where there is soft tissue and bright structures where there is very dense tissue or bone. The reasons for the differing appearance is that soft tissue contains very little that can absorb or scatter X-rays and is considered to be transparent, whereas bone, for example, scatters sufficient radiation to be 'radio-opaque'. Additionally, the high-energy nature of X-rays allows it to interact with matter of all kinds by removing electrons, i.e. giving them enough energy to escape the constraints of their normally associated atoms.

Photosensitisers in Biomedicine Mark Wainwright
© 2009 John Wiley & Sons, Ltd

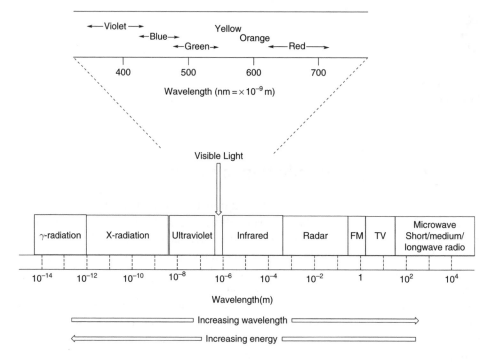

Figure 1.1 The electromagnetic spectrum.

Electromagnetic radiation from X- and gamma-rays up to UV rays is thus considered as 'ionising radiation'. Thus, lead vests/coverings (and the radiographer's remote situation) are necessary when having an X-ray taken.

Many business people, and even some academics these days, wear suits made from blue pinstriped textiles. In daylight or under office lighting, such textiles appear to be dark blue with thin white stripes. Why do the textiles appear dark? What we see depends on the incident radiation. The molecules of blue textile dye absorb long-wavelength (red) light, reflecting the remainder of white light, and minus its red fraction, this appears blue. However, under red light the blue part of the textile would appear black, since all of the light would be absorbed, i.e. none reflected. Ideally, the white pinstripe does not contain any visible light-absorbing structures and so only reflects the incident light.

Microwaves are electromagnetic radiation with much lower energies than visible light. Whereas visible light is associated with the wavelengths in the range of 350–700 nm, microwaves have much longer wavelengths – metres instead of nanometres (Figure 1.1). Wavelengths in this range can be absorbed by water molecules in matter, and the molecular vibration caused by this absorption leads to a heating effect. This is the principle of microwave cooking, and food thus cooked will only radiate heat and not microwaves!

1.2 Radiation and tissue

Now look at the other side – deleterious radiation, and how we can stop it. For X-rays, as mentioned, there is lead shielding. For microwaves, ovens carry an internal metal cage that stops their escape. What of UV light? We are all exposed to UV light at some stage during most days of our lives – no wonder that most of us are quite indifferent to it. However, UV light is short-wavelength/high-energy radiation and as such requires care. Such high-energy radiation is able to remove electrons from biomolecules. Since the skin is (obviously) the first point of contact for UV rays, this is where the electron removal, or *abstraction*, takes place. If this abstraction occurs in molecules involved with genetic information, i.e. deoxyribonucleic acid or DNA, it can have serious consequences, including cancer. In other words, the more often the skin is exposed to the risk, the more likely a cancer-causing event is to occur. However, the skin, and other tissues, can fight back. The colour of a person's skin is due in part to pigmentation with melanin. Again this is an example of light absorption – the absorption spectrum of melanin (Figure 1.2) is very close to that of sunlight output. Melanin can also react with single electron species, or *radicals*, in order to stop their damaging effects on more critical biomolecules such as DNA.

However, it still makes sense to cut down the risks with UV, and rather than avoiding sunlight, many people choose to apply suncreams. Such preparations often have many ingredients, but the essential ones as far as UV protection is concerned are screening compounds and antioxidants. Effectively suncreams can be thought of as externally applied melanin, with the screening compound absorbing the harmful UV and the antioxidant mopping up any radicals formed. So long as there is a sufficient amount of both these components, the suncream should be effective. However – and people do tend to forget this – the cream is only effective as long as the ingredients remain intact, and

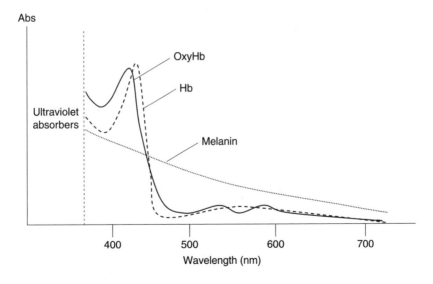

Figure 1.2 Approximate spectra for endogenous visible absorbers.

since one of the results of the interaction between UV light and molecules is chemical degradation, the cream must be re-applied periodically.

It is emphasised at this stage that melanin is not the sole light absorber in the skin. Other, vital biomolecules also have chromophores absorbing in the UV region (for example, the nucleotide bases in DNA/RNA and the aromatic amino acids), while blood in the skin tissue contains haemoglobin (Hb)/oxyhaemoglobin (OxyHb). Both of these contain the iron porphyrin system, which absorbs across the UV and visible regions (Figure 1.2). Endogenous absorption is particularly important when considering the light activation of photosensitisers in practice. As will be seen, a great deal of endeavour is entailed in the production of useful photosensitisers, which avoid the regions of the spectrum covered by these endogenous species. In addition, the involvement of photophysics is essential in determining tissue penetration by light, particularly in the photodynamic therapy of cancer, to ensure that sufficient energy reaches the photosensitiser at its target site. Due to the endogenous absorbers mentioned and light scattering, long-wavelength light is required for tissue penetration. Superficial illumination with red and near infrared light can attain a depth of 5–10 mm, but this depends on the tissue involved, which can be of different types and will also vary in individual cases. Greater volumes of illumination are obtainable by multiple fibre optic implants, again mapped by the photophysicists.

1.3   Light, electrons and molecules

Each of the earlier examples shows us that electromagnetic radiation interacts with matter in ways that depend on the energy of the radiation. In turn, how a material interacts with radiation depends on its chemical make up or, to put it another way, on its electronic structure. In many cases, and certainly where dyes and photosensitisers are concerned, aromatic structures, i.e. conjugated π-systems, are involved, often in conjunction with modifying, peripheral groups. Usually, the key words used in connection with these different moieties are *chromophore* (or *chromogen*) for the aromatic part and *auxochrome* for the relevant peripheral group. Usually, the chromophore alone is colourless or pale yellow in solution and becomes intensely coloured only when joined to the auxochrome. From this it can be understood that there is a strong interaction between chromophore and auxochrome, for example in methylene blue (Figure 1.3).

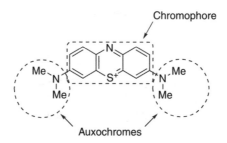

Figure 1.3   Auxochromic groups and chromophore detailed for methylene blue.

It should be noted that auxochromic groups are connected in conjugation with the π-system of the chromophore, enabling their participation in the delocalisation of electrons around the molecule as a whole. This has the effect of lowering the energy of the system in terms of electron transition, as discussed below. The proper attachment of auxochromes, in order to allow electronic delocalisation, depends on correct positioning of the groups. For example, with methylene blue the most efficacious position is at carbon-3 of the phenothiazinium chromophore, since this allows simple communication between the dimethylamino group and the ring sulphur (Figure 1.4). Because of its symmetry, the phenothiazinium ring may be doubly substituted – at C-3 and C-7 – leaving both groups in conjugation (Figure 1.4).

A similar, if less profound, effect is seen on benzenoid aromatic fusion, for example across C-1/C-2 of the phenothiazinium system. The resulting expansion of the π-chromophore again leads to a lowering of the energy of the system and an increase in the magnitude of the longest absorption wavelength (λ_{max}). Such features are applicable across the majority of chromophores covered in this book; thus, benzochlorins absorb at longer wavelengths than chlorins, naphthalocyanines at longer wavelengths than phthalocyanines and so on (see Chapters 6 and 7).

Electrons occupy stratified energy levels in molecules and can be promoted to higher levels by the absorption of the correct amount of energy. This quantified approach explains how the excitation of electrons corresponds specifically with radiation of a certain wavelength. Electrons, in most cases, exist in pairs in specific orbitals. Each electron pair has opposing spins – one clockwise and one anticlockwise, and this is known as the *singlet state*. When one electron of such a pair is promoted to a higher energy level, it keeps the same spin (i.e. singlet ground state → singlet excited state). As mentioned above, the promotion of an electron completely out of its atomic environment (ionisation) is also possible.

Additionally, electrons within organic molecules are usually of one of two types: bonding or non-bonding – in other words, involved in one of the molecule's covalent bonds or existing as a 'lone pair' on an atom other than carbon or hydrogen.

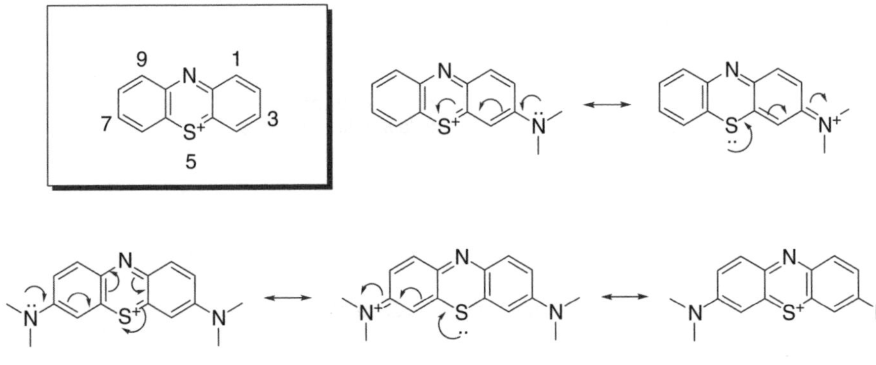

Figure 1.4 Delocalisation in methylene blue-type dyes.

What happens to an excited electron? There are a number of possibilities. It can fall back to its original, preferred (singlet ground) state, either in one go or via molecular vibration (Figure 1.5). Radiation emitted by the former is *fluorescence*. The latter is usually non-radiative. The excited electron may remain in the high-energy state for a more extended period, in which case reversal of its spin is possible, so that this becomes the same as its original partner in the ground state. The promoted electron is now in the excited *triplet* state, which enables reaction. Decay to the singlet ground state is also possible and, if radiative, it is known as *phosphorescence* or delayed fluorescence (Figure 1.5).

From Figure 1.5 it can be seen that the energy involved in absorption is greater than that for the return to the ground state, either from the excited singlet or from the excited triplet states. This is the reason that, normally, both fluorescence and phosphorescence peaks occur at longer wavelengths.

Fluorescence, then, is the manifestation of the rapid return of a singlet-state excited electron to its ground state. However, the fluorescent properties of a given molecule are not always the same. They can change depending on the molecule's environment. While this may not seem logical, it should be remembered that the electronic transitions do not occur in isolation but are dependent on molecular orbitals, and these orbitals feel the effects of environmental change. Such changes may be caused by varying the solvation of the molecule (changing the solvent or pH, for example) or by adsorption behaviour, i.e. binding to substrates such as biomolecules or metal ions, which is of particular

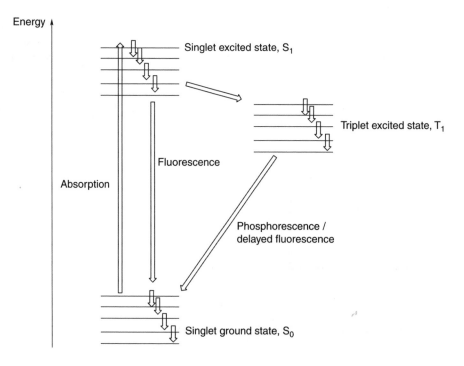

Figure 1.5   Absorption, fluorescence and phosphorescence.

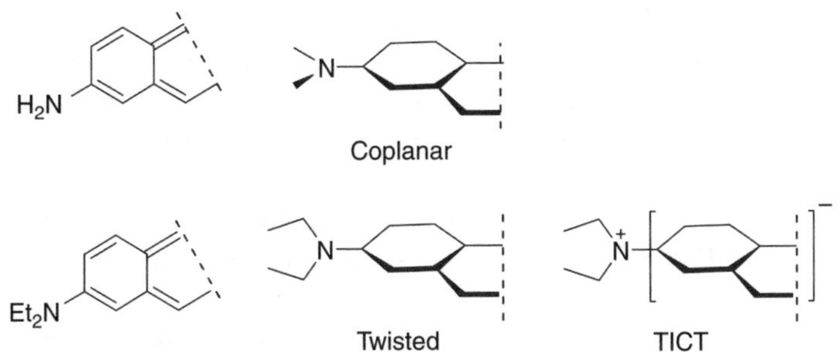

Figure 1.6 Auxochrome twisting leading to TICT formation.

importance when analysing fluorescence staining, and thus also in the burgeoning field of photodiagnosis (Chapter 12). Biomolecular interactions are also important for photosensitisers – some examples only become active when bound.

To give a simplified example, consider the variation in fluorescence of fluorescein itself in methanol and in acetonitrile/benzene. The solvents are quite different – methanol is a polar solvent, whereas a mixture such as acetonitrile/benzene (20/80) is much less polar. The strong dipole present in methanol interacts with the fluorescein dipole, causing a relaxation of the excited state (i.e. moving it to lower energy, as shown in Figure 1.5), a smaller energy gap and thus a lower energy emission (515 nm). This does not occur in the benzene mixture, where emission occurs at 308 nm (Acemioğlu *et al.* 2001). Similar effects are seen in the excitation wavelengths (λ_{max} 456 and 282 nm, respectively).

Clearly, this large variation in excitation and fluorescence reflects the considerable difference in solvent polarity. This is unlikely to be reflected in the excitation of photosensitisers in cells or *in vivo*, but there are different microenvironments that exhibit variations in pH and molecular adsorption. Thus, the laboratory λ_{max} may not be quite the same as that required in practice.

Molecular orbital disposition thus has effects on fluorescence, and these effects may be magnified by functionalisation. For example, conventional fluorescence in amino- ($-NH_2$) substituted molecules is often diminished or lost in higher alkylamino- or dialkylamino-substituted analogues. This is due to the phenomenon of twisted intramolecular charge transfer (TICT), which, as the name suggests, is an alternative route to energy loss caused by the separation of charge between the out-of-plane auxochrome (amine group) and the chromophore itself (Figure 1.6).

Such properties may be utilised in the design of potential photosensitisers where fluorescence is not a requirement, as a way of promoting the Type I/Type II photosensitisation pathway (see Chapter 3).

1.4 Photoreaction

From the preceding section it should be apparent that electrons can be energised by light of a suitable wavelength. This has various applications, one of which is the promotion of chemical reaction. In other words, a non-reactive molecule has its electrons in stable, low-energy orbitals, but this situation can be altered by increasing the energy and disposition of an electron via light absorption. The resulting unstable molecule may either relax back to its stable ground state or react via transfer of the electron, now unpaired.

A good example of this behaviour is given by the basic chemical process of hydrocarbon bromination (Figure 1.7), a stalwart reaction underpinning much of high school organic chemistry.

When discussing the differential reactivity of alkanes and alkenes, i.e. saturated and unsaturated hydrocarbons, respectively, alkene bromination and the resulting decolourisation of bromine (or of bromine water) are given as diagnostic tests: the alkene double bond is reactive enough to brominate at room temperature. Conversely, the relatively inert alkanes do not react under the same test conditions. The bromination of alkanes requires activation by high-energy UV light and occurs via radical (free electron) species. This, or UV chlorination, is probably most students' first experience of photochemistry and photoreaction, although the emphasis is usually on the hydrocarbon reaction rather than the light energy involved.

Further study of organic chemistry, probably at undergraduate degree level, exposes the student to free radical polymer synthesis. Again such reactions may be photoinduced,

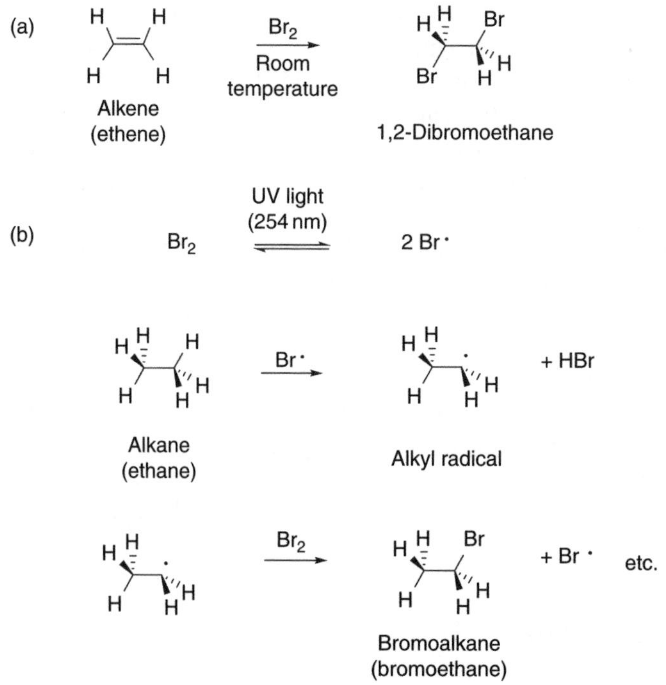

Figure 1.7 Bromination of simple hydrocarbons: (a) alkenes and (b) alkanes.

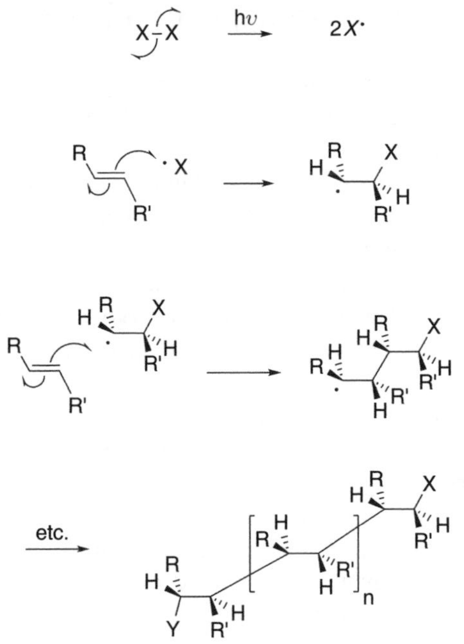

Figure 1.8 Simplified scheme showing photopolymerisation. Initiator X_2 is dissociated (bond homolysis) by high-energy UV light, resulting in free radical formation. Free radicals then react with the olefinic moiety of the monomer, producing further free radicals. Subsequent reaction yields the polymer (Y = quenching group).

UV light providing the means of breakdown of the initiating agent, with the formation of radical species (Figure 1.8). Similar chemistry is entailed in the photoisomerisation or rearrangement of various unsaturated organic compounds.

1.5   Dental/plastics curing

As an application of the above, the UV or blue light curing of plastics is now commonplace. This is a highly useful application of photopolymerisation and has a considerable overlap with the current topic. Local blue light irradiation of a polymer composite to harden it in a tooth cavity gives a safe, targeted method of tooth repair. In many ways, this is similar to a high-energy version of the photo-disinfection of root canals, which is discussed in Chapter 11. Importantly for the current discussion, it also means that there is prior experience among the dental fraternity of using accurate illumination in the oral cavity. This is therefore not a foreign concept and should help in promoting acceptance of the photoantimicrobial application to the oral cavity (Chapter 11).

1.6   Photoaging/photofading/photobleaching

By definition, photosensitisers require light for their activity. As covered in the preceding sections, the absorption of light energy by organic molecules can have a variety of effects following electron promotion.

For the present argument, there is considerable relevance. Studies on the photofading of dyes (particularly textile dyes) carried out over more than 100 years usually point to the importance of the presence of light, oxygen and water in the process. It should therefore not be surprising that photosensitising drugs, working as they do via the production of reactive oxygen species (Chapter 3), should undergo oxidative damage themselves, although it is likely that this is not the only mechanism involved (Bonnett and Martínez 2001).

Clearly, in the textile and paint industries, the usual intention is that the applied colour is as permanent as possible. However, in the present case, where the lifetime of a photosensitiser is short, fading may not be an issue – indeed it may be desirable in limiting side-effect photosensitisation.

The examples given in this chapter are presented to illustrate some possible outcomes of the interaction of light with matter from a molecular viewpoint, since the outcomes are determined at the molecular level. The idea of electron promotion by energy absorption is, of course, important in the understanding of photosensitisers and the photodynamic effect.

References

Acemioğlu B, Arik M, Efeoğlu H, Onganer Y. (2001) Solvent effect on the ground and excited state dipole moments of fluorescein. *Journal of Molecular Structure* **548**: 165–171.

Bonnett R, Martínez G. (2001) Photobleaching of sensitisers used in photodynamic therapy. *Tetrahedron* **57**: 9513–9547.

2
Dyes and stains

2.1 Dye use

Many photosensitisers – although not all – have been derived from molecular types originally used in the dyeing of textiles and the staining of biological samples for microscopy. There are also, of course, several examples of dyes and stains that are photosensitisers in their own right. In order to understand photosensitisers, it is important to grasp the background from which they evolved, since in many cases this provides some explanation of cell selectivity and biomolecular interaction. It is also true to say that many conventional drugs have been derived from dyes and stains.

Dyes and pigments are used to give colour to objects, for either a practical or an aesthetic end use. The main difference between the two types of material lies in solubility. Dyes are generally dissolved at some stage of the colouring process, whereas pigments are usually insoluble. Dyes are often applied to textile substrates as a solution; where solid, if finely ground, pigments are mixed with, for example, polymer melts before forming the requisite structural shape (e.g. plastic furniture). Pigments are also normally more light-stable than dyes – this is an industry requirement in paints and coatings. Textile dye molecules are usually attached to their substrates via chemical bonds; pigment coloration generally utilises size-dependent physical entrapment.

Humans have used colour in their environment for millennia – from cave paintings, earthenware and fabrics to speciality paints, coloured glass and modern décor. Such is the state of technology in which we now reside that it is possible to have almost any object in any colour we wish – a long way, indeed, from Henry Ford's 'Any colour – so long as it's black!' statement of the early twentieth century.

One area in which colour choice has become rather more restricted in recent years is that of food and drink. The toxicological considerations associated with food colourings have meant that only a handful of synthetic colours is now legally sanctioned, whether to avoid juvenile hyperactivity, kidney or liver damage, or the potential for long-term malignancy. Much of the potential damage associated with food colours comes from the presence of aniline fragments within dyes, since these can lead to both cell injury and mutagenicity as consequences of normal human metabolism.

Photosensitisers in Biomedicine Mark Wainwright
© 2009 John Wiley & Sons, Ltd

While human usage of dyes in the environment remains enormous, our use of self-coloration is much more limited. Colour is used now, as in the distant past, to enhance aesthetic qualities (cosmetics and tattooing) or for tribal/religious purposes. The low toxicity of pigments used to such ends in addition to their local application ensures safe usage. The most likely area for toxicity is that of permanent tattooing, where the pigment is introduced beneath the skin (dermis layer), thus relying on care and accuracy with the needle. Temporary tattoos are painted or transferred onto the skin and rely on skin penetration for longevities of days or weeks. Systemic dye use remains in the purview of clinicians and is discussed below.

As will be mentioned in various chapters of this book, dyes employed as biological stains are of great benefit to humans in both the detection/prevention and the treatment of various diseases. Pathology departments routinely use biological stains to differentiate and identify 'non-economic' or 'bad' cells within various types of human samples – blood, tissue, sputum, etc. This is extended to the clinic for a very few stains that can be used inside the living body, normally to delineate malignant tissue at a tumour site or via the lymph to measure the likely route of metastasis. As mentioned, few dyes or stains are considered to be safe enough for such procedures. Fewer still are employed in systemic therapy.

2.2   Textile dyeing and biological staining

Dye chemistry has been investigated thoroughly over the last 150 years or so. Thus, the various interactions occurring between dyes and substrates, and optimum dyeing conditions, are well understood. It would be an oversimplification to suggest – as others have – that photosensitiser–target interactions are the same as those between a dye and its substrate, but there are some similarities nevertheless. Obviously, the main overlap occurs with textile dyeing of natural protein or cellulosic fibres (e.g. wool or cotton, respectively), due to structural commonalities with target biomolecules such as enzymes and cell walls. A progression might be made from the bath dyeing of a piece of wool fabric to the microscopic staining of a cervical cancer smear and the photodynamic treatment of a skin tumour. Each of these involves the interaction between a coloured chemical and a proteinaceous material. An alternative progression might be the dyeing of cotton, microscopic Gram-staining and the treatment of an infected burn wound – here a common interaction occurs between a coloured chemical and polysaccharide-based biomolecules.

However, there are also considerable differences between textile dyes, stains and photosensitisers intended for the clinic. These may be of a chemical nature, for example in the degree of purity or stability required, or may lie in the application protocol or elsewhere. Obviously, there are differences in their interactions with light (Chapters 1 and 3).

As it is more likely that the general reader will be better acquainted with the idea of textile dyeing, this will be employed as the baseline for the following discussion.

Commercial textile dyes are produced usually with the aim of colouring the textile fabric in the desired shade. This colour and shade – again usually – should remain as dyed for the lifetime of the textile; in other words, the dyeing process should fix the dye to the fabric permanently, and, once fixed, the dye should remain chemically unchanged. For most organic dyes this is a vain hope, particularly from the point of view of light-fastness (photostability), and enormous amounts of research have been carried out in order to remedy this. Similarly, fixation of the dye to the fabric should ideally be permanent – wash-fastness is also important, otherwise textiles appear to fade over successive wash cycles. Dye fading/washing out may, of course, be a desirable property as is the case with denim jeans and 'distressed' clothing.

How are textile dyes attached to their substrates? There are several answers here, depending on the substrate type. For protein-based materials such as wool or hair (thus hair dyes would be of similar design), the molecular structure consists of amino acids. There are various sites for dye attachment here, depending on the exact protein constituency, although keratin is the normal model. Binding may be via ionic linkage, depending on pH (amines or carboxylic acid residues), hydrogen bonding (hydroxyls, amines, carboxylic acids and amides) or Van der Waals' or London forces (phenyl/aromatic residues). Reactive dyes, where new covalent bonds are produced to anchor the dye molecule to the substrate, are more akin to cancer chemotherapeutics such as the nitrogen mustards and nitrosoureas than to photosensitisers. Cotton dyes are conventionally planar, being able to lie on the surface of the linear cellulosic polymers constituting the backbone, connecting via hydrogen bonding to the ubiquitous hydroxyl groups of the substrate. The mode of dye attachment to the fabric is important when considering wash-fastness. Generally, a dye attached purely via Van der Waals' forces would not be expected to be as strongly fixed as one connected via ionic or covalent bonds.

The binding of ionic dyes to ionic substrates is relatively straightforward in essence, whether the interaction is that of an acid dye with wool, or a cationic photosensitiser with a biopeptide. Simple ionic attraction governs the initial, i.e. one molecule, attachment. For the acid dye this would be between a negatively charged sulphonate ($-SO_3^-$) group and a cationic ammonium moiety, and for the photosensitiser between a positively charged nitrogen and (usually) a carboxylate residue ($-CO_2^-$). Plainly the state of ionisation of the wool/peptide is important here, and thus the interaction is pH dependent, as shown in Figure 2.1.

Such interactions are not limited to peptides. Indeed, both the outcome of the Gram stain and the generally high activities of cationic photosensitisers against both main types of bacteria are due, at least in part, to the ionic attraction between anionic groups in the bacterial cell wall and the positively charged photosensitiser. Planar, delocalised cations may also aggregate at the original site of attachment, resulting in interactions between the adjacent aromatic (π-) systems and causing an absorption shift to shorter wavelength. This is known as metachromasy (metachromasia), as observed, for example, with planar cationic bacterial stains such as toluidine blue (Chapter 4). The interaction of related methylene blue derivatives with the external membrane of erythrocytes is also known to involve aggregation (Taylor and Jeffree 1969). Both of the preceding examples involve

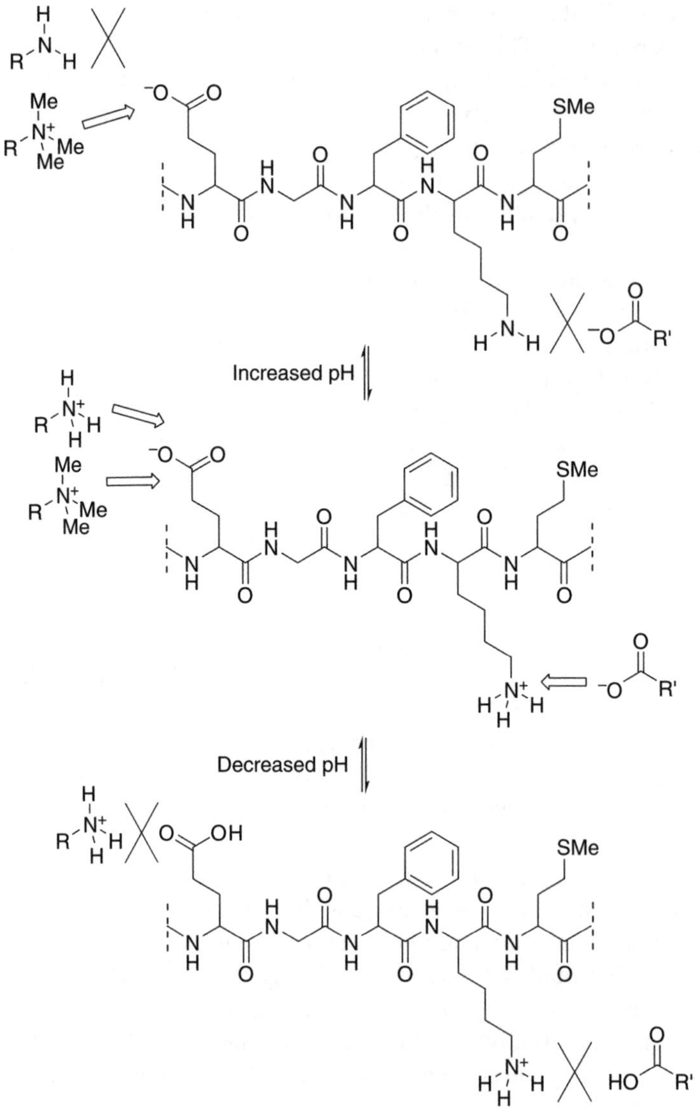

Figure 2.1 Peptide ionisation, pH change and ionic interactions with dyes or photosensitisers (peptide = Glu—Gly—Phe—Lys—Met).

anionic carbohydrate biomolecules (teichoic and sialic acids, respectively, Figure 2.2) rather than peptides *per se*.

The extension of textile dyeing to the staining of microscopy specimens is easily appreciated, especially as the two techniques underwent their greatest periods of growth synchronously in the late nineteenth and early twentieth centuries with the availability of the new aniline dyes.

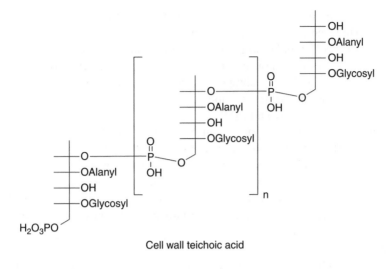

Cell wall teichoic acid

O-Sialic acid

Figure 2.2 Carbohydrates containing anionic groups.

2.3 Biological stains and biomedical photosensitisers

To a certain extent, staining in microscopy, i.e. the scientific areas of histology, histo-chemistry, etc., may be thought of as a half-way house between textile dyeing and the medical use of photosensitisers. In microscopy, cell staining is carried out on both live and dead cells. In both cases, it is necessary to cause coloration without damage to the cell structure and integrity, and this may be extrapolated to photosensitiser use. However, the fixation of dyes with acids or heavy metals is unlikely to be applicable to live cells, far less so when these cells constitute part of a whole organism!

Dyeing or staining of the target cell in photodynamic therapy (PDT), photodynamic antimicrobial chemotherapy (PACT) or photodiagnosis (PD) in the ways discussed above is, in fact, not required. Target binding may be akin to dye fixation on textiles or in microscopy but fixation *per se* is not essential.

Another obvious difference lies in localisation. It is a straightforward matter to apply a staining preparation (addition, fixation and wash-off) to a specimen malignant smear on a microscope slide, to position the cover slip and then to focus the objective lens through the eyepiece. How more complex is the process of achieving the equivalent operation when the aberrant cell population is inside a patient and the colouring agent has to be administered orally or intravenously, to be subject to the normal rules of distribution and metabolism before possibly reaching the target cells? Plainly this is a much more complex scenario and underlines the appeal of topical/local treatment, and this is true for both conventional therapy and PDT.

In both PDT and PACT, an adequate concentration of photosensitiser is required at the target site for a sufficient time in order to produce the requisite amount of reactive oxygen to damage the cell terminally or to initiate the cascade towards this (see Chapter 3). Wash-fastness is (of course) irrelevant, but photostability is required, at least in order to absorb the incoming light and undergo the necessary electronic transitions leading to the production of reactive oxygen species. As mentioned in the previous chapter, photofading after photodynamic action may in fact be advantageous, particularly from the angle of minimising potential toxicity.

To take an alternative viewpoint, what is the similarity between the use of dye-derived photosensitisers and purpose-designed antibacterial agents? As discussed above, textile dyes may be similar to photosensitisers in terms of structure and coloration, but everyday antibacterials – penicillins, fluoroquinolones, etc. – are the obvious comparators in terms of clinical drugs, since this is one of the areas in which compounds are being developed as clinical photosensitisers (i.e. photoantimicrobials; see Chapter 11).

Both antibacterial classes mentioned above are quite dissimilar in action from dyes or stains, as they act on specific enzymes and at specific parts of the bacterial cell. Thus, penicillins exert their action against an enzyme in the cell wall responsible for forming the strengthening peptide cross-links necessary to contain the high internal pressure of the growing bacterial cell. Penicillins are able to do this by mimicking part of the peptide chain involved.

Fluoroquinolones such as ciprofloxacin ('Cipro') are equally specific for a different cellular enzyme, this time one responsible for stabilising the bacterial DNA helix as it unwinds during replication. The planar fluoroquinolone nucleus aids in this process, the molecule being able to mediate between the enzyme and the planar base pairs in the nucleic acid.

Natural penicillins are products of evolution, the cell wall specificity being developed over millions of years by the producing moulds as a method of defence against bacterial ingress. Fluoroquinolones are completely synthetic, and the specific anti-DNA activity was discovered by accident.

Photoantimicrobials, on the other hand, are not specifics. Typical examples, such as methylene blue or toluidine blue (Figure 2.3), will stain the bacterial cell wall similarly to crystal violet in the Gram stain. This staining is akin to dyeing, since there is an ionic attraction between negatively charged residues in the cell wall and the cationic phenothiazinium (or triarylmethane) chromophores. However, this staining is not selective. Over time, both methylene blue and toluidine blue move into the bacterial cell, and this can be judged by the subsequent photodamage observed to DNA and ribosomes. Crystal violet is also photoantibacterial.

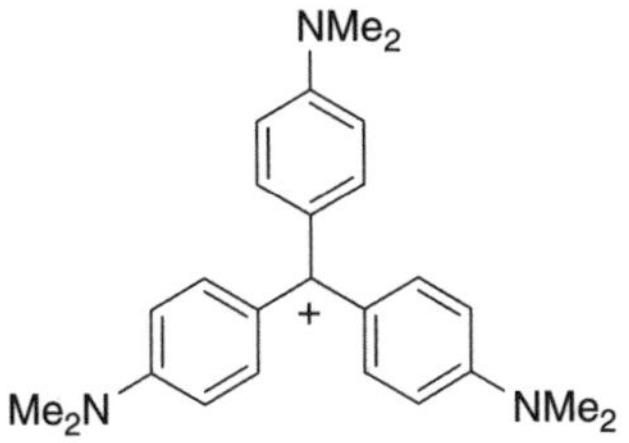

Crystal violet - cell wall stain

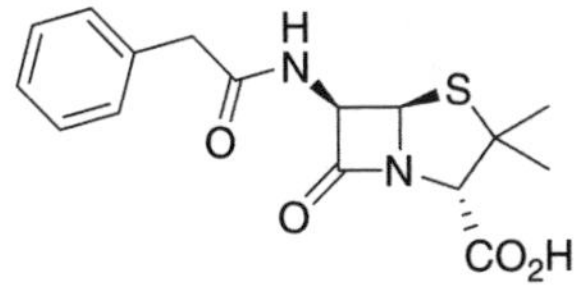

Benzylpenicillin

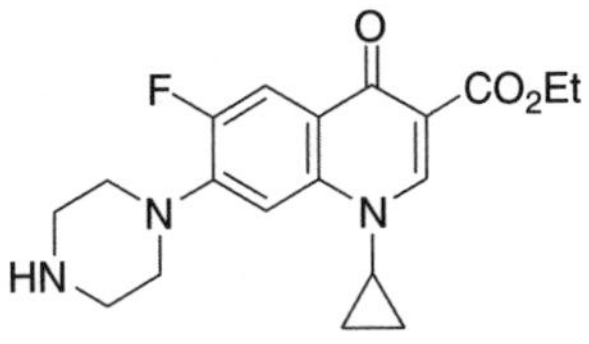

Ciprofloxacin

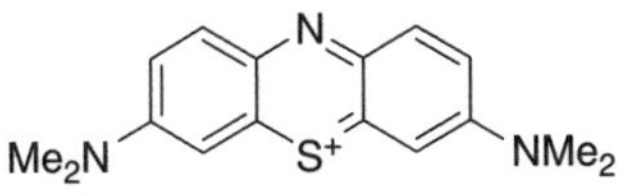

Methylene blue

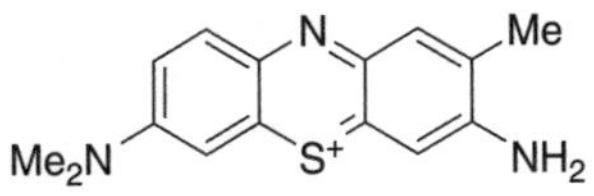

Toluidine blue

Photoantimicrobials

Figure 2.3 Bacterial-specific stains, antibacterials and photoantimicrobials.

Another difference here is the antimicrobial nature of the cationic photosensitisers, compared to conventional drugs. Using benzylpenicillin again as an example, this is only antibacterial; it is neither antifungal nor antiviral. However, both methylene blue and toluidine blue are photoactive against bacteria, fungi, viruses and protozoa (see Chapter 11). This reflects less specific binding at a cellular level.

2.4 The human factor

Dyes are often rejected summarily for human use due to concerns over toxicity, and this is obviously of relevance to the development of photosensitisers for the clinic. There are various reasons for the perceived toxicity. Firstly, some dyes are toxic and some are mutagenic – benzidine derivatives are a particular class in point, i.e. dyes such as Congo red and trypan blue (Figure 2.4). The use of these as histological stains or indicators should not be problematic if good laboratory practice is followed. However, trypan blue

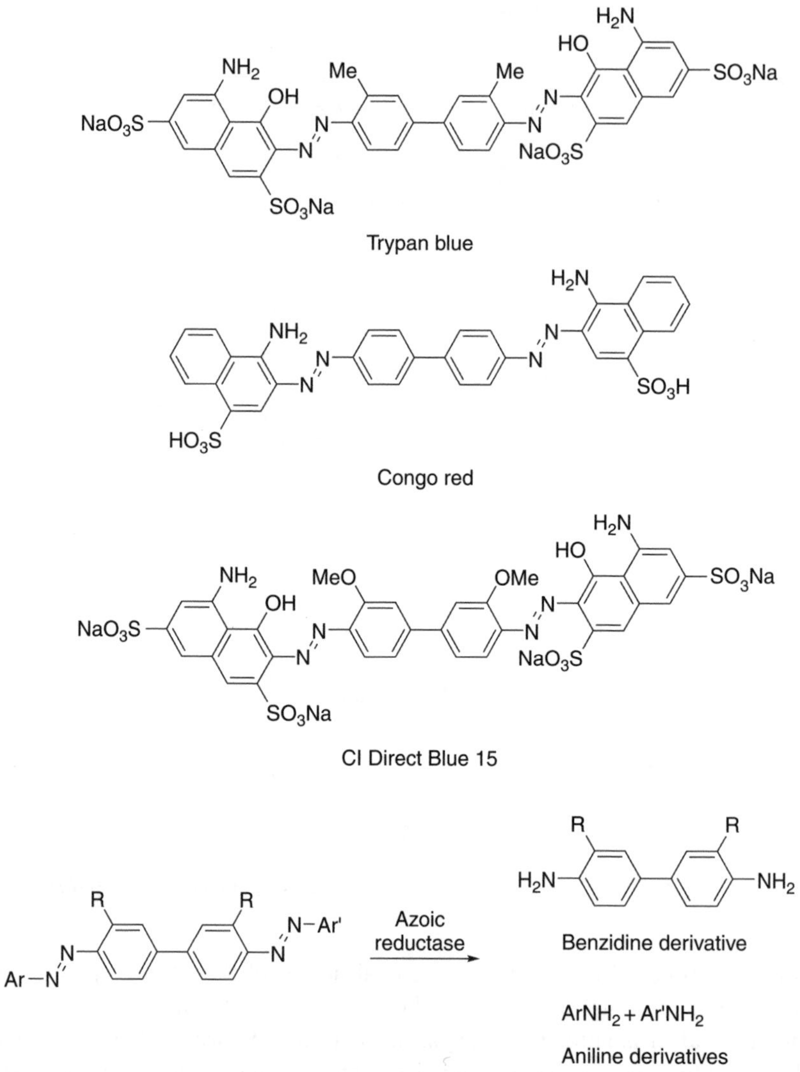

Trypan blue

Congo red

CI Direct Blue 15

Benzidine derivative

$ArNH_2 + Ar'NH_2$

Aniline derivatives

Figure 2.4 Production of benzidine derivatives via metabolic reduction of corresponding bisazo dyes.

and several related derivatives were tested as trypanocidal agents in animals and, to a far lesser extent, in humans.

The knowledge accrued to date is that such derivatives are metabolised by mammals to simple benzidine derivatives (Figure 2.4), thus being potentially activated carcinogens (Morgan *et al.* 1994).

Mammalian enteric reduction of the azoic group ($-N=N-$) has been known almost since the introduction of Prontosil as an antibacterial (*c.* 1935). This step appears ubiquitous and is, indeed, used to deliver therapeutics to the colon (e.g. sulphasalazine in Crohn's disease). However, standard mutagenicity tests on a range of benzidine-derived dyes have shown that not all give positive results.

Similarly, aniline is known to be toxic to the liver, and this is usually taken to include all aniline derivatives – a ridiculous state of affairs, given that different chemical groups attached to the aniline nucleus can change its properties markedly. For example, both *para*-aminobenzoic acid and sulphanilamide are anilines, containing a carboxylic acid $(-CO_2H)$ and a sulphonamide $(-SO_2NH_2)$ residue, respectively, while paracetamol, which enjoys massive use globally as a painkiller, is merely an *N*-acetylated hydroxyaniline (aminophenol).

Where a dye molecule has a continuous planar surface area that approximates to at least the equivalent of three benzene rings and is positively charged, there is a strong possibility that it will intercalate with nucleic acid in buffer, say, in a test tube. Given that the buffer solution is required to keep the shape of the nucleic acid, the only significant components here are the nucleic acid and the dye. The question that is rarely asked is: 'Where *else* could the dye go?' The extrapolation of this *in vitro* finding is often that the dye is (1) nucleic acid specific and (2) thus (!) mutagenic in the human host. While this may, indeed, be the case, it should not be a foregone conclusion. Herring sperm DNA in a test-tube is an easy target for an intercalating dye. To cause mutagenicity in humans, the dye must reach the cell concerned, cross into its interior, seek out and enter the nucleus, *then* intercalate into the helix, before binding sufficiently strongly to, or reacting

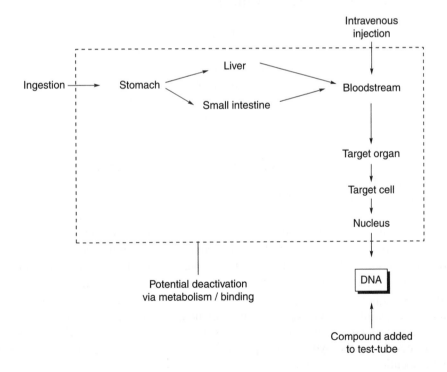

Figure 2.5 Contrasting routes to DNA intercalation.

chemically with, one or more of the nucleotide bases. Even this ignores the possibility of DNA repair mechanisms. Quite plainly, the two scenarios are not equivalent (Figure 2.5) and, in the author's opinion, the former should not be used as a predictor of the latter.

Another factor, not normally considered, is the half-life of the dye in the human system. This is governed, in the first instance, by both the route of administration and the lipophilic/hydrophilic character of the dye. It is well established that hydrophilic species are rapidly cleared but that lipophilics are metabolised in order to produce more hydrophilic derivatives. Plainly, a more lipophilic compound will remain for a longer period within the system, thus increasing the chances of a deleterious event. Indeed, highly lipophilic molecules tend to be concentrated in tissue containing high levels of lipid – this is the basis of general anaesthesia.

In the 21st century, dye chemistry is well understood, enabling the production of many derivatives if desired. The interaction of dyes with the human body is also well appreciated because of drug metabolism studies – in other words a dye having a similar structure to a certain drug will probably interact with the human metabolism in the same way, given similar administration routes and dosages.

2.5 Dyes and drug development

The medical use of dyes was developed from the hypothesis that non-economic cells (i.e. not useful to humans, typically microbial) could be stained *selectively* in the presence of human host cells in clinical samples. While Koch used the technique in the identification of the tuberculosis microbe (*Mycobacterium tuberculosis*) and allowed simple determination of the clinical progress of the major killer of the age, Ehrlich's endeavours led him to the development of dyes as selective agents for pathogenic microbes.

Biological stains were originally developed from textile dyes without much idea as to which structures were likely to stain which substrates. Given the nascence of the contemporary chemical industry, Ehrlich employed the specific chemistry understood at the time – for example the 'affinity' of arsenic for sulphur in trypanosomes, which is now known be the manifestation of the arsenical binding of cysteine residues in trypanothione. Similar explanations also underpin the antibacterial activities of arsenic- and mercury-based drugs, such as the effects of Salvarsan in syphilis. However, while the spirochaete implicated in syphilis (*Treponema pallidum*) contains many sulphur moieties, so does the human patient, as attested by the many records of arm loss due to gross necrosis stemming from the injection site.

While it is unfair to expect high selectivity in the early therapeutics derived from dyes, gross hypotheses concerning oil and water solubility in dyes did provide a basis for modern thought on the hydrophilic/lipophilic balance and ionisation in both histology and therapeutics.

Ehrlich's work ultimately furnished a range of front line (conventional) therapeutic agents. From his use of methylene blue in antimalarial therapy at the end of the nineteenth century, chloroquine and several other aminoquinoline antimalarials were eventually derived. Moreover, the Romanovsky stain (containing methylene blue or one of the azure stains) is still a standard indicator of malarial blood pathology. Bacterial staining using acridine dyes such as acriflavine led to the use of this in local antibacterial therapy (see Chapter 11). Furthermore, the inheritors of Ehrlich's research, such as Browning and Roehl, encouraged the further testing and screening of dyes and stains, leading to Domagk's discovery of the antibacterial azoic dye Prontosil and thus the sulphonamide class. Other classes, such as the heterocyclic neuroleptics (antipsychotics), may also be traced back to dyes (Figure 2.6).

Cancer chemotherapy too is indebted to dye research. Again via histology and histo-pathology, the understanding of structure–function relationships among the dye classes has led to dye-derived chemotherapeutics such as the acridines Nitroakridin and Amsacrine (Ferlin *et al.* 2000).

Thus, it should be remembered that both the idea of selective toxicity and the development of modern chemotherapy came directly from the use of synthetic dyes in clinical medicine during the early part of the twentieth century. However, widespread penicillin availability from the mid-1940s made the treatment of many infectious diseases almost trivial, effectively making dye therapy an archaic curiosity. This triviality, of course, led to overuse and misuse of the penicillins and their descendants, to such a degree that many bacterial types now enjoy almost a pre-antibiotic existence.

The modern pharmaceutical industry – whether anticancer or antimicrobial – shows little interest in its dye-based roots. Indeed it is doubtful that new starters in the industry will be familiar with sulphanilamide, never mind its gestation in azoic dyestuffs, and very few chemistry or pharmacy undergraduates have encountered Gerhard Domagk. Today's new therapeutics – synthetic or semi-synthetic – are usually colourless compounds, and *not* dyes.

The fact remains, however, that there are various areas of pharmaceutical endeavour where widespread success is absent – notably in the field of antibacterial chemotherapy where, as mentioned above, we have attacked friendly and pathogenic bacteria alike to such an extent as to guarantee the evolution of strains that are now quite unaffected by antibiotics. *Staphylococcus aureus*, once summarily dismissed by simple benzylpenicillin is now, as methicillin-resistant *S. aureus* (MRSA), resistant to all β-lactams. While this is undoubtedly a massive tragedy, those stalwarts involved in bacterial assaying will very simply demonstrate that both *S. aureus* and MRSA are Gram-positive organisms. Both are still stained in the same way with crystal violet, iodine and safranin or another co-stain, as demonstrated by Christian Gram a long time before Fleming's dirty Petri dishes!

As will be discussed at length in Chapter 11, the ability to stain both drug-susceptible and drug-resistant cells with dyes, which are also photosensitisers, must now be seen as a great advantage.

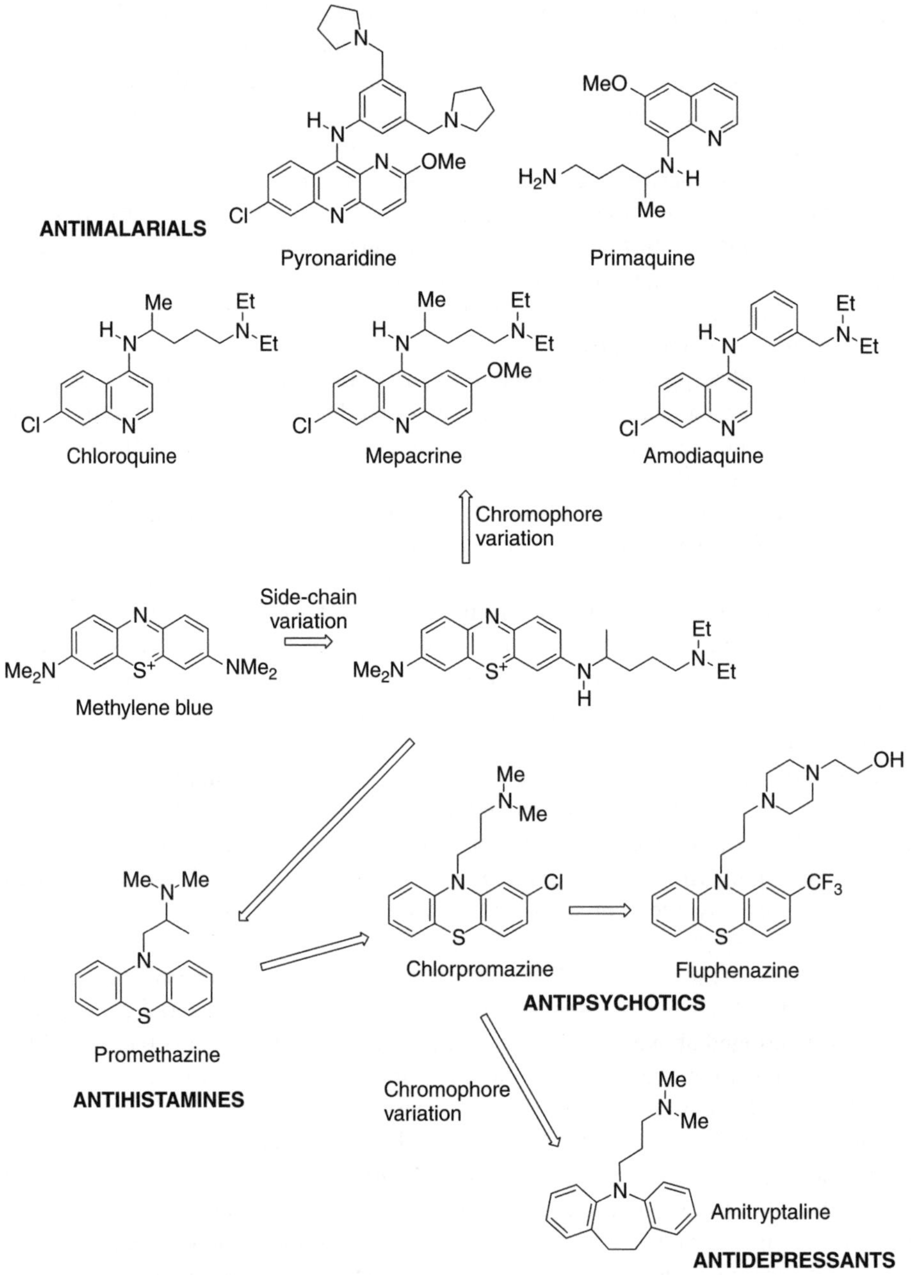

ANTIMALARIALS

Pyronaridine

Primaquine

Chloroquine

Mepacrine

Amodiaquine

Chromophore variation

Side-chain variation

Methylene blue

Promethazine

ANTIHISTAMINES

Chlorpromazine

ANTIPSYCHOTICS

Fluphenazine

Chromophore variation

Amitryptaline

ANTIDEPRESSANTS

Figure 2.6   Drugs derived from methylene blue.

2.6 Dyes and stains and photosensitiser design

Much has been made of the various structural mixtures constituting the premier clinical photosensitiser haematoporphyrin derivative (HpD, Chapter 6). Early PDT meetings had multiple lectures dedicated to the chemical constitution of the active molecules within the mixture, and how the various chemical steps produced these. As has been stated elsewhere, HpD, porfimer sodium, etc. still remain mixtures. However, the basis for the original choice of these porphyrins was based on clinical observations of *in vivo* staining behaviour. It is perhaps surprising that more is not made of this approach in the search for other photosensitisers.

The use of synthetic dyes as stains, as covered above, has a long history – almost since the production of the first synthetic dyes, in fact. Consequently, there is a considerable body of evidence covering vital staining, which might be of use in identifying other structural types for photosensitiser development. The discovery of HpD arose from clinical fluorescence, but the principle is the same. The only significant class otherwise identified in this way is the Nile blue derivatives (NBDs).

Here, histological findings by Lewis in the 1940s demonstrated the selective staining of tumour sections by the benzo[*a*]phenoxazine Nile blue (Lewis, Goland and Sloviter 1949). Advances were made on this structure by Crossley *et al.* in the mid-1950s in their search for more specific compounds for use as conventional anticancers (Crossley *et al.* 1952). This work was used as a basis for photosensitiser discovery by Foley *et al.*, by use of the heavy atom effect (*q.v.*), and this has furnished candidates for anticancer, antimicrobial and fluorescence imaging applications (Cincotta, Foley and Cincotta 1993; Foley *et al.* 2006; Song, Kassaye and Foley 2008; see also Chapter 4).

The use of staining behaviour to provide routes to putative photosensitisers is appealingly simple, although it might upset the complex modellers among us. There is a certain resonance here also, as this is undoubtedly the way in which Ehrlich tested and developed his candidate magic bullets. Moreover, this should underline rather than contradict the widespread use among PDT groups of Photofrin and related porphyrins (Chapter 6), which has evolved from clinical observations of porphyrin fluorescence in tumour tissue. Both routes use the demonstration of tissue selectivity, and ultimately this must be the underpinning principle of drug discovery and development.

References

Cincotta L, Foley JW, Cincotta AH. (1993) Phototoxicity, redox behaviour and pharmacokinetics of benzophenoxazine analogues inEMT-6 murine sarcoma cells. *Cancer Research* **53**: 2571–2580.

Crossley ML, Dreisbach PF, Hofmann CM, Parker RP. (1952) Chemotherapeutic dyes. I. 5-Aralkylamino-9-alkylaminobenzo[*a*]phenoxazines. *Journal of the American Chemical Society* **74**: 573–578.

Ferlin MG, Marzano C, Chiarelotto G, Baccichetti F, Bordin F. (2000) Synthesis and antiproliferative activity of some variously substituted acridine and azacridine derivatives. *European Journal of Medicinal Chemistry* **35**: 827–837.

Foley JW, Song X, Demidova TN, Jilal F, Hamblin MR. (2006) Synthesis and properties of benzo[*a*]phenoxazinium chalcogen analogues as novel broad-spectrum antimicrobial photosensitizers. *Journal of Medicinal Chemistry* **49**: 5291–5299.

Lewis MR, Goland PP, Sloviter HA. (1949) The action of oxazine dyes on tumors in mice. *Cancer Research* **9**: 736–740.

Morgan DL, Dunnick JK, Goehl T, *et al.* (1994) Summary of the National Toxicology Program Benzidine Dye Initiative. *Environmental Health Perspectives* **102**: 63–78.

Song X, Kassaye DS, Foley JW. (2008) 5,9-Diaminodibenzo[*a*]phenoxazinium chloride: a rediscovered efficient long wavelength fluorescent dye. *Journal of Fluorescence* **18**: 513–518.

Taylor KB, Jeffree GM. (1969) A new basic metachromatic dye, 1:9-dimethyl methylene blue. *Histochemical Journal 1*: 199–204.

3

Photosensitisers and photosensitisation

What does the term *photosensitisation* mean? In the current context it is used to describe the transfer of light energy via a molecule (the photosensitiser) into either chemical reaction – the transfer of electrons – or across to another molecule, oxygen. The intracellular cascade of further reactions emanating from these events, which are the endpoints of the application, i.e. the killing of uneconomic cells, tend to be included also (Figure 3.1a).

However, there are other related uses of the term, and it may be as well to clear up any possible misunderstandings at this stage. Skin photosensitisation is a fairly common side effect of conventional drug therapy, for example with sulphonamides or phenothiazines. The concentration of the drugs in the skin allows the absorption of daylight by the drugs *in situ*, again promoting radical reactions, cell damage and concomitant syndromes of the skin, for example erythema, blistering, etc. With chronic use of such drugs there may be particular problems caused by the photosensitisation of various structure of the eye. Skin photosensitisation is also a side effect of systemic photosensitiser administration, i.e. in photodynamic therapy, where the photosensitiser is hydrophobic and tends to compartmentalise into the skin layers. We might thus think in terms of positive (advantageous) and negative (deleterious) drug-mediated photosensitisation.

Until quite recently, photosensitisers were also widely used in the photographics industry, in this case being the means of transferring energy and electrons to pigments at certain wavelengths for colour film development. A large number of cyanine dyes (see Chapter 8) were developed from the 1920s onwards for this purpose.

The chemical industry also uses photosensitisers, often immobilised on inert polymers, as oxidation catalysts for large-scale processes. Rose Bengal (Chapter 5) and methylene blue (Chapter 4) have found considerable use in this respect (Figure 3.1b). Photodynamic materials with desired applications, for example in infection control, are usually photosensitisers immobilised within polymer matrices and so may be considered to be a development of the oxidation catalyst idea.

Another word that occurs in much of the literature associated with photosensitisers is *photosensitive*. For example, the phrase *photosensitive dye* is used by some

Photosensitisers in Biomedicine   Mark Wainwright
© 2009 John Wiley & Sons, Ltd

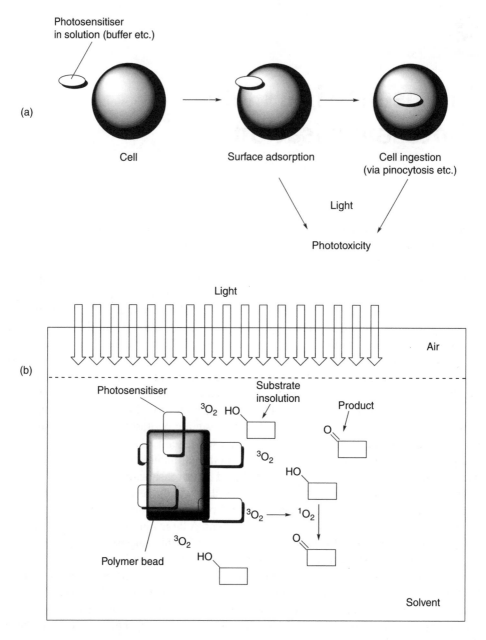

Figure 3.1 Comparison of (a) basic cellular and (b) chemical photosensitisation.

authors when they actually mean *photosensitising dye*. There is quite a difference, since the term *photosensitive* infers sensitivity or reaction to light and thus includes the breakdown of the dye chromophore (photofading/photobleaching) and photochromism – chemical isomerism on light absorption. Clearly, none of these is the same as photosensitisation as described above, although some photosensitisers do also exhibit such properties.

As discussed in Chapter 1, photosensitisation depends on the absorption of light of the correct wavelength by a chromophore, which can exist in the excited state for sufficient time to undergo electron spin inversion and achieve the excited triplet state. In addition, the yield of conversion to the triplet also has to be high enough so that meaningful quantities of, for example, singlet oxygen are formed. Typical quantum yields, i.e. the fraction or percentage of absorption events leading to singlet oxygen production, for useful photosensitisers such as methylene blue are around 0.5 (50%).

3.1 Photosensitiser action

As mentioned above, dye molecules undergo transient electronic excitation. Conversely, efficient photosensitiser action depends on the initial singlet electronic excited state of the molecule having sufficient longevity to allow intersystem crossing and electron spin inversion to yield the triplet excited state from which interaction with the local environment is effected, via either electron transfer (Type I photosensitisation) or energy transfer and singlet oxygen generation (Type II). It should also be noted that some photosensitisers, such as the psoralens, react at the initial singlet excited state. These processes are shown in Figure 3.2 (also compare Figure 1.5, Chapter 1). There are further differences in molecular behaviour on light absorption, the psoralens mainly reacting via photoaddition, i.e. photochemical attack of the excited molecule on the biomolecular target (see Chapter 9).

Plainly, the dependence on a long-lived triplet excited state has encouraged synthesis in order to promote this, usually via the heavy atom effect, i.e. the attachment of heavy

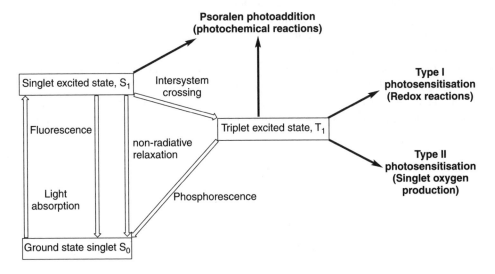

Figure 3.2 Electronic pathways leading to photosensitiser action (see also Figure 1.5).

halogens to the photosensitiser molecule, or the replacement of chromophoric atoms with those from a lower period (e.g. sulphur for oxygen). In electronically excited molecules, intersystem crossing (Figure 3.2) is facilitated by spin–orbit coupling, allowing otherwise forbidden changes in the spin state (i.e. singlet–triplet). As the spin–orbit coupling constant is proportional to the fourth power of the atomic number of the element concerned (Z^4), the presence of an aforementioned heavy atom in a molecule increases the degree of spin–orbit coupling. In terms of the compounds discussed here, this should lead to an increase in the triplet yield and improved photosensitising activity.

Improved photosensitising potency is normally measured as an increase in the singlet oxygen yield, i.e. the percentage efficiency of singlet oxygen production resulting from quantum light absorption. As mentioned, useful photosensitisers normally have associated quantum yields in the range 0.2–0.7 (i.e. 20–70%).

Singlet oxygen generation may be assayed in several ways. Direct measurement of singlet oxygen luminescence at 1270 nm is possible (Krasnovsky 2008). More commonly it is measured via its interaction with singlet oxygen quenching agents, either using electron spin resonance or, more simply, spectrophotometrically, via the oxidation and consequent fading of sensitive indicator dyes such as diphenylisobenzofuran or tetraphenylcyclopentadienone (TPCPD, Figure 3.3) in organic solvents or nitrosoaniline derivatives in aqueous media. Direct measurement can give access to quantum yield data or yields may be measured relative to a known standard (rose Bengal, methylene blue, etc.).

The main difference between conventional textile dyes and photosensitisers occurs here. The majority of textile dyes – like the dye in the blue pin-striped suit mentioned in Chapter 1 – undergo excitation and then non-radiative relaxation from the singlet excited state. Photosensitisers possess longer-lived singlet states, which thus allow significant population of the excited triplet and thence reaction with molecules in the immediate environment. This is the basis of photodynamic action.

It should perhaps be mentioned here that occasionally textile dyers get it wrong. There is a phenomenon known as *bikini dermatitis* that occurs due to the photosensitising action of some (typically) blue anthraquinone dyes sometimes used in swimwear fabrics. A combination of light absorption by the dye and water in the environment of the fabric causes slight leaching of the dye and the production of radical species *in situ*, leading to cell damage and, ultimately, to minor skin irritation.

Anticancer and antimicrobial photosensitisers constitute the two main classes in terms of therapeutic (and potentially therapeutic) applications in this area. Conventional drugs

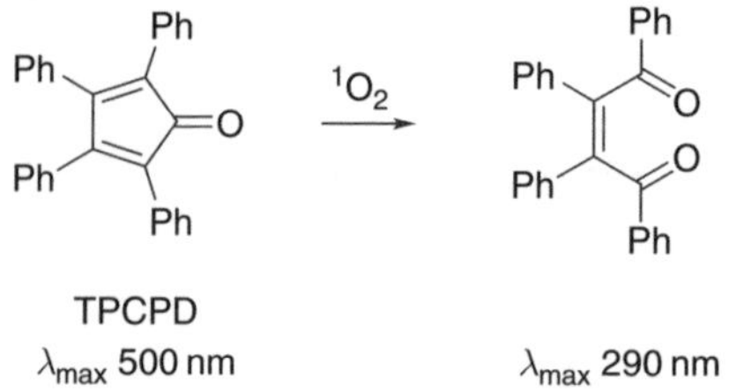

TPCPD
λ_{max} 500 nm λ_{max} 290 nm

Figure 3.3 Spectrophotometric singlet oxygen assay.

here are clearly split – most anticancer chemotherapy agents are too toxic in humans to be used in anything less serious than malignant disease. Conversely, most antibacterial agents, for example, have little or no effect on mammalian cells, including tumour cells. With photosensitisers, however, the boundaries are less well defined. Since the cytotoxic effects of photosensitisers depend on the production of reactive oxygen species in the cellular environment, this might be applicable to whichever non-economic cells (microbial, tumour, etc.) are involved. Thus, while mammalian cells are usually less sensitive to photosensitisers than are simpler microbial species, anticancer photosensitisers have been proposed as photoantimicrobials without significant attempts at drug design or rationale. However, it is now accepted that, for example, broad-spectrum photoantibacterial agents should bear a positive charge.

Although this topic is covered in proper depth in Chapter 11, it should be emphasised at this point that protoporphyrin IX (PPIX, Chapter 6) is an anionic photosensitiser that *is active* against some Gram-negative bacteria, contradictory to the final statement in the preceding paragraph. However, this represents the efficacy of PPIX as an endogenous (internally supplied) rather than an exogenous (externally applied) photosensitiser, the porphyrin being produced by the metabolism of the cell either from endogenous or from exogenous γ-aminolaevulinic acid.

In addition to the overall charge of the molecule, shape is also an important factor in antimicrobial activity, mainly from the aspect of site of action. For example, planar molecules might be expected to act against microbial DNA. Since most of the molecules under consideration here are light absorbing, it follows that they should have a significant π-system, which in turn endows a significant degree of molecular planarity. However, interaction with DNA is governed by the degree of planarity *in combination with molecular charge*. Thus, cations with sufficient planar area may intercalate into the helix between adjacent base pairs (Wainwright 2001), whereas planar anions cannot, due to repulsion by anionic phosphate groups. Thus, cationic aminoacridines, such as proflavine, are effective intercalators but phthalocyaninetetrasulphonic acids are not.

3.2   The photodynamic effect and cell death

Cell killing by photosensitisers may vary depending on the photosensitiser, the cell type and the cellular target or targets. In tumour cells there are two major routes, necrosis and apoptosis, neither of which is a unique consequence of photodynamic treatment. The former normally describes gross damage to a cellular structure, whereas in the latter the damage may be subacute but leads to a cascade mechanism involving various aspects of the cell machinery and signalling. For example, hydrolytic enzymes (proteases), called *cathepsins* and *caspases*, appear to be intimately involved, both in the signalling and in the actual degradative pathway, while mitochondria are heavily implicated in the processing of the apoptotic event (Oleinick, Morris and Belichenko 2002). Among 'standard' targets, photodamage to the membrane, endoplasmic reticulum, lysosomes, mitochondria or DNA may induce apoptosis (Figure 3.4).

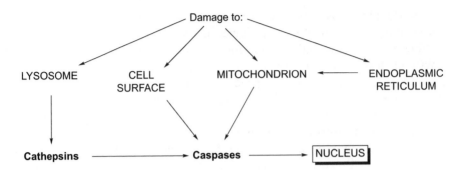

Figure 3.4 Simple apoptotic pathways.

Whether the killing process is due to apoptosis or necrosis may be influenced by photosensitiser localisation and the time of illumination, for example illumination of high levels of photosensitiser in the plasma membrane is likely to cause sufficient damage to lead to cell lysis, before apoptosis takes effect. However, it is likely that cell death is often a mixture of both processes.

Simpler versions of apoptosis may be present in bacteria, reflecting the far more basic unicellular structures involved, in comparison to mammalian tumours (Klarsfield and Revah 2004). A more direct necrotic mechanism is likely with simple unicellular organisms, depending on the majority localisation of the photosensitiser involved.

The multifactorial aspect of photodynamic cell damage is an important positive point in arguments for the clinical use of photosensitisers, particularly where the circumvention of resistance mechanisms is required. This is perhaps more applicable in the antibacterial arena, where clinical drug resistance is considerably more of a problem, but has been demonstrated by several groups working on drug-resistant tumour cell lines also. However, as will be covered in Chapter 11, the 'multiple sites of attack' paradigm not only circumvents established resistance but also constitutes a sound basis for avoiding resistance development *per se*.

3.3 Photosensitising drug discovery

The various research groups around the world involved in photosensitiser development need to supply a requirement for candidate materials for test. Where do these candidates come from? How is a structure decided upon? What are the screening requirements to produce a clinical photosensitiser?

As far as lead compounds are concerned, these can be of either a natural or a synthetic source, for example haematoporphyrin derivative was originally derived from bovine blood, whereas methylene blue is a synthetic compound first produced in 1876. Haematoporphyrins were originally investigated due to clinical manifestations of tumour localisation, whereas methylene blue has a considerable history in vital staining, both in oncology and in microbial pathology. However, neither is an optimum compound in either cancer photodynamic therapy (PDT) or photodynamic antimicrobial chemotherapy (PACT), and new derivatives have been produced in order to address this.

As with conventional therapeutics, one of the major reasons for new compound synthesis is to improve selectivity, and thus therapeutic efficacy. New compounds may have altered hydrophilic/lipophilic balance, since this may improve cellular uptake. However, there may also be effects on both the photosensitising properties and the pharmacology/pharmacokinetic profiles of the new compounds.

As can be seen from Figure 3.5, the pathway of photosensitiser discovery/development is very similar to that for conventional therapeutics. The major difference lies, unsurprisingly, in the incorporation of photoproperty assays into the development schedule. Among these, however, the value of a singlet oxygen assay *per se* is questionable, since the ability of a compound to produce singlet oxygen is dependent, to varying degrees, upon its environment or it may act mainly via a Type I mechanism. For example, the triphenylmethane derivatives crystal violet and Victoria blue BO produce no measurable singlet oxygen in standard assays, and yet both are efficient photosensitisers in cellular media (Wilson, Dobson and Harvey 1992; Burrow *et al.* 1999). Conversely, the benzo[*a*]phenoxazinium lead compound Nile blue produces no singlet oxygen in assays *and* is ineffective as a photosensitiser in cell culture. Plainly, if the singlet oxygen assay were used as a basis for screening, none of these compounds would progress. Similarly, a photosensitiser that exhibits high cellular uptake is more likely to exert a significant photocytotoxic effect than one of similar singlet oxygen yield that exhibits poorer uptake. Thus, although it is unlikely that a candidate that produces measurable singlet oxygen in the assay will fail to do so in the cellular challenge (Figure 3.5), it is sensible to use the cellular challenge stage directly from synthesis (obviously after purification) or formulation. The singlet oxygen generation assay should be treated as merely informative. More important are the associated photoproperties, typically λ_{max} and ε_{max}, i.e. where the candidate absorbs light and how strongly. Plainly, these parameters have a considerable effect on how the photosensitiser may be used, due to the presence of endogenous light absorbers such as haem and melanin (see Chapter 1).

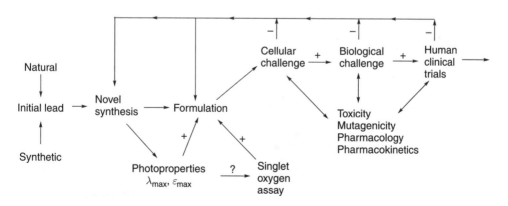

Figure 3.5 Photosensitiser discovery pattern (+, positive result; −, negative result).

As shown in Figure 3.5, feedback from the various assays is used to drive new compound synthesis.

It is possible to synthesise series of compounds from many different lead compounds, and this is the normal way of things in photosensitising drug discovery programmes. There is little evidence of *de novo* synthesis, and certainly not the computer-aided drug design associated with conventional pharmaceutical research – usually, there is a basis in the reported activity of one or other dye, stain or biomolecule. A simple view of the derivation of chromophoric photosensitiser types from a known biologically active example is shown in Figure 3.6 for malachite green. Each of the derived candidate structures given either is a photosensitiser or may be converted into one via, for example the heavy atom effect.

For series of compounds specifically aimed at the production of effective biological photosensitisation, there is little requirement for computational involvement, other than in the modelling of photoproperties, for example in increasing the λ_{max}. As will be covered in Chapter 11, one of the main reasons for drug resistance in bacteria is the use of antibacterial drugs that have only one mode/site of action. This encourages the

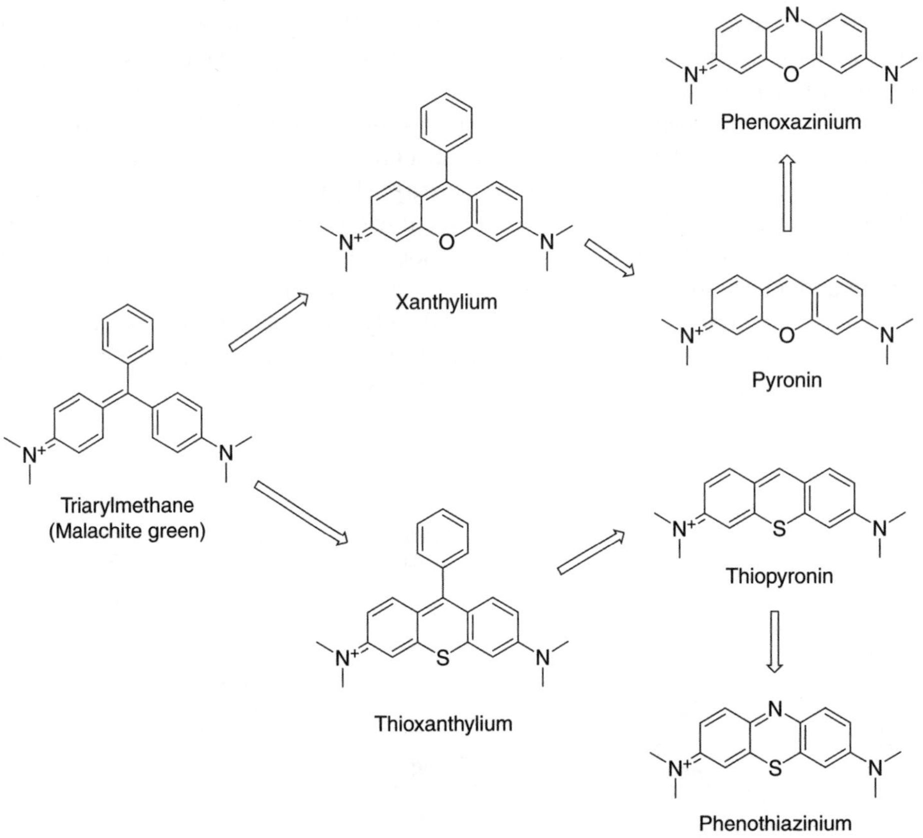

Figure 3.6 Close structural relationships between some different cationic photosensitiser classes.

selection of resistant strains, whereas photobactericidal agents have multiple sites of attack and there is growing evidence to support the absence of resistance selection. The design of a photosensitiser specific for, say, part of the bacterial ribosome could thus be counterproductive.

Much has been written by groups reporting their latest discoveries of 'novel' photo-sensitisers and the advantages of these as potential anticancer or antimicrobial therapeutics. This is, of course, as it should be. However, the search for the ultimate photosensitiser for a particular application remains a conservative endeavour. How many groups are willing to try something different? It is true that the inventive step is normally provided by the chemist (or in rare cases, the chemistry *section*), but often this discovery has to be 'sold' to the remainder of the group. There are even tales of chemists being asked to prove that their new creation is a possible entity. It is unfortunate that there are usually financial considerations that determine the 'safe' choice of photosensitiser moving forward from the laboratory. This lack of adventure is of particular relevance to those involved in photosensitiser research based on a chromophore–auxochrome system.

A reasonable example of this is provided by the phenothiazinium photosensitisers (Chapter 4). Here there were two lead compounds: Methylene blue had been known and used as a therapeutic and in biological staining since the late nineteenth century; toluidine blue O is a related compound, again widely used in biological staining. Both compounds have significant associated photosensitising activities and had been shown to be powerfully photoantimicrobial by 1930. However, there are very few examples of derived phenothiazinium photosensitisers where more than auxochromic variation has been attempted (Figure 3.7). Admittedly it is far simpler to vary the amino moieties at positions 3 and 7 of the heterocyclic ring system than it is to alter the chromophore itself, but this should not be a barrier to drug discovery.

As can be seen from Figure 3.7, novel derivatives are centred on extension or alteration of the amino function. However, toluidine blue O has a chromophoric methyl group, and

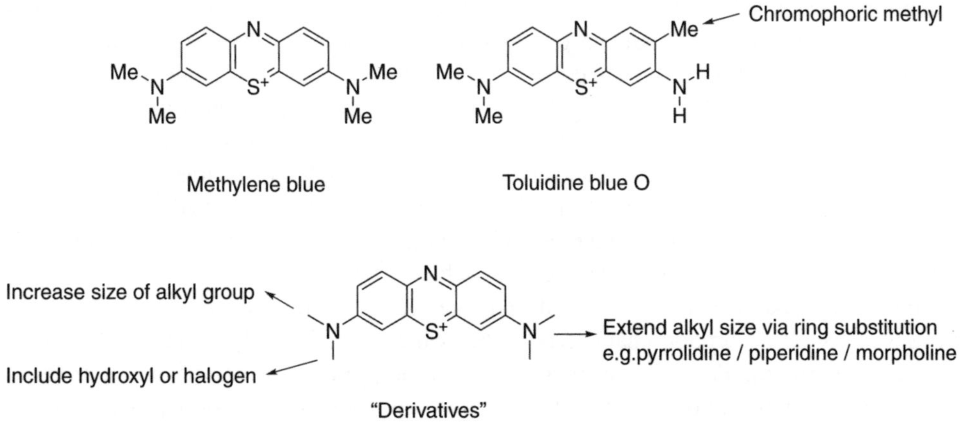

Figure 3.7 Phenothiazinium leads and reported derivatives.

new methylene blue and Taylor's blue (Chapter 4) have two each. Furthermore, methylene green (4-nitro) and iodomethylene blue (4-iodo) have functionalised chromophores and are active photosensitisers, tested for PDT use. Yet there is still a widespread dependence on auxochromic variation to provide new photosensitisers and the only clinically approved derivative remains methylene blue.

The lack of chromophore functionalisation work in this situation would not be countenanced in conventional drug research. The optimum molecular structure for a particular application, by definition, must evolve from consideration of the molecule as a whole (i.e. the holistic approach). It may be that an auxochromic variant is the answer to a particular problem, but without parallel chromophoric derivation this cannot be established on a rational basis.

Holistic considerations are important in drug design. In connection with this, the use of molecular functionalisation aimed at a particular effect is apt also to have effects on other aspects of behaviour, and this can impinge on overall performance.

For example, a traditional method of increasing aqueous solubility – obviously important in drug formulation – is the inclusion of sulphonic acid groups in the drug structure. While this is generally successful in endowing solubility, another result is that an anionic molecule is produced. In turn, this affects the cellular interaction/localisation behaviour of the resultant candidate. Similarly, chromophoric iodination can convert a non-photosensitiser into a useful one via the heavy atom effect, but the lipophilic character of the resulting molecule is considerably increased, which will alter aqueous solubility, aggregation behaviour and, again, cellular localisation. Proper analogue design should take account of such ramifications, and synthetic chemists involved in photosensitiser research over a period of time should have a good idea of the effects of their manipulations on molecular properties. This knowledge should also form part of the decision process in designing or proposing new candidates.

3.4 Fitness for purpose

Supposing that much of the work covered in preceding sections has been carried out efficiently, a functioning candidate photosensitiser should result. Given that the process outlined in Figure 3.5 is included in the programme of work, a successful outcome should be guaranteed, at least insofar as moving the candidate into the clinic is concerned. This overall process has been completed for very few candidate photosensitisers – porfimer sodium, temoporfin and 5-aminolaevulinic acid are the obvious examples at the time of writing – and this is not due to a single factor. While there are always monetary considerations in drug development – conventional or photodynamic – the search for a new compound suitable for clinical use often loses sight of the word 'suitable' in the daily laboratory toil. Thus, for example, a candidate may be produced that is an efficient photosensitiser with long-wavelength absorption characteristics but which has solubility properties that have worsened during the development process due to the presence of one of the formulation excipients. In short, the product is not fit for purpose since there are now problems in its administration to the patient. A similar effect might be seen in series where the early derivatives are easily soluble but might have unimpressive photosensitising activities, whereas subsequent, iodinated derivatives are better photosensitisers in the cellular challenge but now require considerable efforts to ensure solubility in mammalian systems.

Most authors of reviews concerning photosensitisers, whether from a synthesis or an application viewpoint, include, in the initial stages of the article, a list of the properties associated with the ideal photosensitiser for the area under discussion. This list is usually along the lines of the following.

Essential characteristics

- Pure chemical compound, not a mixture
- Efficient photosensitiser *in situ*
- Chemically stable
- Photostable
- Selective for target
- Intense, long-wavelength light absorption
- Non-toxic to host – rapidly excreted?
- Non-mutagenic in host

Desirable characteristics

- Aqueous solubility
- Easily formulated
- Short drug–light interval
- Straightforward synthesis/ease of scale-up

As can be seen, there are essential and desirable criteria. These may change, depending on the photosensitiser application in question. However, these criteria should be examined before embarking on a drug discovery programme.

Chemical purity

In employing a therapeutic agent (photodynamic or conventional) against a target, it is essential to know which chemical entity is being administered. Drug molecules are subject to the vagaries of human metabolism and are thus altered. However, if the original chemical identity is known, the metabolism, pharmacokinetics, etc. can be measured and understood, and any side effects explained. The administration of a mixture makes the situation much more complicated, due to the required identification of useful and deleterious constituents.

Efficiency of photosensitisation *in situ*

As mentioned earlier in this chapter, a lack of singlet oxygen production *in vitro* does not necessarily translate into the clinic. The laboratory chemical testing of candidates has them in a quite different environment to that encountered in the cellular milieu. As suggested, it is sensible to use a cellular challenge as the deciding screen for further work.

There is a conventional parallel here in the discovery of sulphonamide antibacterials in the 1930s: If Domagk's bacterial screen for his azoic dyes had been carried out only in test-tubes and petri dishes, the reductive liberation of sulphonamide would not have occurred, giving a negative result. Positive data were gained from mouse tests, where the metabolism of the dyes (e.g. Prontosil) produced sulphanilamide, which overcame the bacterial challenge.

Chemical stability

In order to act in an effective manner, the photosensitiser molecule should be relatively inert. The metabolism and the immune system exist to detoxify xenobiotics, and the administered photosensitiser will be subject to such forces to a greater or lesser degree, given the modern availability of 'stealth' delivery systems (Chapter 10). However, it is important that the photosensitiser is not altered via facile processes such as hydrolysis or the actions of esterase or protease enzymes. Generally, photosensitiser chromophores are stable entities, but side chains may be altered via oxidative routes, for example N- or O-dealkylation.

Photostability

Given that photosensitisers rely on illumination for activity, it is important that this ability is not diminished before the target site is reached. Photodegradation as a result of target illumination and the photosensitisation process is perfectly acceptable – given the absence of toxic degradants – and would help in the minimisation of post-treatment photoeffects.

Target selectivity

In terms of the ideal photosensitiser, selectivity for the target would be total – this is, after all, the *magic bullet* paradigm. Such a situation is difficult with any systemically administered therapeutic agent, but is somewhat simplified by topical application and is in the present case given dual selectivity by accurate illumination. However, it is necessary to design the photosensitiser so that a significant therapeutic ratio is obtained, rather than to rely on the illumination of a select tissue volume. Early porphyrin-based PDT, for example, was carried out with very low ratios between tumour and surrounding tissue (skin, muscle, etc.), and there remain problems with post-treatment photosensitivity.

The issue of dark toxicity is included here. Again, given the ideally selective photosensitiser, any effects of dark toxicity would be overall beneficial to the therapeutic effect. With photosensitisers lacking selectivity for the target, dark toxicity is effectively systemic toxicity, in a similar way to that experienced with conventional chemotherapeutic agents.

Light absorption

It is usually supposed that photosensitisers must have long-wavelength absorption in order to be useful clinically. This is true if it is desired to avoid endogenous absorbers such as haem and melanin, and so most researchers in the area strive to produce candidate compounds absorbing in the red and near-infrared regions. However, there are situations where this is quite unnecessary, for example in the decontamination of blood plasma and related products (Chapter 11) where absorption above 400 nm is usually effective. The intensity of the absorption is important in that this is related to the effective therapeutic concentration.

Toxicity and mutagenicity

Both criteria here refer to the target selectivity of the candidate. Both phototoxicity and dark toxicity occur when the candidate is concentrated in, for example, non-tumour tissue, usually as a result of the normal metabolism, leading to liver or kidney damage via cellular necrosis or apoptosis as discussed above. Usually, highly hydrophilic compounds are rapidly excreted and *vice versa*.

Mutagenicity is usually a more subtle process, which may of course take years to present as symptoms. It should be noted that mutagenicity occurs via DNA damage and is thus not specific to intercalators, i.e. other cellular processes may end in DNA damage also. As mentioned earlier, intercalation in a test-tube does not necessarily correlate with human mutagenicity – indeed the Ames test, which is based on a mutagenicity assay in *Salmonella typhimurium*, does not do so either. In order to affect human nuclear DNA in the same way, the agent must localise in the cell nucleus. Many intercalators do not exhibit such behaviour. Admittedly, it would be difficult to plead this case with some regulatory authorities.

Solubility and formulation

In order to reach the site of action, the candidate will spend some time in solution, either in the initial administration phase or after absorption from the alimentary canal. Consequently, either aqueous solubility or some methods of formulation using co-solvents, surfactants, liposomes, etc. are required. This is discussed in Chapter 10. The automatic answer here is usually to guarantee solubility via functionalisation (e.g. sulphonation), but this alters other characteristics – perhaps deleteriously – as mentioned above. Certainly, the use of liposomal encapsulation for the tumour delivery of water-insoluble zinc phthalo-cyanine appears to be an excellent alternative to the aqueous mixture of its sulphonic acids (see Chapter 7). Once in the bloodstream, candidates are exposed to the body's defences and the metabolism with little regard for solubility, although highly hydrophilic species generally have a short plasma half-life, and protective formulation (Chapter 10) is a sensible precaution in increasing tumour uptake.

Drug–light interval

Following the administration of the photosensitiser, there is a period over which the drug concentrates at its target site. Plainly, this is different across a patient population, even given the same tumour type/site. Logically, light is applied to the tumour at the maximum tumour/surrounding tissue concentration ratio. This is true whether for PDT or for PACT, although it would be expected that the drug–light interval for local disinfection, for example in the oral cavity (Chapter 11), would be considerably shorter. It is likely that a contact time (photosensitiser–bacteria) of less than a minute would be preferable in such cases, compared with hours or days for distribution/accumulation pre-PDT.

Straightforward synthesis

This is a desirable rather than an essential criterion, since the more important criterion of drug efficacy is the driver. Having established that the candidate represents a sufficient advance, it is usually 'up to the chemists' to provide a ready supply. Indeed, it makes little sense to reverse the arrangement and much interesting chemistry has been developed as a consequence. However, scale-up is often fraught with problems, particularly where large quantities of drug are required to be prepared under conditions of good manufacturing practice, for example for a clinical trial.

3.5 Organic *versus* inorganic photosensitisers

Although this book is concerned primarily with organic photosensitisers, it should be remembered that there are also inorganic examples, including titanium dioxide, zinc oxide and cadmium sulphide. However, it is the first of these that has been the subject of the majority of investigations in this area.

The major differences between organic photosensitisers and titanium dioxide lie in mechanism of action and colour. The organic compounds rely on promotion of an electron by light absorption form a ground state to an excited state orbital (Figure 3.2), whereas the light absorption step in TiO_2 causes the excitation of an electron from a valence to a conduction band. The resulting excited species are thus different. In addition, organic photosensitisers cover the whole spectrum of visible light absorption, whereas the inorganic oxide is white, absorbing UV-A light (approximately 320–400 nm). As a *prima facie* example of a pigment, titanium dioxide is also insoluble in water. While there are similarly insoluble organic photosensitisers (e.g. unfunctionalised phthalocyanines), the normal approach is to solubilise these via sulphonation, etc. The development of nanotechnology, of course, encompasses both forms.

The photoexcitation step in TiO_2 leads to the formation of an electron–hole pair, i.e. the negative electron is promoted, leaving a corresponding positive 'hole'. The electron in the conduction band may act as a reducing agent, the positive (valence) hole as an oxidiser. Irradiated titania in its anatase form has been shown to produce both singlet oxygen and the superoxide anion on illumination (Konaka *et al.* 1999). Correctly delivered, this activity can thus be targeted at non-economic cells in the same way as the other photosensitisers covered here, and TiO_2 cell photodestruction has been demonstrated for both tumour and microbial cells. The drawback of UV-A excitation is a considerable one, although this may be circumvented by doping the lattice with non-metals such as sulphur, phosphorus, etc. (Chen and Zhang 2008). However, the utility of such an arrangement in cell photokilling, especially *in vivo*, appears uncertain.

The main advantage of TiO_2 photosensitisation appears to be in the action of various 'self-cleaning' materials meant for more robust environments than the human body. Thus, glass and structural features may be coated with anatase in order to provide non-fouling properties. However, inorganic photosensitisers have received comparatively scant attention when considered beside the wide range of organic chromophores available.

References

Burrow SM, Phoenix DA, Wainwright M, Waring JJ. (1999). Uptake and cell-killing activities of Victoria blue derivatives in a mouse mammary tumour cell line. *Cytotechnology* **29**: 35–43.

Chen X, Zhang X. (2008) Preparation of visible-light responsive P-F codoped TiO_2 nanotubes. *Applied Surface Science* **254** : 6693–6696.

Klarsfield A, Revah F. (2004) *The Biology of Death: Origins of Mortality*, Cornell University Press, 136.

Konaka R, Kasahara E, Dunlap WC, Yamamoto Y, Chien KC, Inoue M. (1999) Irradiation of titanium dioxide generates both singlet oxygen and superoxide anion. *Free Radical Biology and Medicine* **27**: 294–300.

Krasnovsky AA. (2008) Luminescence and photochemical studies of singlet oxygen photonics. *Journal of Photochemistry and Photobiology A: Chemistry* **196**: 210–218.

Oleinick NL, Morris RL, Belichenko I. (2002) The role of apoptosis in response to photodynamic therapy: what, where and how? *Photochemical and Photobiological Sciences* **1**: 1–21.

Wainwright M. (2001) Acridine – a forgotten antimicrobial chromophore. *Journal of Antimicrobial Chemotherapy* **47**: 1–13.

Wilson M, Dobson J, Harvey W. (1992) Sensitization of oral bacteria to killing by low-power laser radiation. *Current Microbiology* **25**: 77–81.

PART II

Chemistry of photosensitisers

4

Azines

Among the synthetic classes of photosensitiser, the azines are almost as commonly encountered as the phthalocyanines, although there is more exemplified variety in the former group in terms of heteroatom inclusion, since this defines the subclass of chromophore (Figure 4.1).

The grouping of various different chromophores under the heading of *azines* is, for the most part, correct – ox*azine*, phen*azine*, etc. However, it may be thought that the heading has been stretched somewhat by the inclusion of the acridines here. Although these have been classified as azaanthracenes in the past, it remains the oxidised carb*azines* (Figure 4.2), which exhibit the closest similarity (Goldstein and Kopp 1928).

Nevertheless, consideration of phenazinium neutral red, phenothiazinium toluidine blue O and acridinium flavicid (Figure 4.3) shows that these three differ in only their central heteroatoms. A similar situation pertains with acridine orange (AO) and methylene blue (MB), and with proflavine and thionin. Certainly, in terms of chemical properties and biological behaviour there is sufficient commonality.

Using the *photosensitisers-from-biological-stains* paradigm, the azine class covers most of the famous examples, including MB, Nile blue, neutral red and AO. Activity here covers both the histopathology and the vital staining of malignant and infectious disease, and this has been translated into anticancer and antimicrobial effects, e.g. AO in photodynamic therapy (PDT) (Chapter 10) and MB in photodynamic antimicrobial chemotherapy (PACT) (Chapter 11). MB and the acridines in particular also have a long history in conventional antibacterial chemotherapy, which obviously supplies a sound rationale for photoantimicrobial development.

One result of the structural similarity is that many examples interact strongly with nucleic acid. This is due to the common three- or four-ring cationic heteroaromatic (planar) system. However, this might be seen as a negative aspect due to the potential – if not proven – ramifications in mutagenicity for the human organism. This is discussed in some detail below.

As photosensitisers, the premier group among the azine derivatives is that containing the MB derivatives, the phenothiaziniums. However, it should be remembered – or recognised – that there are many other related structures and, as such, these should be

Photosensitisers in Biomedicine   Mark Wainwright
© 2009 John Wiley & Sons, Ltd

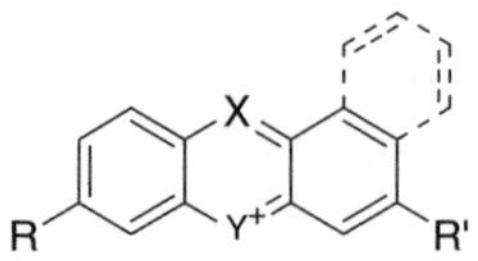

X	Y	
N	NH	Phenazinium
N	O	Phenoxazinium
N	S	Phenothiazinium
N	Se	Phenoselenazinium
CH	NR"	Acridinium

Figure 4.1 Structural range of azine and benzo[a]-fused photosensitisers (R, R' = auxochromes; R'' = H or alkyl, etc.).

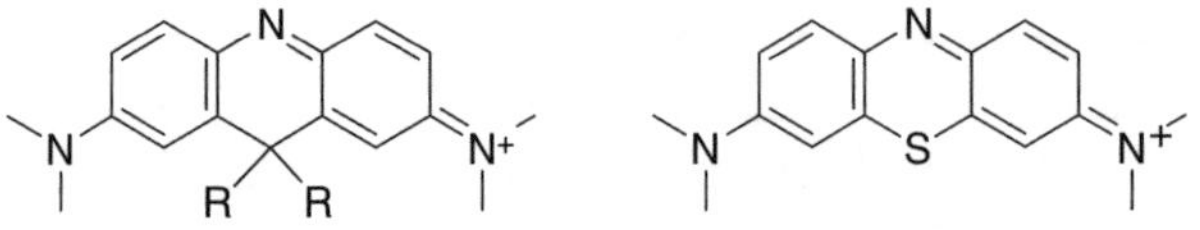

Carbazine   Methylene blue

Figure 4.2 Similarity between carbazine/carbazinium (R = alkyl and aryl) and phenothiazinium structures.

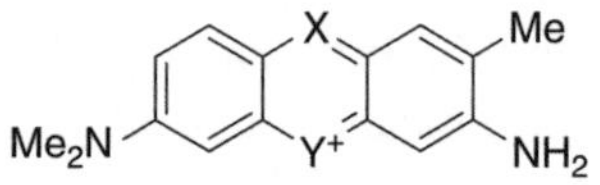

X	Y	
N	NH	Neutral red (phenazinium)
CH	NMe	Flavicid (acridinium)
N	O	Brilliant cresyl blue (phenoxazinium)
N	S	Toluidine blue O (phenothiazinium)

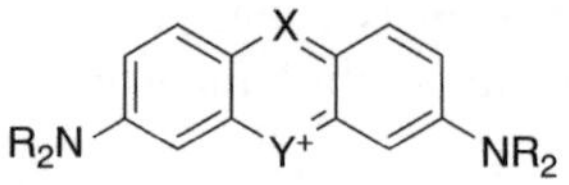

R	X	Y	
Me	CH	NH	Acridine orange (acridinium)
Me	N	S	Methylene blue (phenothiazinium)
H	CH	N	Proflavine (acridinium)
H	N	S	Thionin (phenothiazinium)

Figure 4.3 Commonality in azine-derived structures.

included from the drug discovery angle. Small differences in similar molecules can be magnified significantly with regards to drug action.

4.1 Acridines

In many reviews concerned with photodynamic therapy and/or photosensitisers, mention is made of the early – probably the earliest – experimentation carried out by Oskar Raab in Germany at the turn of the twentieth century, dealing with the lethal photosensitisation of paramecia (Raab 1900). These experiments reportedly employed two photosensitisers: eosin and acridine.

It is rather ironic that acridine is neither an efficient generator of singlet oxygen under normal screening conditions nor is it particularly water soluble, and yet Raab continues to be widely cited, if not lionised, for this early breakthrough. Admittedly, the use of synthetic, dye-based photosensitisers in an antimicrobial setting is appealing to the non-porphyrin lobby! It is also ironic that the doyenne of acridine chemistry, Adrien Albert, stated in *The Acridines* (which remains the acridine researcher's bible) that the use of acridines in 'photodynamy' would have little effect. Subsequent workers in the field have reported quite the opposite, both in the anticancer and in the antimicrobial application.

As with the phenothiazinium and triarylmethane classes, there is much to recommend the use of acridines as photoantimicrobials, particularly given the number of acridine antibacterial agents that have seen conventional clinical use. Indeed, the clinical history of acridine antibacterials is very closely related to that of both MB and crystal violet. The negative side of acridine therapy – at least the perception for many clinicians – is that nucleic acid targeting implies too great a potential for mutagenic sequelae in patients.

Browning's introduction of topical proflavine (3,6-diaminoacridine) into wound therapy in 1916/1917 (Browning *et al.* 1917) undoubtedly saved many lives that would otherwise have been lost to sepsis. This demonstration of efficacy also led to further investigation of synthetic dye-based molecules in the inter-war period, supporting the large-scale screening of dyes aimed at the therapy of both bacterial infection and tropical disease. The evolution of the sulphonamide antibacterials as well as the acridine and quinoline antimalarials can thus be traced back to dye screening (Wainwright 2003). In addition, while proflavine was again employed in wound therapy during World War II, Albert (1944) had also developed less colouring derivatives, such as aminacrine (9-aminoacridine) and salacrin (9-amino-4-methylacridine) for wound use. A full list of clinically used acridine antibacterial agents and their photosensitising potential is given in Table 4.1.

Although the emergence of penicillin by 1944 eclipsed the acridine antibacterials, they remained in pharmacopoeias for decades afterwards. The binding of aminoacridines to nucleic acids was reported in 1963 (Woese *et al.* 1963), and this was accepted as the antibacterial mode of action, since it explained the requirement for significant chromophore cationic charge and minimum planar molecular surface area exhibited by the many effective acridines derivatives reported by Albert himself in the 1930s and 1940s (Albert *et al.* 1945).

Table 4.1 Clinically used antibacterial acridines

	R^2	R^3	R^4	R^6	R^9	R^{10}
Proflavine*	H	NH_2	H	NH_2	H	H
Euflavine*†	H	NH_2	H	NH_2	H	Me
Diflavine	NH_2	H	H	NH_2	H	H
Sinflavin	H	MeO	H	MeO	H	Me
Flavicid	Me	NH_2	H	Me_2N	H	Me
Ethacridine*	EtO	H	H	NH_2	NH_2	H
Aminacrine*	H	H	H	H	NH_2	H
Salacrin*	H	H	Me	H	NH_2	H

*Photobactericidal according to Wainwright *et al.* (1997a).
† Purified acriflavine.

Logically, as acridines have been reported to be photosensitisers at various times since Raab and as there exists a catalogue of effective aminoacridine antibacterials, it should be possible to design effective photoantibacterials based on the acridine chromophore. In fact, the inherent antibacterial nature of the aminoacridine drugs alluded to above *is* increased by illumination with white light. Thus, for example, the minimum bactericidal concentration (MBC) for ethacridine (rivanol; 3,9-diamino-7-ethoxyacridine) against *Escherichia coli* decreased from 100 µM in dark conditions to 10 µM under white light at a light dose of 6.3 J/cm². In the same circumstances, AO, tested against *Staphylococcus aureus* had a dark MBC of 25 µM and this decreased to 1 µM on illumination (Wainwright *et al.* 1997a).

The aminoacridines have mainly been used as topical antibacterials – indeed the remaining example in UK hospitals is proflavine cream, used in burn wound treatment. Earlier systemic use (i.v.) of acriflavine, aminacrine and ethacridine demonstrated rapid elimination (Wainwright 2001). However, topical application would be the favoured approach in conjunction with illumination.

As mentioned above, Albert introduced aminacrine as a non-colouring topical agent in order to replace the yellow-staining proflavine. The highly active diflavine (2,6-diaminoacridine) also remained unapproved for clinical use due to its red colour. However, in terms of the photodynamic use of such compounds, colour intensity would be a positive factor and would permit the consideration of acridine-based stains such as acridine yellow and AO as lead compounds in drug design.

The negative reception associated with acridine drugs is based on the easily demonstrable interaction of these agents with bacterial DNA. Indeed, the affinity of aminocridines and the isomeric aminophenanthridines (such as ethidium bromide) for DNA, while providing superb service in fluorescent labelling, is truly a double-edged sword. Mutagenic effects have been reported in both bacteria and yeasts (Ito 1973; Hass and Webb 1981). However, whether these effects are transferable to the human subject is still unclear.

It is unlikely that there is another group of compounds as roundly vilified as the acridines, particularly given that there is such a dearth of human evidence. There are compounds that are established to cause cancer, for example 1-naphthylamine within the dyes milieu. However, the case against acridines and phenanthridines (Figure 4.4) is derived mainly from experimental data and only for a small range of compounds. While it is obviously impossible to trial such compounds for side effects in humans, there exists a considerable group of individuals who were treated with aminoacridines in the period 1917 to c. 1950. As noted, the use of proflavine, acriflavine and aminacrine (Table 4.1) was significant in the area of antisepsis both in the military and in the civilian populations, while the antimalarial mepacrine was widely employed in the South-East Asian/Pacific theatre (Sweeney 2003). If the potential for oncogenesis is so great, where are the data showing increased incidence of cancer among these groups? However, the fact remains that it is very difficult to interest drug developers in molecules that contain either an acridine or a phenanthridine moiety, unless it is for anticancer purposes.

The basis of the DNA-intercalation/tumour production argument, although not proved clinically, obviously requires intercalation. However, the idea that intercalation requires a cationic, planar molecule of approximately acridine size is merely a rule of thumb. Important data in this area were reported by Zimmerman in the mid-1980s (Zimmermann 1986). The data showed that simple cationic character is insufficient, since hydrogen bonding is also required. Plainly this may be 'designed out' of a molecule quite simply. Thus, although proflavine and acriflavine (euflavine) satisfy both the cationic/planarity criteria, each molecule contains free amino groups with hydrogen available for H-bonding. Methylation of these amino moieties, as in N-methyl AO (acridine structure, Figure 4.4, R = R' = Me) produces molecules that no longer intercalate. The related ethidium cation behaved similarly, and it was also found that the replacement of one of the free amine moieties with hydroxyl allowed H-bonding.

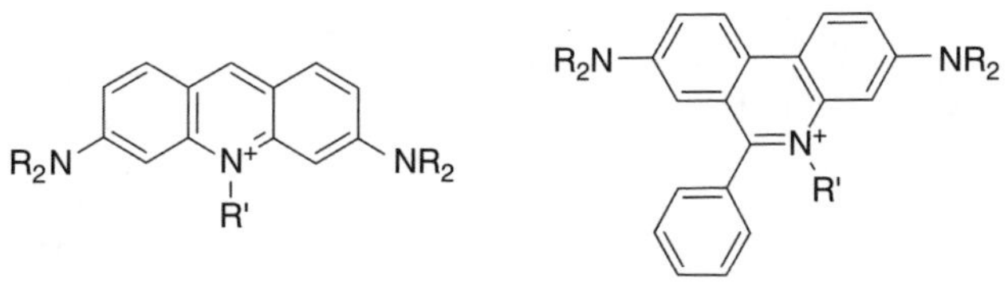

Figure 4.4 Acridinium and phenanthridinium intercalators (R = H; R' = H and alkyl) and non-intercalators (R = R' = alkyl).

4.2 Acridine synthesis

The acridine chromophore has been known since the end of the nineteenth century, and the synthesis of analogues is mainly carried out along well-established lines, as detailed in Figures 4.5–4.7, for symmetrical and non-symmetrical compounds. Given the fundamental requirement for activity of cationic ionisation, it is not surprising that the majority of published acridine structures contain either an amino functionality or a quaternised ring nitrogen. Delocalisation of the positive charge is important in activity, and this dictates where the amine functionality may be placed, i.e. at positions 3, 6 or 9.

However, it is possible to have a wide variety of other substituents present in addition to an amine group in one or more of these positions (Figure 4.6). This obviously allows the development of structure–activity relationships over a range of physicochemical properties and has resulted in successful conventional drugs such as ethacridine (Figure 4.7).

In terms of photosensitiser development from the acridine chromophore, this may use conventional aminoacridines. Indeed, their bactericidal potency has been reported against

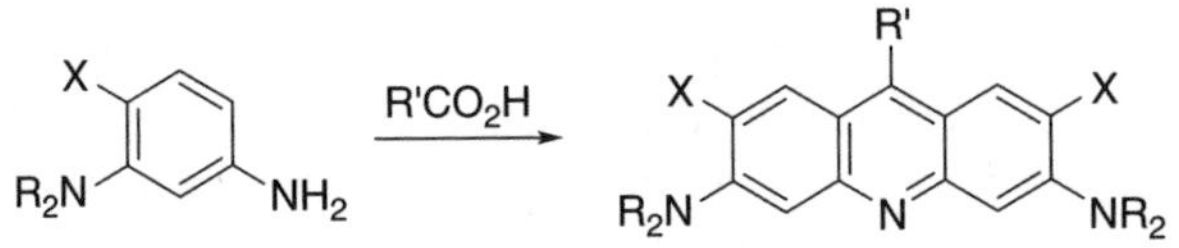

X	R	R′	Acridine
H	H	H	Proflavine
H	Me	H	Acridine orange
Me	H	H	Acridine yellow
H	H	Ph	Benzoflavine

Figure 4.5   3,6-Diaminoacridine synthesis.

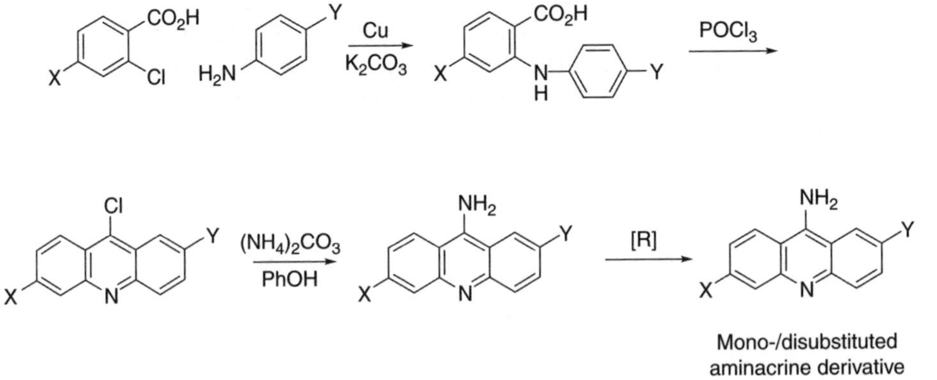

Mono-/disubstituted
aminacrine derivative

Figure 4.6   Classical Bernsthen synthesis of isomerically pure mono- or disubstituted 9-aminoacridine (this synthetic route may be used to produce aminoacridines bearing a heavy atom).

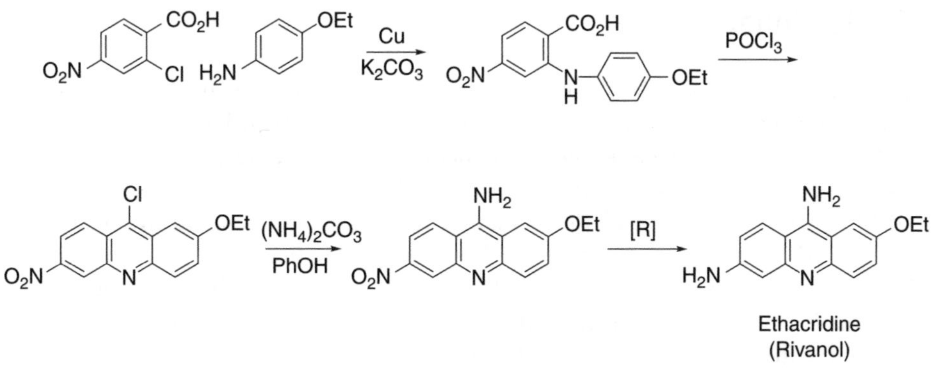

Figure 4.7 Synthesis of ethacridine via the Bernsthen route.

both Gram-positive and Gram-negative pathogens (Wainwright *et al.* 1997a). The main drawback of such compounds as photosensitisers in the biological milieu lies in the short wavelengths of absorption exhibited – typically these are in the region 400–500 nm.

The absorption situation can be improved by the use of styryl or cyanine formation (Figure 4.8), although the former at least appear to be of little utility as photosensitisers. In addition, the availability of precursor diphenylamines is not high.

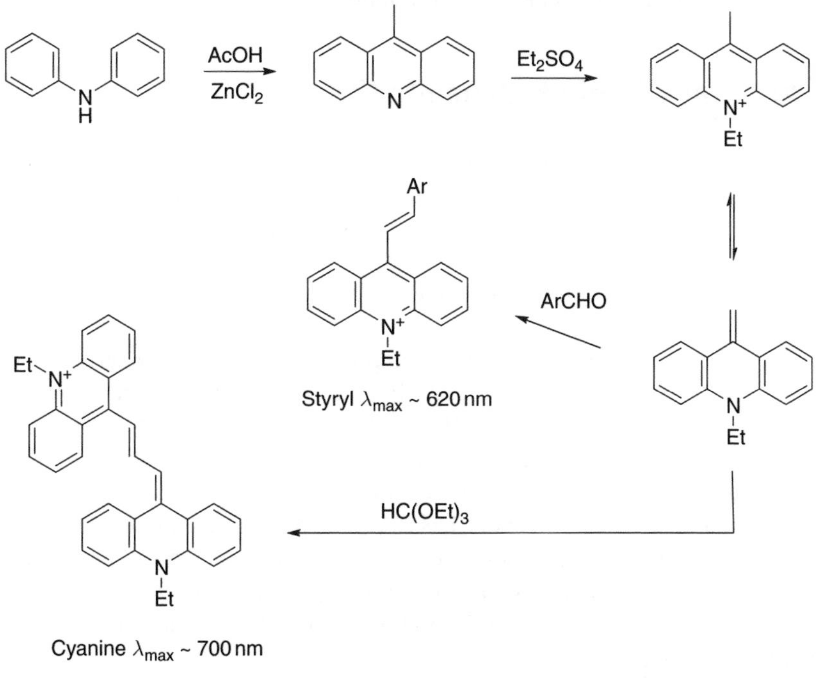

Figure 4.8 Synthesis of long-wavelength absorbing acridine–styrenes and acridine–cyanines from diphenylamine.

4.3   Rationale

The biological activities of the simple aminoacridines, particularly in the field of antisepsis/antibacterial action, are well known. Since the photodamage caused to yeasts in mutagenicity studies is also well documented; the hypothesis that the aminoacridine antibacterials would also be photobactericidal was a short, logical step. Indeed, had this knowledge been available to Browning during World War I, much more rapid cures would presumably have been effected, given that the acridines used were applied topically.

Thus, there are several examples of widely used but obsolete antibacterial agents in this class, which have shown significant increase in efficacy *in vitro* on illumination. The potential for their use in topical disinfection should be recognised, but the shadow of potential mutagenicity represents a considerable obstacle to progress.

Interestingly, a simple iodinated aminoacridine derivative (3-amino-6-iodoacridine) has been shown to interact, predictably, with DNA *in vitro* but to cause phage photodamage via two different routes, depending on the wavelength of illumination. Long-wave UV light caused photolysis of the iodine–carbon bond and free-radical formation *in situ*, leading to further sequelae, this reaction being independent of oxygen. Conversely, under identical conditions, except that longer wavelength light at 420 nm was used, typical oxygen-dependent photosensitisation was observed (Chen *et al.* 1996). This example is interesting, given the reliance placed on heavy atoms for the realisation of photodynamic effects in a range of chromophoric types.

4.4   Acridines in photodynamic therapy

AO is the main acridine employed in anticancer PDT research. It is reported to localise in the lysosomes of normal mammalian cells, but this can be altered by ring *N*-alkylation (= quaternisation), as indicated by the mitochondrial localisation of the *N*-nonyl derivative, which may have an affinity for cardiolipin (Mileykovskaya *et al.* 2001). However, the *N*-hexyl derivative has been shown to be more selective for mitochondria (Rodriguez *et al.* 2008).

There are differences in behaviour between proflavine and the AO derivatives (albeit that AO may be thought of as tetramethyl proflavine). The main differences are due to the presence or absence of *N*-alkyl groups. The binding of proflavine to DNA is intercalative and strengthened by the hydrogen bonding possible from the auxochromic amino moieties. Dimethylation of these groups, as in AO, negates such bonding. Additionally, both proflavine and AO, as nitrogenous bases, can be protonated (Figure 4.9), furnishing neutral–cationic equilibria, which are important in cellular uptake and compartmentalisation. The quaternary derivatives such as nonyl AO are, by definition, 100% cationic and so would be expected to exhibit slightly different biological properties to the non-quaternised derivatives.

It is well established that AO localises in the acidic food vacuole of the malarial parasite (*Plasmodium* spp.) and has been used to investigate changes in pH related to antimalarial resistance (Dzekunov, Ursos and Roepe 2000). Related to this use, AO

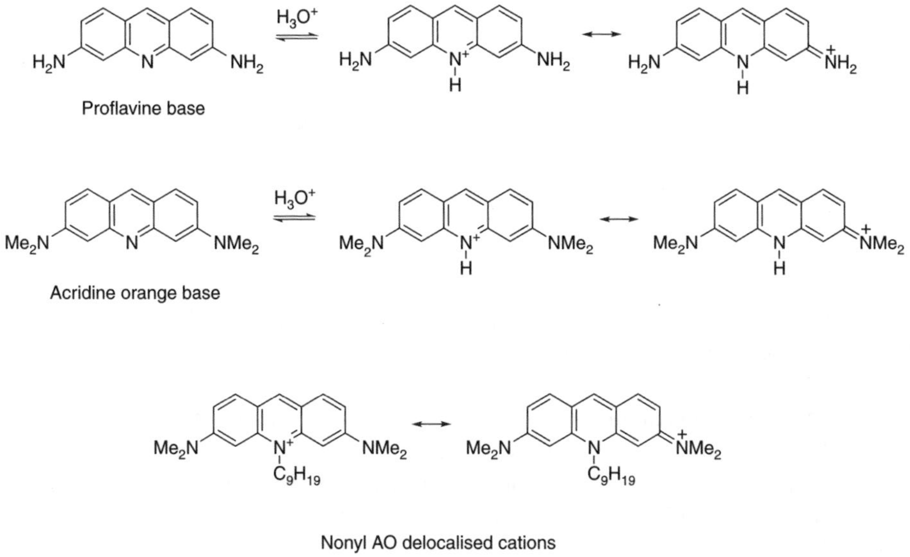

Proflavine base

Acridine orange base

Nonyl AO delocalised cations

Figure 4.9 Proflavine and acridine orange forms.

causes bleb formation from acidic vesicles in the mammalian cell membrane, post-illumination, for example in malignant melanoma cells. Singlet oxygen production is involved in AO-mediated cell killing (Hiruma *et al.* 2007).

Acridines, mainly proflavine and AO, have been used for virus inactivation, nucleic acid binding being a deciding factor. For example, both acridines were shown to inactivate poliovirus (Crowther and Melnick 1961; Wallis and Melnick 1963). Proflavine was also used clinically in the treatment of herpes lesions in the 1970s (Moore *et al.* 1972).

AO does not stain DNA in live mammalian cells, although this situation may alter after cell illumination due to the release of lysosomal constituents (Delic *et al.* 1991). Since this means that the cell would be in a dying or at least apoptotic state, the DNA interaction – cell studies have shown strand breaks post-illumination (Uggla and Sundell-Bergman 1990) – is probably irrelevant. Once again, this should encourage researchers to produce novel analogues that are safe, in terms of DNA damage, but promote an apoptotic mechanism from the lysosome, mitochondrion, etc.

Effective AO-PDT activity has been demonstrated in implanted stomach tumours in rats, via intraperitoneal injection of the photosensitiser, followed by direct tumour illumination after 2 hours with 488 nm light (argon laser) (Tatsuta *et al.* 1984). Similarly, successful results were reported as early as 1976 for skin tumours in an animal model (Tomson 1976).

4.5 Phenanthridines

The phenanthridine ring system is isomeric with that of the acridines, being an angular arrangement of the three aromatic nuclei, rather than linear (Figure 4.10). Consequently,

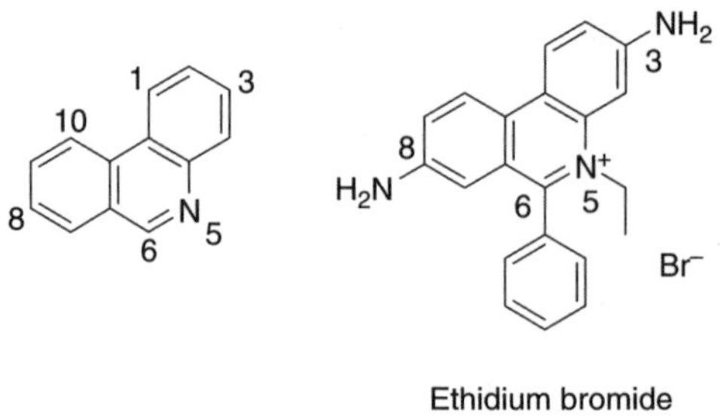

Ethidium bromide

Figure 4.10 Numbering in the phenanthridine chromophore and ethidium bromide (3,8-diamino-5-ethyl-6-phenylphenanthridinium bromide).

suitably substituted phenanthridines exhibit similar properties to the acridines covered above, for example amino derivatives where the amino moieties are conjugated with the ring nitrogen are usually highly cationic in nature. The high nucleic acid affinity of derivatives such as ethidium bromide can thus be explained in terms of planar area and cationic charge.

As with several drugs derived from dyes, early phenanthridines were developed for intentional use in tropical medicine, for example in Carl Browning's work on human African trypanosomiasis (sleeping sickness) after World War I (Browning *et al.* 1938). Ethidium bromide and its later derivatives are established trypanocidal agents (although mostly now used in the veterinary milieu) and are known to target the DNA of the target parasite – indeed, this may be plainly seen using fluorescence microscopy (Boibessot *et al.* 2002).

However, again in common with the acridines, this nucleic acid specificity constitutes a double-edged sword due to mutagenicity concerns.

In terms of the present work, again as with the acridines, the common ethidium bromide is an effective cellular photosensitiser, but this effect has been exhaustively demonstrated in yeasts, such as *Saccharomyces cerevisiae*, to cause mutation (Ferguson and Borstel 1992). However, such findings do not necessarily pertain to structurally altered phenanthridinium derivatives, for example, the 3- or 8-desamino or the 3,8-didesamino were all significantly less effective at petite induction in growing *S. cerevisiae* and, unlike ethidium bromide, were non-mutagenic in the resting phase (Fukunaga *et al.* 1984). Thus, it is obviously important to have both amino groups present to achieve efficient mutagenicity, in line with the findings reported by Zimmerman, discussed above.

4.6 Phenaziniums

In the phenazinium chromophore, C-9 in the acridine nucleus is replaced by nitrogen (Figure 4.11). The cationic nature of the chromophore is provided either by protonation of the lower nitrogen (*N*-5) of the central ring or, more commonly, by alkylation/arylation. Clearly protonation is reversible and allows equilibria to be set up in the cellular

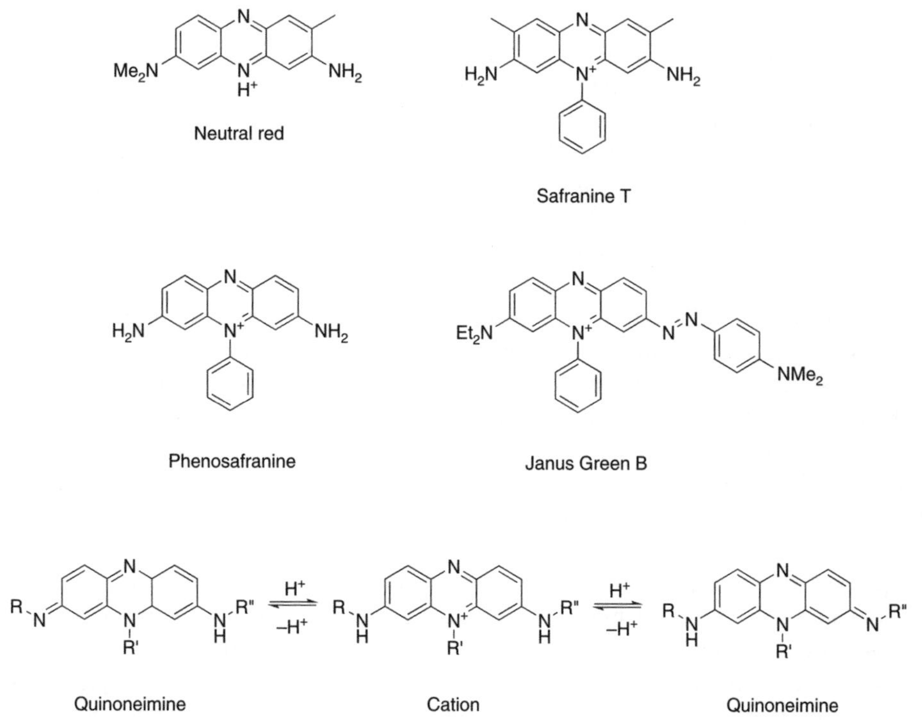

Figure 4.11 Standard phenazinium derivatives and quinoneimine formation.

environment. For example, neutral red can exist in a neutral or protonated form. As with other azine-type chromophores, phenaziniums having suitably conjugated primary or secondary amino groups may also form neutral quinoneimines (Figure 4.11).

Also in common with similar azines, the phenazinium chromophore has properties normally associated with typical DNA-intercalating behaviour, i.e. sufficient molecular planar area and cationic charge. Accordingly, *in vitro* intercalating behaviour has been reported for neutral red (Wang *et al.* 1999), methylene violet (MV) (Chuan, Yu-xia and Yan-li 2005; Li *et al.* 2007), Safranine T (Cao and He 1998) and phenosafranine (Das and Kumar 2008). The phenaziniums provide a good example of the possible confusion caused in *in vitro* DNA testing.

Neutral red was among the group of cationic dyes, screened in the 1940s and 1950s, which were discovered to inhibit tumour growth in mice and to be tumour cell selective stains (Riley 1948). Neutral red is often employed as a viability stain for mammalian cells, i.e. it is taken up by live cells. In addition, it reportedly gives a positive Comet assay, inferring DNA damage, for example in *S. aureus* (Phoenix and Harris 2003).

Interestingly, in the aforementioned study, which provided the Comet data, the phenazinium exhibited high photoactivity against *S. aureus* but produced no measurable singlet oxygen during *in vitro* testing. This suggests a type I photosensitisation pathway. Similarly, the related Janus green has been shown to be a photosensitiser in mammalian cells, its efficacy concomitant with use as a mitochondrial stain (Van Duijn 1961). The fact that this photosensitiser presumably also acts via a type I mechanism is emphasised by the singlet oxygen quenching activity of the azoic linkage in the auxochromic group at position 7 of the ring system (Figure 4.11). Neutral red was also involved in early antiviral photosensitiser research (Wallis and Melnick 1964), which eventually led to the topical photovirucidal programme of the 1960s/1970s aimed mainly at genital herpes treatment (Kaufman *et al.* 1973).

Two other phenazine-containing molecules might also prove fruitful as lead compounds for photosensitiser research. Pyocyanin (Figure 4.12) is a well-known, if very simple, zwitterionic derivative of phenazine itself (i.e. bearing both a positive and a negative charge), while clofazimine (Figure 4.12) was originally synthesised by Barry *et al.* (1948) in the search for antileprosy drugs. Pyocyanin appears to be photoactive, probably via a type I mechanism (Gardner 1996), and is produced, typically by *Pseudomonas aeruginosa*, as a defensive chemical against other microbes (Hassan and Fridovich 1980; Okuda, Ito and Ito 1987). Given the blue colour of the pigment, it is strange that no organised search for new photosensitisers based on pyocyanin has been reported. However, there is a series of publications by McIlwain concerning the synthesis and properties of analogues from the 1930s (Dickens and McIlwain 1938).

Clofazimine is a purple/damson-coloured compound that contains the phenazine chromophore as a quinoneimine. Although it was originally aimed at Hansen's bacillus (*Mycobacterium leprae*) (Lockwood 2005), its activity against the opportunistic HIV-associated pathogen *Mycobacterium avium* has given it a use outside tropical medicine (Arbiser and Moschella 1995). It is also used in the treatment of drug-resistant tuberculosis (Shah, Bhagat and Panchal 1993) and Crohn's disease (Selby *et al.* 2007). It has also been shown to reverse multi-drug resistance in cancer cell lines (Van Rensburg *et al.* 1994). Clofazimine has been shown to be a photosensitiser *in vitro* and *in vivo* (Arutla *et al.* 1998; Bennet 2007).

There are thus a number of structures available for further exploration based around the phenazine/phenazinium chromophore. From a synthetic standpoint, the original, related

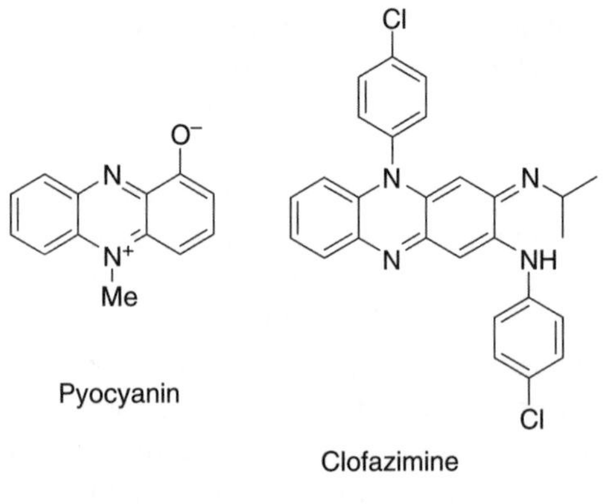

Pyocyanin

Clofazimine

Figure 4.12   Pyocyanin (phenazinium) and clofazimine (quinoneimine).

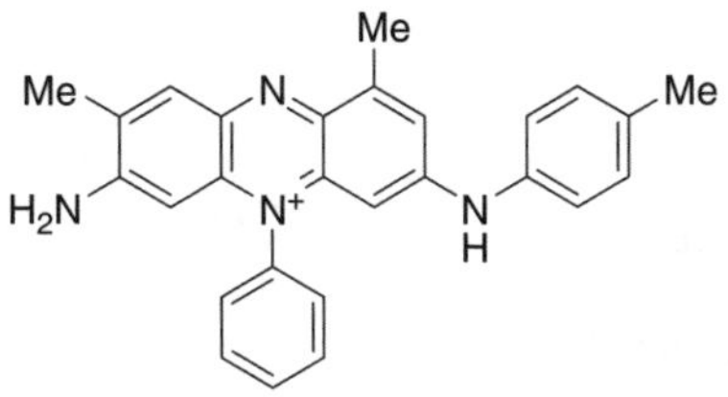

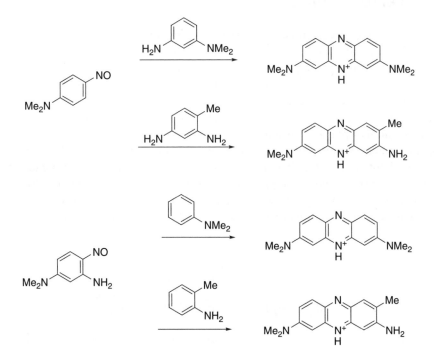

Figure 4.13 Structure of mauveine, after Meth-Cohn and Smith (1994).

dyestuffs were produced via the oxidation of anilines – the most famous example obviously being that of Perkin's mauve (Figure 4.13). The danger here is that oxidation is difficult to control unless the reagents involved are carefully selected. However, unlike Perkin, 21st century chemists know that the structure they are trying to produce is a phenazine, rather than a quinoline! It should also be appreciated that peripheral substituents may not stand up to treatment with strong oxidants (see *Phenothiaziniums* section of this chapter).

As with other related azinium-type photosensitisers, the synthesis of neutral red and its derivatives is oxidative. However, the simple derivatives may be prepared quite simply, usually from a nitroso-containing starting material (Figure 4.14). Plainly, there is considerable scope here for derivatisation (yet again mainly unexplored!), such as

Figure 4.14 Classical synthesis of phenazinium dyes.

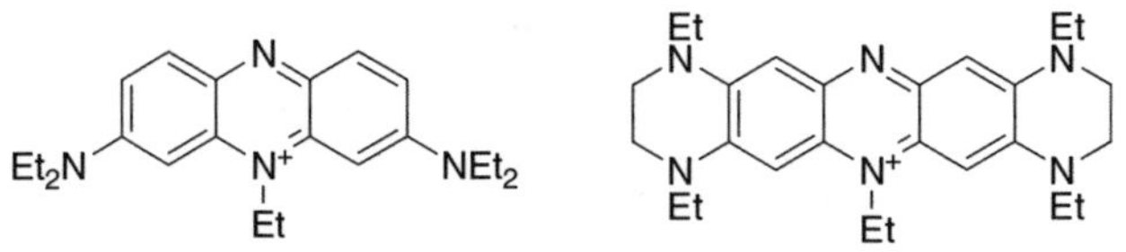

λ_{max} 569 nm
θ_Δ 0.04

λ_{max} 606 nm
θ_Δ 0.36

Figure 4.15 Increased wavelength and singlet oxygen quantum yield in pentacyclic phenazinium derivatives.

increasing the hydrophilic nature of the molecule via hydroxylation of the auxochromic alkyl groups (Proevska, Ignatova-Avramova and Pojarlieff 1993).

One of the problems with established phenazinium compounds is that, in most cases, the associated λ_{max} values lie outside (short) of the therapeutic window normally expected for photodynamic applications. However, both extension of the chromophore and rigidification of the amine auxochromic groups result in increased λ_{max} values and singlet oxygen yields, for example the neutral red analogues shown in Figure 4.15 (Gloster, Cincotta and Foley 1999).

4.7 Phenoxaziniums

Within this group, closely related to the MB series, there are two examples that are particularly important in relation to anticancer and antimicrobial action. These are Capri blue and Nile blue (Figure 4.16), somewhat exotic tags denoting the brilliant intensity of the associated blue solutions. However, neither compound is a significant photosensitiser, and the heavy atom effect must be relied on in order to produce such behaviour. Plainly, ring heteroatom substitution of sulphur or selenium for oxygen furnishes the phenothiazinium or phenoselenazinium families, which are dealt with elsewhere.

The importance of compounds Capri blue and Nile blue, and closely related analogues such as brilliant cresyl blue, lies in their use as biological stains (thus demonstrating biological selectivity). For example, Nile blue was shown to be selective for animal tumours around 60 years ago (Lewis, Goland and Sloviter 1949) and was subsequently used as a model also for antimalarial (Eveland and Allen 1972) and antituberculosis drug

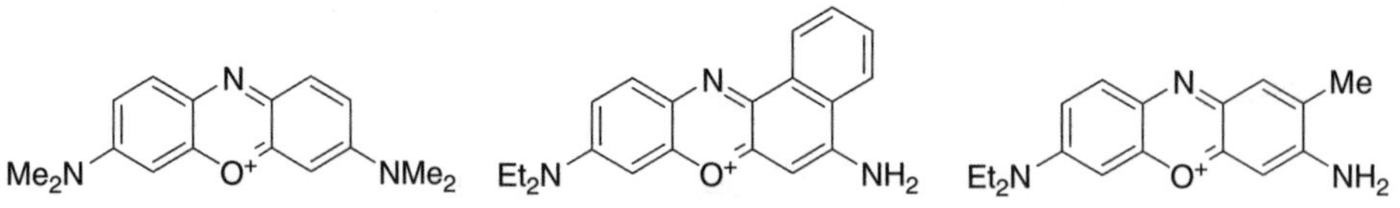

Capri blue   Nile blue   Brilliant cresyl blue

Figure 4.16 Lead phenoxazinium compounds.

research (Crossley *et al.* 1952a1952a, 1972b). Renewed interest in photosensitising compounds built on the earlier work and the use of the heavy atom effect, as mentioned, led to highly effective PDT agents (Cincotta, Foley and Cincotta 1987). Recent work has extended the active spectrum to include photobactericidals (O'Riordan *et al.* 2007).

Various substitution patterns were tested with both bromine and iodine as the heavy atom (Figure 4.17), the resulting derivatives offering limited increase in singlet oxygen yield unless disubstituted at positions 6 and 8 (Cincotta, Foley and Cincotta 1987). Interestingly, iodination in the benzo-fused ring had very little influence, actually decreasing the effect of 6-iodination.

There exists a sound understanding concerning important structure–light absorption relationships in the benzannelated derivatives based on the work of Crossley *et al.* (mentioned above) in the early 1950s, this being an investigation of mainly auxochromic

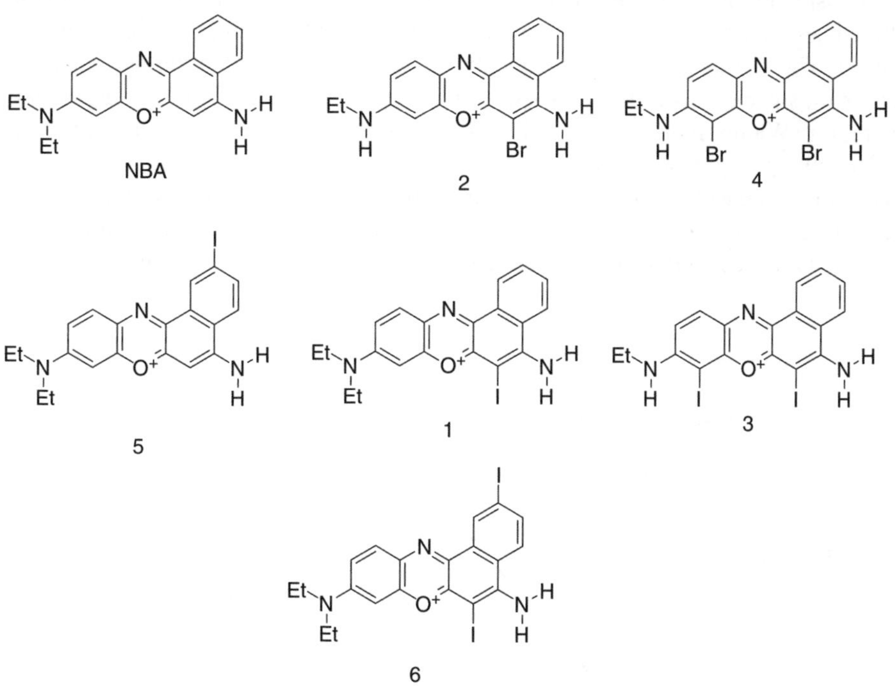

	Halogen	$^1O_2$ yield	λ_{max} (nm)	ε_{max}
NBA	–	0.005	623	75 300
1	6-I	0.036	642	79 000
2	6-Br	0.007	643	80 600
3	6,8-I$_2$	0.500	620	61 000
4	6,8-Br$_2$	0.082	621	67 000
5	2-I	0.008	633	83 200
6	2,6-I$_2$	0.034	650	83 000

Figure 4.17   Halogenated Nile blue derivative photosensitisers.

variants as conventional chemotherapeutics for tuberculosis (Crossley, Hofmann and Dreisbach 1952; Clapp *et al.* 1952; Crossley *et al.* 1952).

As expected, increasing alkyl chain length in the auxochromic moiety was associated with slight increments in maximum absorption wavelength and concomitant increases in lipophilicity. Interestingly, the use of *N*-aryl groups at position 5 of the benzo[*a*]phenoxazine nucleus provided a number of highly active antimycobacterial compounds, having equal or greater activity against *Mycobacterium bovis* when compared to streptomycin. Moreover, a simple plot of side chain carbon content against antimycobacterial activity (relative to that of streptomycin) demonstrates a normal, bell-shaped activity/log*P* model (Figure 4.18).

As mentioned, this work was extended to the field of photosensitiser development by Foley and co-workers. Recourse to heavy atom inclusion was required for measurable photosensitising behaviour in the new derivatives produced, typically using iodination, although sulphur and selenium analogues were also synthesised (see *Phenothiaziniums* section). The use of *N*-arylamino substitution as an auxochromic variant appears less successful, since this is usually associated with the disappearance of any singlet oxygen generation (Wainwright, Grice and Pye 1999). The drug development potential of the benzylamino group should not be ignored, however.

The holistic consideration of photosensitiser molecules produced by such manipulation is essential, since the inclusion of a heavy atom is often associated with increased lipophilicity in the resultant derivatives. Thus, where a phenoxazinium or benzo[*a*]phenox-

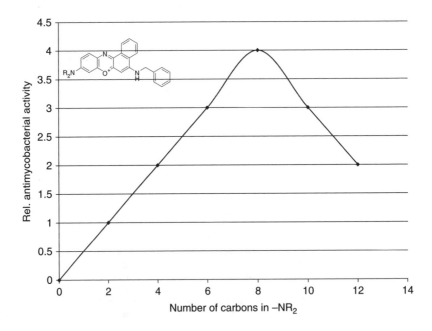

Figure 4.18 Lipophilicity (auxochromic carbon content) *versus* activity* in some Nile blue derivatives (*activity versus *Mycobacterium bovis* relative to streptomycin = 1.0; Crossley *et al.* 1952a).

azinium derivative may be active, for example as an antimycobacterial *in vitro*, the uptake and localisation of its 6-iodo analogue may be considerably different due to increased lipophilicity. Such iodination may also affect amino function basicity (though not normally to such an extent), which in turn may affect drug concentration in the target cell.

Synthesis

Phenoxazinium compounds (Figure 4.19) and their benzannelated derivatives (Figure 4.20) are much more straightforward in synthesis than their sulphur or selenium congeners. Generally, this is due to the availability of suitably functionalised aminophenols, i.e. the oxygen function does not require introduction *de novo*, as with the sulphur of selenium analogues. In addition, the nitrosation of aminophenols and naphthols is relatively simple. Iodination or bromination may be carried out after chromophore synthesis (Foley and Cincotta 1990).

A considerable amount of work carried out by Moura and co-workers has realised a number of novel derivatives in this area. Based on the benzo[*a*]phenoxazinium nucleus,

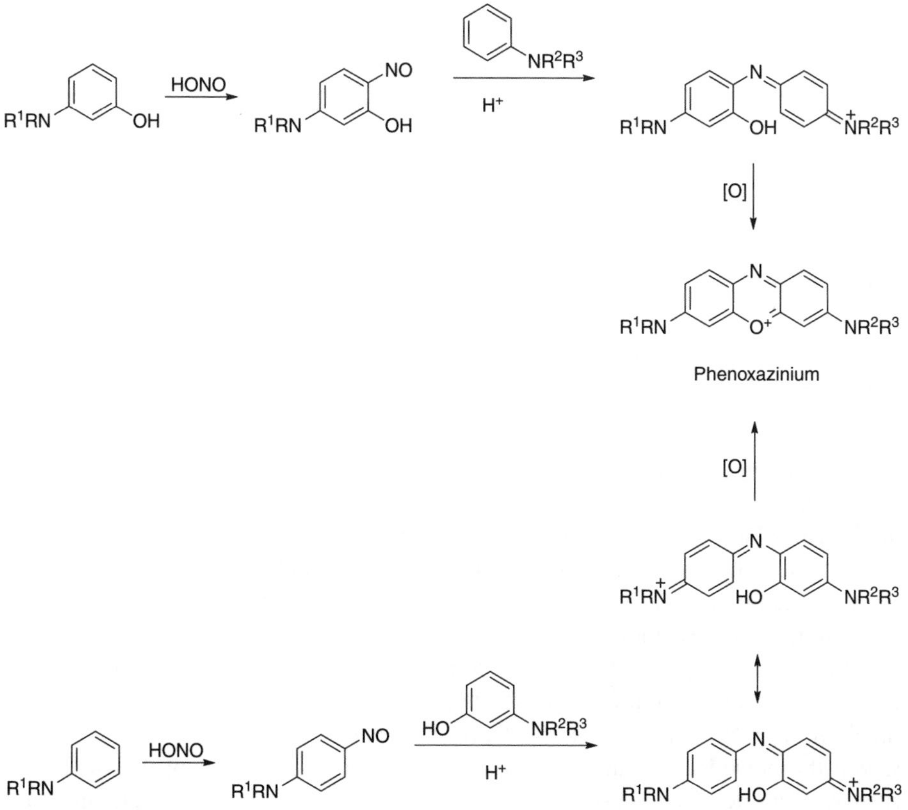

Figure 4.19 Phenoxazinium synthesis via nitrosation of either the aminophenol or the aniline precursor.

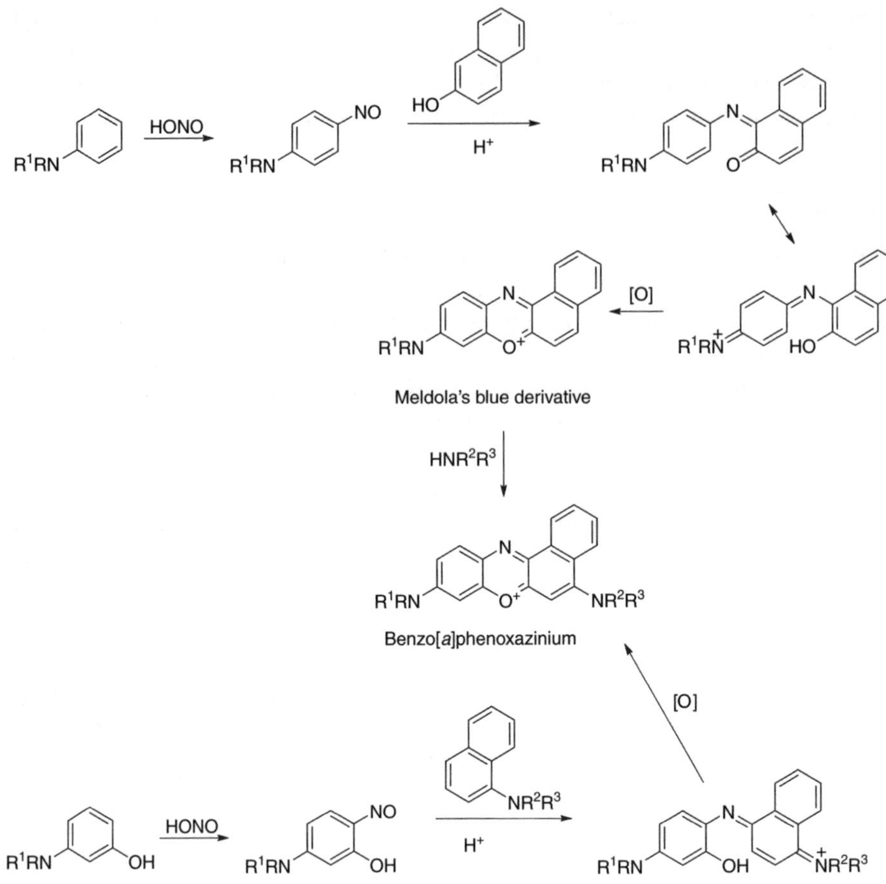

Figure 4.20 Benzo[a]phenoxazinium synthesis via nitrosation of either the aminophenol or the aniline precursor with intermediacy of the Meldola's blue derivative.

the derivatives are not photosensitisers but have been designed as probes and derivatising agents, utilising the high fluorescence yields associated with this chromophore (Frade *et al.* 2007a). Fluorescence maxima are usually in the region 640–660 nm, the molecules being constructed via the standard arylnitroso/1-aminonaphthalene derivative condensation route (Figure 4.20). Probe targeting may be achieved by the inclusion of amino acid residues in the naphthyl moiety.

The fluorescence λ_{max} may be increased by the replacement of the aminonaphthalene derivative by an aminoanthracene analogue (Figure 4.21). Resulting derivatives exhibit a bathochromic shift of approximately 20 nm (Frade *et al.* 2008).

Given the considerable antimycobacterial activities of the benzo[a]phenoxazinium derivatives reported by Crossley *et al.* (mentioned above), it is not surprising that the amino acid derivatives are also antimicrobial (Frade *et al.* 2007b). The results from both of these studies should encourage photoantimicrobial synthesis.

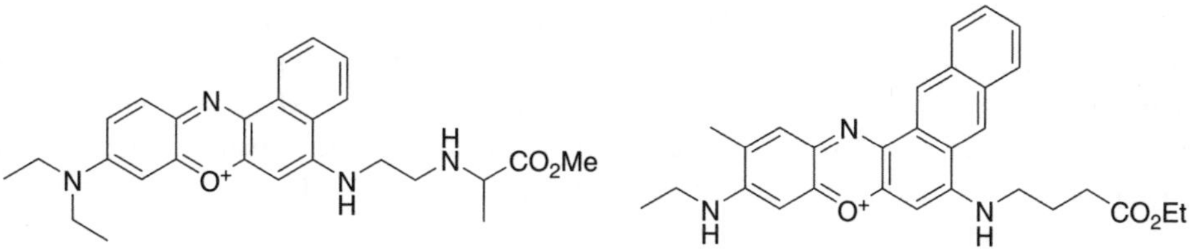

Figure 4.21 Fluorescent probes based on the benzo[*a*]phenoxazinium and naphtho[2,3-*a*]phenoxazinium chromophores.

4.8 Phenothiazinium derivatives

The original phenothiazinium dyes, thionin and MB, were first reported in the late nineteenth century. Both Caro (MB) and Lauth (thionin, thus *Lauth's violet*) are key figures in the discovery of aniline dyes. In both examples, the synthesis entailed the oxidation of an intended substance that was in fact impure – the incidence of sulphur in coal tar often made this inevitable. Thus, oxidation of aniline in the presence of sulphur gave rise to coloured species, among them the first crude phenothiazinium dyes.

The contemporary, flourishing growth of organic chemistry was mirrored in various areas of what is now known as biomedicine, via the involvement of scientists such as Romanovsky, Koch and Ehrlich, and the use of phenothiazinium dyes in specimen staining for microscopy. Ehrlich developed the idea of structure–activity relationships in stains in this era and thereby laid the foundations for chemotherapy and modern medicinal chemistry.

Ehrlich developed his ideas on the selective uptake of dyes. He used dyes as therapeutics and actually achieved clinical success in the cure of malaria, although this involved only two patients, using MB (Guttmann and Ehrlich 1891). As mentioned elsewhere in this book, Ehrlich's efforts led to many breakthroughs in drug discovery, and, such is the versatility of MB, his favoured dye is still employed in medicine and often as a lead compound in research.

Due to Ehrlich's research, MB was undoubtedly tested in a much wider range of disease than (perhaps) it would merit today. However, in the early years of the last century, there were few safe materials for drug use. Possibly as a consequence, phenothiazinium salts have been shown to offer more in terms of the therapy of disease states than other dye types – and certainly more so than other photosensitisers. MB and toluidine blue have been lead compounds in drug research against bacterial infection – local (Deeds, Stockton and Thomas 1939; Zeina *et al.* 2001), tuberculosis (DeWitt 1913) and rickettsial (Peterson and Fox 1947), tropical disease – trypanosomiasis (T'ung 1938) and malaria (Guttmann and Ehrlich 1891; Vennerstrom *et al.* 1995) – yeast infection (Wilson and Mia 1994), viral blood colonisation (Wainwright 2002) and cancer (Williams *et al.* 1989; Link 1999). In addition, the reduced-phenothiazine drugs, again developed from observations on MB, remain an important part of neuroleptic (antipsychotic) therapy (Kristiansen 1989). Among dye types, only the acridines approach such versatility (Wainwright 2001).

The increasing acceptance of PDT and the more recent rediscovery of useful photo-sensitiser activity in the antimicrobial milieu have led to greater legitimacy in expanding the research effort in terms of photosensitiser design. For example, MB is currently in use in several regions for blood plasma photodecontamination. In the expanded field of total blood product disinfection, an ideal candidate photosensitiser would need to be effective in the inactivation of bacteria, viruses, yeasts and protozoa, while being essentially non-toxic and non-mutagenic in a human recipient. It is hardly surprising that none of the currently available agents (including MB) is completely satisfactory against all of these criteria.

Concerning photosensitisers for both PDT and the related antimicrobial approach, where phenothiazinium derivatives have been employed, there has been a reliance on commercial phenothiazinium stains (Table 4.2), rather than novel molecules. This demonstrates the lack of involvement of chemists as contemporary researchers in photosensitiser development. Many PDT groups have investigated the behaviour of MB or occasionally toluidine blue, but where novel analogues have been prepared these are usually auxochromic side-chain derivatives (at positions 3 and 7 of the ring). This approach to photosensitising drug discovery does not reflect conventional research in the pharmaceutical industry: lead structure optimisation requires consideration of the whole molecule, not merely selected, straightforward variations. Cellular uptake is determined by a combination of charge type/distribution and lipophilicity, both of which characteristics may be controlled by informed synthesis.

The long-wavelength absorption of phenothiazinium dyes generally occurs in the region 590–660 nm, depending on structure. Significant factors in shifting this to the longer wavelength include amino auxochrome alkylation, chromophore heteroatom

MV

Table 4.2 Established, commercially available phenothiazine-based photosensitisers

	R^1	R^2	R^3	R^4	R^7	R^8	R^9
Thionin	H	H	NH_2	H	NH_2	H	H
Azure A	H	H	NH_2	H	NMe_2	H	H
Azure B	H	H	NHMe	H	NMe_2	H	H
Azure C	H	H	NHMe	H	NH_2	H	H
Methylene blue	H	H	NMe_2	H	NMe_2	H	H
New methylene blue	H	Me	NHEt	H	NHEt	Me	H
Dimethyl methylene blue	Me	H	NMe_2	H	NMe_2	H	Me
Toluidine blue O	H	Me	NMe_2	H	NH_2	H	H
Methylene green	H	H	NMe_2	NO_2	NMe_2	H	H
Methylene violet							

substitution and benzannelation. Thus, for example, thionin having no methyl groups on its auxochromes absorbs at 590 nm, whereas the tetramethylated MB absorbs at 660 nm, seleno-Nile blue (658 nm) absorbs at 26 nm longer than the parent molecule and the thionated Nile blue derivative EtNBS absorbs at 652 nm, compared to Azure B at 648 nm. As examples of the odd-alternant system, small-wavelength changes can also be made in the phenothiazinium series by ring substitution.

In addition to the suitability of light absorption behaviour, photosensitising potential (Table 4.3) and significant historical use, the phenothiazinium class has other associated facets that may be adapted to use in the biological milieu.

4.9 Reduction–oxidation activity

Photosensitisation involves the promotion of electrons to excited states and, in the case of the type I mechanism, involves electron transfer reactions with other molecules. The phenothiazinium chromophore is well suited to this behaviour, and there is a considerable body of evidence regarding the electrochemistry of MB and its congeners (Tani, Thomson and Butt 2001). However, the major cell killing agent in both PDT and PACT appears to be singlet oxygen.

It should be noted that facile redox capability may be something of a double-edged sword. The ease of reduction of some phenothiazinium salts, while useful, e.g. in decolorisation tests for bacteria, may be deleterious to photodynamic action, i.e. the reduced (leuco) form is generally colourless and a poor photosensitiser. This may be inhibited by the steric crowding of N-10 in the structure, e.g. methyl groups at positions 1 and/or 9 appear to inhibit reduction (Wainwright *et al.* 1997b). However, the reduced (10H-phenothiazine) form, being neutral and lipophilic, may be more easily absorbed by cells and may in turn undergo simple oxidation once inside. Such a scenario has been suggested for the uptake of MB by the malarial parasite (Buckholz *et al.* 2008).

4.10 Hydrophilicity–lipophilicity

The most commonly used phenothiazinium photosensitiser, MB, is highly water soluble (hydrophilic) and this is reflected in a short pharmacological half-life. Such hydrophilic nature is a barrier to lipid partitioning and explains the poor cellular uptake of this and similar compounds. In quantitative laboratory terms, hydrophilic character is associated with poor distribution from water into octanol, the logarithm of the distribution coefficient (log$P$, Table 4.3) for compounds such as MB and toluidine blue having a negative value. Lipophilic character is associated with compounds having log$P > +1.5$, with species in the intermediate range being considered amphiphilic. From a practical viewpoint, highly lipophilic photosensitisers can be problematic in that they do not easily dissolve in aqueous media (i.e. for administration), although this problem may be solved via the use of benign co-solvents or liposomal formulation.

Table 4.3 Photoproperties and log*P* values for commercially available phenothiazine-based photosensitisers

	λ_{max} (nm, aq.)	Relative $^1O_2$ yield	log P
Methylene blue	656	1.00	−0.10
Toluidine blue	625	0.86	−0.21
New methylene blue	630	1.35	+1.20
Dimethyl methylene blue	648	1.21	+1.01

Simple changes such as increasing the alkyl content can endow the phenothiazinium structure (and others) with greater lipophilicity. This may be facilitated either by chromophoric or by auxochromic alteration. However, derivatives with long alkyl chains are often insoluble in water. For example, in a homologous series based on MB reported by Mellish *et al.*, the dimethylamino auxochromes were replaced by diethylamino, dipropylamino, etc., up to dihexylamino. Aqueous solubility was noticeably poor above the dibutylamino analogue, requiring the use of a co-solvent in testing protocols (Mellish *et al.* 2002).

As might be expected, the inclusion of increased aromatic character, whether fused or side chain, generally leads to an increase in lipophilicity. For example, the benzo[*a*]phenothiazinium analogues were reported to have log*P* values in the region of +3 (Figure 4.24, c.f. MB = −0.1), while bis(arylamino)phenothiazinium salts were too insoluble in water to attempt meaningful log*P* measurement (Wainwright, Grice and Pye 1999).

For useful biological activity, a balance of hydrophilic/lipophilic character is usually required (i.e. *amphiphilicity*). For example, Bennetto *et al.* (1981) have shown that the chromophore reduction of commercially available phenothiazinium dyes by *Escherichia coli* was closely related to the hydrocarbon character of the auxochromic moieties. Although none of the commercial phenothiazinium dyes is lipophilic, this study underlines the increased uptake of the series with increasing log*P*, since bacterial reduction relies on entry of the dye into the cell.

4.11   Molecular planarity

The phenothiazinium chromophore is planar, regardless of the substitution pattern of the derivative, whereas the reduced 10*H*-phenothiazine system has a dihedral angle of 115°. In targeting nucleic acid, e.g. in blood disinfection programmes, planarity is obviously an advantage (Dardare and Platz 2002). However, the opposite construction may be applied where potential mutagenicity in transfusion recipients is concerned (see the discussion on potential acridine mutagenicity, above). A representation of the possible overlap of MB, as a typical example, with a cytosine–guanosine base pair is given in Figure 4.22.

Although this may seem to be an impasse, a compromise may be made possible by carefully directed synthesis, since it has been reported that increased alkyl character in the auxochromic moieties of MB furnishes photosensitisers that do not localise in the cell nucleus (Mellish *et al.* 2002).

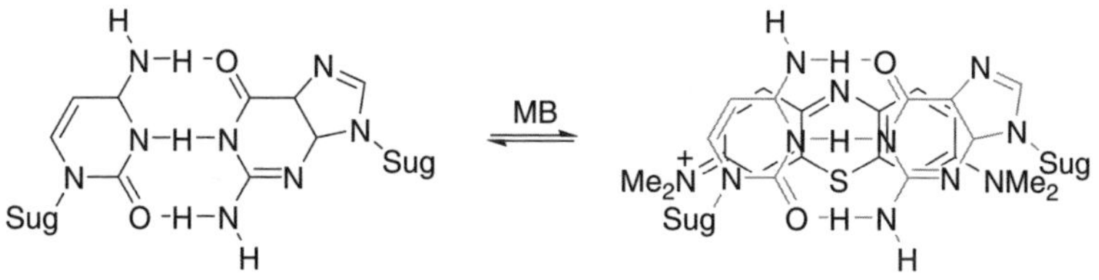

Figure 4.22   Base-pair overlap between cytosine–guanine and methylene blue.

The inclusion of electronically inert (alkyl) groups into the ring system may affect the intercalative properties of the phenothiazinium chromophore. However, this would require the use of relatively large groups, probably C_4, although changes in the inter-calating behaviour of the related aminoacridine chromophore were brought about by bromination alone (Tomosaka *et al.* 1997).

4.12   Benzannelated derivatives

As mentioned in the phenoxazines section above, much has been made of the biological activities of Nile blue, from both the anticancer and the antimicrobial aspects – typically for the azine group – from original observations relating to biological staining. While it is well established that the phenoxazinium chromophore and its benzannelated derivatives are non-photosensitisers unless suitably altered with heavy atoms, this is not the case with the sulphur and selenium analogues, *viz.* the phenothiaziniums, phenoselenaziniums and their respective benzologues. A simple demonstration of this is given by compounds shown in Figure 4.23 as synthesised by the indefatigable Foley (Cincotta, Foley and Cincotta 1987).

As may be noted, the use of mono-iodination of Nile blue A (NBA) provides similar photoproperties in the resulting derivative as those associated with the sulphur analogue, NBS. Iodination of NBS provides a potentially useful photosensitiser. Selenium analogues offer further significant improvement, particularly in the *in vitro* singlet oxygen yield (Figure 4.24), although the value of such comparative data in their extension to mammalian systems is uncertain.

The potential for improved photosensitising properties not to translate *in vivo* usually reflects differences in localisation, binding, etc. As mentioned previously, this may be due to alteration in the hydrophilic–lipophilic balance of the molecule caused by heavy atom introduction. However, similar non-translation has been reported in terms of inherent biological activities. For example, in the Nile blue antimycobacterial project cited earlier, active compounds in the benzo[*a*]phenoxazine group were up to four times as active as streptomycin in culture. (Note; This was inherent antibacterial, rather than photodynamic, activity.) No such activity was seen for the direct sulphur analogues, and this decrease is unlikely to be related to change in lipophilicity, as such analogues are similar in this respect (see log*P*, Figure 4.24). The decreased performance may be due to the greater ease of reduction of the sulphur-containing chromophore.

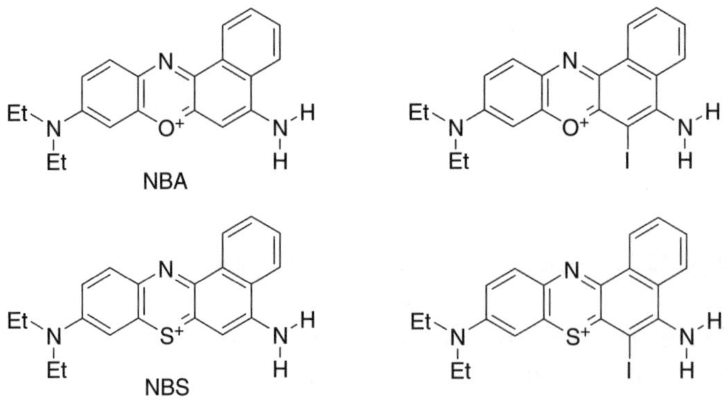

Figure 4.23 Improved photoproperties of benzo[a]phenothiazinium derivatives compared with benzo[a]phenoxazinium congeners.

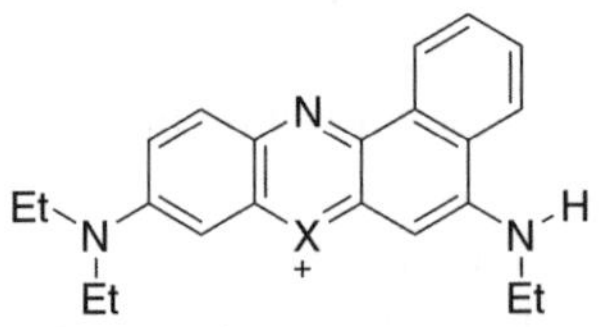

Halogen	$^1O_2$ yield	λ_{max} (nm)	ε_{max}	O/S
NBA	0.005	623	75 300	O
6-I	0.036	642	79 000	O
NBS	0.024	645	67 500	S
6-I	0.170	660	72 600	S

X	$^1O_2$ yield	λ_{max} (nm)	ε_{max}	logP
O	0.005	632	65 800	+2.69
S (EtNBS)	0.025	652	68 600	+2.76
Se	0.650	659	81 900	+2.08

Figure 4.24 Improved singlet oxygen yields *in vitro* for sulphur and selenium analogues of EtNBA (Cincotta, Foley and Cincotta 1987).

Such observations apart, benzo[a]phenothiazinium derivatives developed by Cincotta, Foley and co-workers have proved to be effective in both PDT and PACT applications (Cincotta *et al.* 1994; O'Riordan *et al.* 2007). The subcellular (lysosomal) localisation of EtNBS (Figure 4.24) in tumour cells led to its investigation in a small animal tumour model as a synergist with a benzoporphyrin derivative, i.e. providing a two-pronged attack on the tumour (Cincotta *et al.* 1996). Most recently, the selenium analogues have been reported as effective photobactericidal agents (Foley *et al.* 2006).

4.13 Synthetic approaches

The original, dye industry, synthesis of phenothiazinium dyes employed dichromate oxidation of anilines containing a source of sulphur (Fiertz-David and Blangey 1949). For example, MB itself can be prepared from *N,N*-dimethylaniline, sodium thiosulphate via acid dichromate and then copper(II) oxidation (Figure 4.25). This synthetic approach is a fairly robust one, requiring considerable aqueous solubility in both starting materials and products. Other methods may be required (see below).

Researchers working in this particular area in recent times have used an alternative route to symmetrical MB analogues, utilising the oxidation of 10*H*-phenothiazine with either bromine or iodine. Both routes yield phenothiazinium salts that are active electrophiles at ring positions 3 and 7. These sites undergo facile attack by amines, yielding symmetrical MB derivatives (Figure 4.26). The iodine route also allows the production of asymmetrical derivatives via a stoichiometry-driven 3-*N,N*-dialkylamino intermediate, which is isolated and may be purified before further reaction with a second (different) amine at position 7 (Figure 4.26).

Phenothiazinium drug design requires particular attention when chromophore alteration is desired. Using the oxidative thiosulphonic acid route, it is possible to place substituents at various sites in the ring, and the situation of these depends on the identity of the precursor aniline isomers.

Due to the relative size of any dialkylamino group, it is not possible to utilise *N,N*-dialkylanilines in the synthesis that have groups at the adjacent positions. Adjacent substitution forces the amino group out of co-planarity with the aromatic ring and precludes any reaction at position 4 of the ring. Consequently, direct MB analogues have no substituents at positions 2/4 or 6/8. This leaves only positions 1 and 9 of the ring available, as in Taylor's blue (1,9-dimethyl methylene blue, DMMB). It is noticeable that other directly synthesised phenothiaziniums with substituted rings do not have two dialkylamino functions: toluidine blue has a dialkylamino group at C-3 and an amino function at C-7, next to a methyl group and new MB has two ethylamino groups next

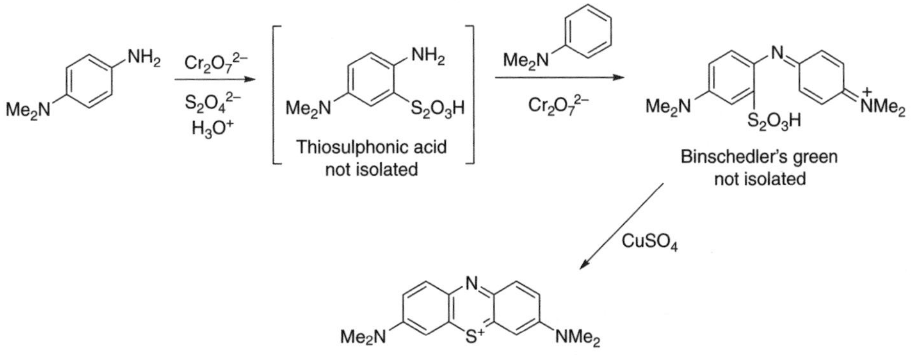

Figure 4.25 Classical methylene blue synthesis.

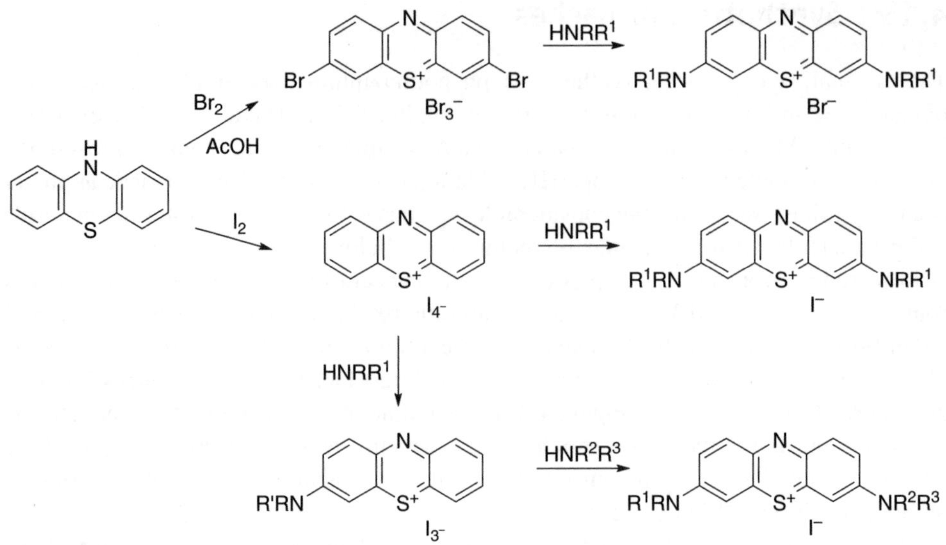

Figure 4.26 Phenothiazinium synthesis via bromine or iodine oxidation.

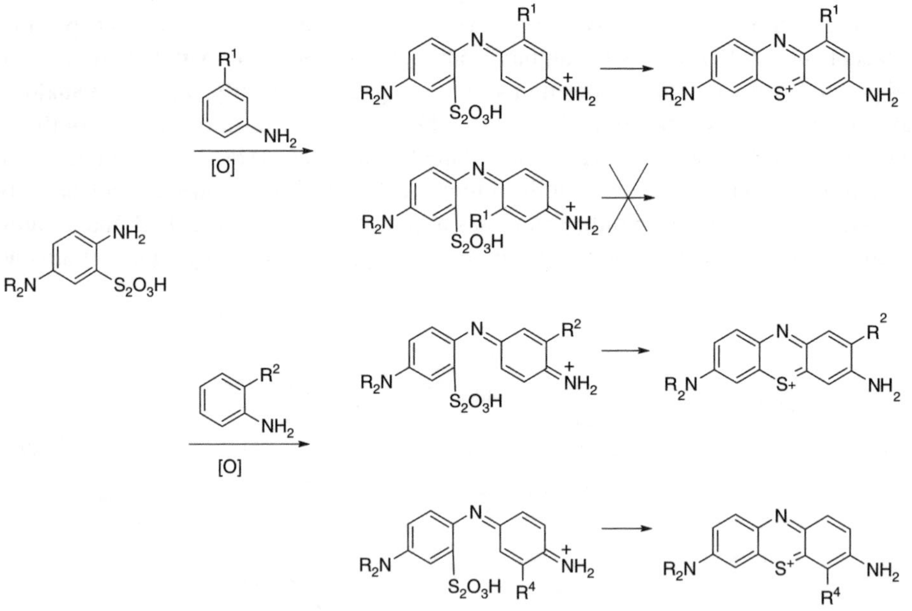

Figure 4.27 Isomerism in phenothiazinium synthesis.

to methyl groups. However, functionalisation of MB itself, as in methylene green (4-nitro MB), is possible.

For the aniline that is coupled to the thiosulphonic acid, again there are limitations. The initial oxidative coupling must occur at position 4 relative to the amine function, and so this must be unsubstituted (Figure 4.27). A 3-substituted aniline would leave the substituent next to the bridging nitrogen with two orientations, only one of which allows the required sulphur ring closure and product formation.

For anilines with substitution adjacent to the amine function, again two orientations are possible in the coupled intermediate. However, in this case, both of these may lead to a ring-closed product, being either a 2- or a 4-substituted phenothiazinium. A standard example here is that of toluidine blue. The aromatics employed are *N,N*-dimethylaniline and 2-methylaniline, and the oxidative synthesis yields a mixture of the 2- and 4-methyl isomers, which may be separated chromatographically. The usual, given structure for toluidine blue is, of course, the 2-methyl isomer.

Another cause of product impurity is the potential for side reactions associated with the generally oxidative nature of the synthesis. Historically, as mentioned, much use has been made of dichromate as an oxidant, but this is potentially damaging to other oxidisable groups present in the aniline precursors. Methyl substitution is an obvious example, since these and other alkyl groups may be oxidised to carboxylic acid, and further cleaved as carbon dioxide, with dichromate. Consequently, if investigating chromophoric functionalisation, it is often preferable to use weaker oxidants, such as silver(I) (Wainwright and Giddens 2003).

Other approaches to the oxidised ring synthesis include the thionation of Bindschedler's green derivatives (without the thiosulphonic acid moiety) with hydrogen sulphide and ferric chloride (Figure 4.28).

The above methodologies allow the synthesis of a wide range of analogues, but simple amino derivatives are not produced in this way. Although it is perhaps surprising that these compounds are still required in photosensitiser research, examples such as thionin are under current investigation for the disinfection of blood plasma (Mohr 2001). While phenothiaziniums having one amino group may be synthesised via the thiosulphate/ oxidation route employing aniline or a substituted derivative, thionin may be produced via the direct nitration of 10*H*-phenothiazine. The 3,7-dinitro derivative is then reduced to the diamino compound conventionally (Figure 4.29), and this is easily oxidised to the product (Fiedeldei 1994).

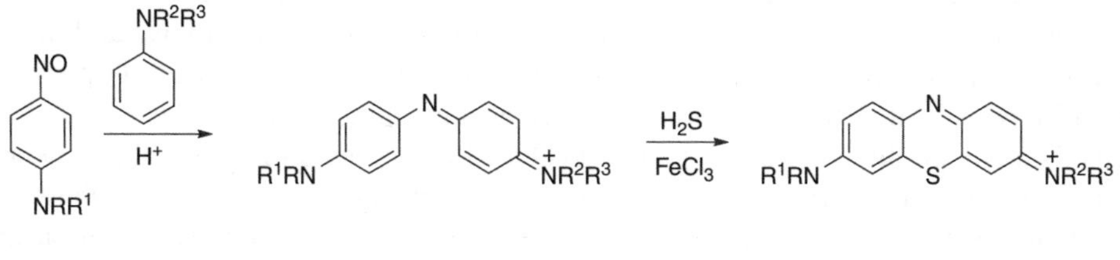

Figure 4.28 Phenothiazinium synthesis from Bindschedler's derivatives.

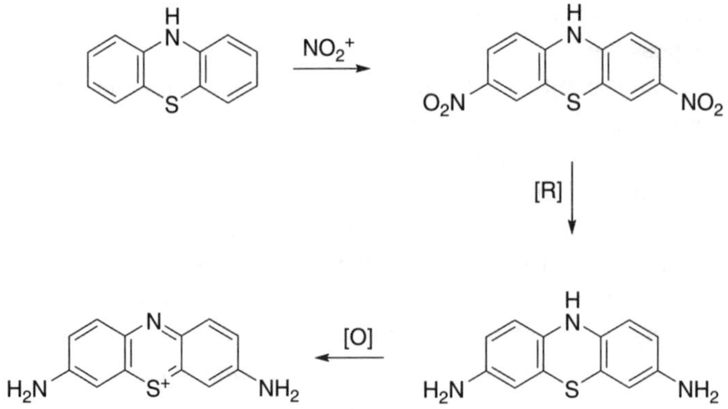

Figure 4.29 Thionin synthesis.

Proper consideration of the aqueous chemistry of potential photosensitising drugs is a necessary part of the development process, as would be the case elsewhere, since all drug molecules have a solution phase at some stage of their usage. Candidates based on the phenothiazinium and related chromophores, while normally considered as cations, often have other possible forms, as a consequence of either metabolism or pH change.

In fact, the overall charge on the molecule may be cationic, anionic or neutral, depending on ring substitution. For example, anionic groups ($-CO_2H$ and $-SO_3H$) may be included in the auxochromic amino functions to give an overall negative charge (Moura, Oliveira-Campos and Griffiths 1997). While this would be useful in preventing nucleic acid intercalation (and thus potential mutagenicity), it would also alter cellular uptake and activity. This approach could also be employed to furnish an overall neutral (zwitterionic) molecule and also has the advantage of conferring negative charge without significant – and probably deleterious – effects on the chromophore, as would be the case with, for example, direct ring sulphonation.

The presence of a delocalised cation as the chromophore can also promote the formation of neutral species where the auxochromic (i.e. resonance-linked) nitrogen has an N–H bond (Figure 4.30). Although there are more examples of commercial phenothiazinium dyes with this feature (e.g. the azures, TBO, thionin, etc.) than without, there is scant reference to the formation of the neutral quinoneimine in the literature (Wainwright 2000). Cellular uptake for neutral species is simpler than for charged species.

A related commercially available derivative here is MV (Table 4.2). This is a neutral species having a double-bonded oxygen in place of one of the auxochromic moieties in MB. Indeed, alkaline hydrolysis of MB yields MV (Adamcikova, Paylikova and Sevcik 2000). A useful demonstration of the improved cellular uptake of neutral species is provided by MV, this being far more effective against intracellular viruses in red blood cells than is MB, the latter being mainly excluded from the cell interior (Skripchenko, Robinette and Wagner 1997). However, this advantage is balanced by the inhibition of action of this photosensitiser (and its halogenated analogues) in the presence of plasma proteins (Houghtaling et al. 2000). Greater aqueous solubility has been attained via the

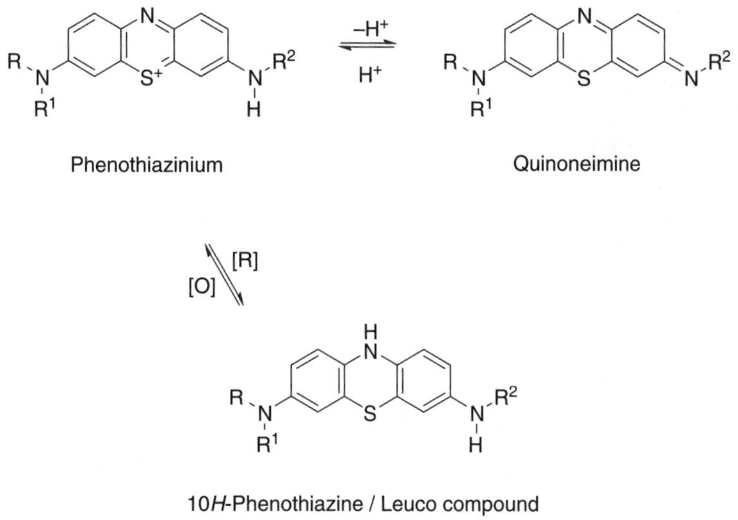

Phenothiazinium   Quinoneimine

10H-Phenothiazine / Leuco compound

Figure 4.30 Neutral species formed by phenothiazinium salts. For the reduction step, H may be an alkyl group.

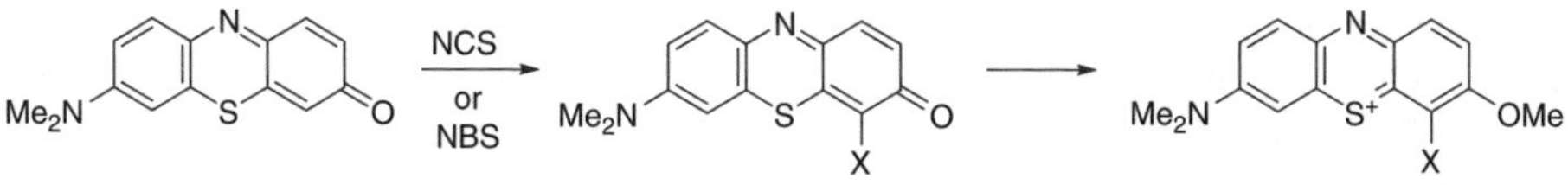

Figure 4.31 Methylene violet and water-soluble/halogenated derivatives (X = Cl, Br or I).

conversion of the neutral oxo-function to alkoxy, regenerating a phenothiazinium salt (Figure 4.31), and halogenated versions exhibited both similar singlet oxygen yields to MB in combination with lowered protein binding compared with the parent MV.

Although the majority of synthetic work in the phenothiazinium field has been aimed at producing improved photosensitisers *per se*, the preparation of macromolecule-linked examples has also been carried out in related areas of biomedical science and is similar to that covered in the porphyrins section of this book (Chapter 6).

The synthesis of phenothiazinium bioconjugates, as with other photosensitiser types, depends on the presence in the side chain of a reactive group (usually carboxylic acid or amino), which will react with the biomolecule under the mild conditions required to maintain biomolecular integrity (ambient temperature, neutral pH and aqueous media). To this end, usually the synthesis of phenothiazinium salts having suitable functionality in the auxochromic side chain is carried out, although attachment through a chromophoric 4-amino group (normally derived from methylene green) is also possible (Moller, Schubert and Cech 1995). Protein attachment via a carboxylic ester function remote from the chromophore has been reported, but the phenothiazinium bearing the carboxylic acid group was synthesised from a suitably substituted aniline via the thiosulphonic acid derivative as described above rather than the carboxyl-bearing analogues made possible via the 10*H*-phenothiazine/iodine method (Motsenbocker *et al.* 1993).

Several of the azine derivatives discussed in the current chapter have been established and widely used biomedical dyes for more than a century. It is thus surprising that there are such a low number of novel derivatives of this class and that published research usually refers to derivatives having auxochromic (side chain) rather than chromophoric elaboration. In consequence, much remains to be discovered about the properties and performance of more complex photosensitisers. This is especially surprising concerning candidates based on the phenothiazinium chromophore. Moreover, it is doubly mystifying that such a situation should pertain, given the relative ease of synthesis of the phenothiazinium and related systems and the range of possible end uses for new photosensitisers. It is to be fervently hoped that new avenues of research emanating from cancer PDT, for example pathogen inactivation in blood products, will continue to encourage the small amount of pioneering work begun in this area in the recent past.

References

Adamcikova L, Paylikova K, Sevcik P. (2000) The decay of methylene blue in alkaline solution. *Reaction Kinetics and Catalysis Letters* **69**: 91–94.

Albert A. (1944) Cationic chemotherapy, with special reference to the acridines. *Medical Journal of Australia* **31**: 245–248.

Albert A, Rubbo SD, Goldacre RJ, Davey ME, Stone JD. (1945) The influence of chemical constitution on antibacterial activity. Part II: a general survey of the acridine series. *British Journal of Experimental Pathology* **26**: 160–192.

Arbiser JL, Moschella SL. (1995) Clofazimine: a review of its medical uses and mechanisms of action. *Journal of the American Academy of Dermatology* **32**: 241–247.

Arutla S, Arra GS, Prabhakar CM, Krishna DR. (1998) Pro- and anti-oxidant effects of some antileprotic drugs *in vitro* and their influence on surperoxide dismutase activity. *Arzneimittel Forschung* **48**: 1024–1027.

Barry VC, Belton JG, Conalty ML, Twomey D. (1948) Anti-tubercular activity of oxidation products of substituted *o*-phenylenediamines. *Nature* **162**: 622–623.

Bennet SL. (2007) Photosensitisation in a cat. *Australian Veterinary Journal* **85**: 375–380.

Bennetto HP, Dew ME, Stirling JL, Tanaka K. (1981) Rates of reduction of phenothiazine redox dyes by *E. coli*. *Chemistry & Industry (London)* **21**: 776–778.

Boibessot I, TurnerCMR, Watson DG, *et al.* (2002) Metabolism and distribution of phenanthridine trypanocides in *Trypanosoma brucei*. *Acta Tropica* **84**: 219–228.

Browning CH, Gulbransen R, Kennaway EL, ThorntonLHD. (1917) Flavine and Brilliant Green, powerful antiseptics with low toxicity to the tissues: their use in the treatment of infected wounds. *British Medical Journal* **i**: 73–78.

Browning CH, Morgan GT, Robb JVM, Walls LP. (1938) The trypanocidal action of certain phenanthridinium compounds. *Journal of Pathology and Bacteriology* **46**: 203–204.

Buckholz K, Schirmer RH, Eubel JK, *et al.* (2008) Interactions of methylene blue with human disulfide reductases and their orthologues from *Plasmodium falciparum*. *Antimicrobial Agents and Chemotherapy* **52**: 183–191.

Cao Y, He X. (1998) Studies of interaction between Safranine T and double helix DNA by spectral methods. *Spectrochimica Acta Part A: Molecular and Biomolecular Spectroscopy* **54**: 883–892.

Chen T, Voelk E, Platz MS, Goodrich RP. (1996) Photochemical and photophysical studies of 3-amino-6-iodoacridine and the inactivation of λ phage. *Photochemistry and Photobiology* **64**: 622–631.

Chuan D, Yu-xia W, Yan-li W. (2005) Study on the interaction between methylene violet and calf thymus DNA by molecular spectroscopy. *Journal of Photochemistry and Photobiology A: Chemistry* **174**: 15–22.

Cincotta L, Foley JW, Cincotta AH. (1987) Novel red absorbing benzo[a]phenoxazinium and benzo[a]phenothiazinium photosensitizers: *in vitro* evaluation. *Photochemistry and Photobiology* **46**: 751–758.

Cincotta L, Foley JW, MacEachern T, Lampros E, Cincotta AH. (1994) Novel photodynamic effect of a benzophenothiazine on two different murine sarcomas. *Cancer Research* **54**: 1249–1258.

Cincotta L, Szeto D, Lampros E, Hasan T, Cincotta AH. (1996). Benzophenothiazine and benzoporphyrin derivative combination phototherapy effectively eradicates large murine sarcomas. *Photochemistry and Photobiology* **63**: 229–237.

Clapp RC, Clark JH, English JP, Fellows CE, Grotz RE, Shepherd RG. (1952) Chemotherapeutic dyes. IV. Phenoxazines and benzo[a]phenoxazines. *Journal of the American Chemical Society* **74**: 1989–1993.

Crossley ML, Hofmann CM, Dreisbach PF. (1952) Chemotherapeutic dyes. III. 5-Heterocyclicamino-9-dialkylaminobenzo[a]phenoxazines. *Journal of the American Chemical Society* **74**: 584–586.

Crossley ML, Dreisbach PF, Hofmann CM, Parker RP. (1952a) Chemotherapeutic dyes. I. 5-Aralkylamino-9-alkylaminobenzo[a]phenoxazines. *Journal of the American Chemical Society* **74**: 573–578.

Crossley ML, Turner RJ, Hofmann CM, Dreisbach PF, Parker RP. (1952b) Chemotherapeutic dyes. II. 5-Arylamino-9-dialkylaminobenzo[a]phenoxazines. *Journal of the American Chemical Society* **74**: 578–584.

Crowther D, Melnick JL. (1961) The incorporation of neutral red and acridine orange into developing poliovirus particles making them photosensitive. *Virology* **14**: 11–21.

Dardare N, Platz MS. (2002) Binding affinities of commonly employed sensitizers of viral inactivation. *Photochemistry and Photobiology* **75**: 561–564.

Das S, Kumar GS. (2008) Molecular aspects on the interaction of phenosafranine to deoxyribonucleic acid: model for intercalative drug – DNA binding. *Journal of Molecular Structure* **872**: 56–63.

Deeds F, Stockton AB, Thomas JO. (1939) Studies on phenothiazine VIII. Antiseptic value of phenothiazine in urinary tract infections. *Journal of Pharmacology and Experimental Therapeutics* **65**: 353–371.

Delic J, Coppey J, Magdelenat H, Coppey-Moisan M. (1991) Impossibility of acridine orange intercalation in nuclear DNA of the living cell. *Experimental Cell Research* **194**: 147–153.

DeWitt LM. (1913) Preliminary report of experiments in the vital staining of tubercles. *Journal of Infectious Diseases* **12**: 68–92.

Dickens F, McIlwain H. (1938) Phenazine compounds as carriers in the hexosemonophosphate system. *Biochemical Journal* **32**: 1615–1625.

Dzekunov SM, UrsosLMB, Roepe PD. (2000) Digestive vacuolar pH of intact intraerythrocytic *P. falciparum* either sensitive or resistant to chloroquine. *Molecular and Biochemical Parasitology* **110**: 107–124.

Eveland LK, Allen EG. (1972) Nile blue stain: *Plasmodium berghei* and uninfected erythrocytes. *Transactions of the Royal Society of Tropical Medicine and Hygiene* **66**: 512–513.

Ferguson LR, Borstel RC. (1992) Induction of the cytoplasmic 'petite' mutation by chemical and physical agents in *Saccharomyces cerevisiae*. *Mutation Research* **265**: 103–148.

Fiedeldei U. (1994) 4,302,013.

Fiertz-David HE, Blangey L. (1949) *Fundamental Processes of Dye Chemistry*, Interscience Publishers, New York, 311.

Foley JW, Cincotta L. (1990) US Patent 4962197.

Foley JW, Song X, Demidova TN, Jilal F, Hamblin MR. (2006) Synthesis and properties of benzo[*a*]phenoxazinium chalcogen analogues as novel broad-spectrum antimicrobial photosensitizers. *Journal of Medicinal Chemistry* **49**: 5291–5299.

Frade VHJ, Barros SA, Moura JCVP, Gonçalves MST. (2007a) Fluorescence derivatisation of amino acids in short and long-wavelengths. *Tetrahedron Letters* **48**: 3403–3407.

Frade VHJ, Sousa MJ, Moura JCVP, Gonçalves MST. (2007b) Synthesis, characterisation and antimicrobial activity of new benzo[*a*]phenoxazine based fluorophores. *Tetrahedron Letters* **48**: 8347–8352.

Frade VHJ, Sousa MJ, Moura JCVP, Gonçalves MST. (2008) Synthesis of naphtho[2,3-a]phenoxazinium chlorides: structure–activity relationships of these heterocycles and benzo[*a*]phenoxazinium chlorides as new antimicrobials. *Bioorganic & Medicinal Chemistry* **16**: 3274–3282.

Fukunaga M, Mizuguchi Y, Yielding LW, Yielding KL. (1984) Petite induction in *Saccharomyces cerevisiae* by ethidium analogs action on mitochondrial genome. *Mutation Research* **127**: 15–21.

Gardner PR. (1996) Superoxide production by the mycobacterial and pseudomonad quinoid pigments phthiocol and pyocyanin in human lung cells. *Archives of Biochemistry and Biophysics* **333**: 267–274.

Gloster DF, Cincotta L, Foley JW. (1999) Design, synthesis, and photophysical characterization of novel pentacyclic red shifted azine dyes. *Journal of Heterocyclic Chemistry* **36**: 25–32.

Goldstein H, Kopp W. (1928) Syntheses in the class of carbazines. VI. Some derivatives of *C*-diethylcarbazine. *Helvetica Chimica Acta* **11**: 486–489.

Guttmann P, Ehrlich P. (1891) Ueber die wirkung des methylenblau bei malaria. *Berliner Klinische Wochenschrift* **39**: 953–956.

Hassan HM, Fridovich I. (1980) Mechanism of the antibiotic action pyocyanine. *Journal of Bacteriology* **141**: 156–163.

Hass BS, Webb RB. (1981) Photodynamic effects of dyes on bacteria: IV. Lethal effects of acridine orange and 460- or 500-nm monochromatic light in strain of *Escherichia coli* that differ in repair capability. *Mutation Research/Fundamental and Molecular Mechanisms of Mutagenesis* **3**: 277–285.

Hiruma H, Katakura T, Takenami T, *et al.* (2007) Vesicle disruption, plasma membrane bleb formation, and acute cell death caused by illumination with blue light in acridine orange-loaded malignant melanoma cells. *Journal of Photochemistry and Photobiology B: Biology* **86**: 1–8.

Houghtaling MA, Perera R, Owen KE, Wagner S, Kuhn RJ, Morrison H. (2000) Photobiological properties of positively charged methylene violet analogs. *Photochemistry and Photobiology* **71**: 20–28.

Ito T. (1973) Some differences between 470 nm and 510 nm in the acridine-orange-sensitized photodynamic actions on yeast cells. *Mutation Research/Fundamental and Molecular Mechanisms of Mutagenesis* **20**: 201–206.

Kaufman RH, Gardner HL, Brown D, Wallis C, Rawls WE, Melnick JL. (1973) Herpes genitalis treated by photodynamic inactivation of virus. *American Journal of Obstetrics and Gynecology* **117**: 1144–1146.

Kristiansen JE. (1989). Dyes, antipsychotic drugs and antimicrobial activity. *Danish Medical Bulletin* **36**: 178–185.

Lewis MR, Goland PP, Sloviter HA. (1949) The action of oxazine dyes on tumors in mice. *Cancer Research* **9**: 736–740.

Li J, Wei YX, Wei YL, Dong C. (2007) Study on the spectral behavior of four fluorescent dyes and their interactions with nucleic acid by the luminescence method. *Journal of Luminescence* **124**: 143–150.

Link EM. (1999) Targeting melanoma with 211 At/131 I-methylene blue: preclinical and clinical experience. *Hybridoma* **18**: 77–82.

Lockwood DNJ. (2005) Leprosy. *Medicine* **33**: 26–29.

Mellish KJ, Cox RD, Vernon DI, Griffiths J, Brown SB. (2002) *In vitro* photodynamic activity of a series of methylene blue analogues. *Photochemistry and Photobiology* **75**: 392–397.

Meth-Cohn O, Smith M. (1994). What did W.H. Perkin actually make when he oxidised aniline to obtain mauveine? *Journal of the Chemical Society, Perkin Transactions* 1, 5–7.

Mileykovskaya E, Dowhan W, Birke RL, Zheng D, Lutterodt L, Haines TH. (2001) Cardiolipin binds nonyl acridine orange by aggregating the dye at exposed hydrophobic domains on bilayer surfaces. *FEBS Letters* **507**: 187–190.

Mohr H. (2001) Methylene blue and thionine in pathogen inactivation of plasma and platelet concentrates. *Transfusion and Apheresis Science* **25**: 183–184.

Moller U, Schubert F, Cech D. (1995) Versatile procedure of multiple introduction of 8-aminomethylene blue into oligonucleotides. *Bioconjugate Chemistry* **6**: 174–178.

Moore C, Wallis C, Melnick JL, Kuns MD. (1972) Photodynamic treatment of herpes keratitis. *Infection & Immunity* **5**: 169–171.

Motsenbocker M, Masuya H, Shimazu H, Miyawaki T, Ichimori Y, Sugawara T. (1993) Photoactive methylene blue dye derivatives suitable for coupling to protein. *Photochemistry and Photobiology* **58**: 648–652.

Moura JCVP, Oliveira-Campos AMF, Griffiths J. (1997) Synthesis and evaluation of phenothiazine singlet oxygen sensitising dyes for application in cancer phototherapy. *Phosphorus, Sulfur and Silicon and the Related Elements* **120–121**: 459–460.

Okuda S, Ito T, Ito K. (1987) Pyocyanine as a potent inhibitor for the growth of *Rhodopseudomonas sphaeroides* B5. *Current Microbiology* **16**: 167–169.

O'Riordan K, Akilov OE, Chang SK, Foley JW, Hasan T. (2007) Real-time fluorescence monitoring of phenothiazinium photosensitizers and their anti-mycobacterial photodynamic activity against *Mycobacterium bovis* BCG in *in vitro* and *in vivo* models of localized infection. *Photochemical and Photobiological Sciences* **6**: 1117–1123.

Peterson OL, Fox JP. (1947) The antirickettsial effect of thionine dyes: I. The use of methylene blue and toluidine blue to combat experimental Tsutsugamushi disease (scrub typhus). *Journal of Experimental Medicine* **85**: 543–558.

Phoenix DA, Harris F. (2003) Phenothiazinium-based photosensitizers: antibacterials of the future? *Trends in Molecular Medicine* **9**: 283–285.

Proevska LI, Ignatova-Avramova EP, Pojarlieff IG. (1993) *N,N*-dialkylsafranines with hydroxyl groups in the alkyl substituents. *Dyes and Pigments* **21**: 13–21.

Raab OZ. (1900) Uber die wirking fluoreszierender stoffe auf infusorien. *Zeitschrift Biologie* **39**: 524–546.

Riley JF. (1948) Retardation of growth of a transplantable carcinoma in mice fed basic metachromic dyes. *Cancer Research* **8**: 183–188.

Rodriguez ME, Azizuddin K, Zhang P, *et al.* (2008) Targeting of mitochondria by 10-N-alkyl acridine orange analogues: role of alkyl chain length in determining cellular uptake and localization. *Mitochondrion* **8**: 237–246.

Selby W, Pavli P, Crotty B, *et al.* (2007) Two-year combination antibiotic therapy with clarithromycin, rifabutin, and clofazimine for Crohn's disease. *Gastroenterology* **132**: 2313–2319.

Shah A, Bhagat R, Panchal N. (1993) Resistant tuberculosis: successful treatment with amikacin, ofloxacin, clofazimine and PAS. *Tubercle and Lung Disease* **74**: 64–65.

Skripchenko A, Robinette D, Wagner SJ. (1997) Comparison of methylene blue and methylene violet for photoinactivation of intracellular and extracellular virus in red cell suspensions. *Photochemistry and Photobiology* **65**: 451–455.

Sweeney AW. (2003) *Malaria Frontline. Australian Army Research During World War II,* Melbourne University Press.

Tani A, Thomson AJ, Butt JN. (2001) Methylene blue as an electrochemical discriminator of single- and double-stranded oligonucleotides immobilized on gold substrates. *Analyst* **126**: 1756–1759.

Tatsuta M, Yamamura H, Yamamoto R, *et al.* (1984) Destruction of implanted gastric tumors in rats by acridine orange photoactivation with an argon laser. *European Journal of Cancer and Clinical Oncology* **20**: 543–552.

Tomosaka H, Omata S, Hasegawa E, Anzai K. (1997) The effects of substituents introduced into 9-aminoacridine on frameshift mutagenicity and DNA binding affinity. *Bioscience, Biotechnology and Biochemistry* **61**: 1121–1125.

Tomson S. (1976) Laser destruction of photosensitized superficial malignant tumours. *Optics & Laser Technology* **8**: 81–84.

T'ung T. (1938) *In vitro* photodynamic action of methylene blue on *Trypanosoma brucei. Proceedings of the Society of Experimental Biology and Medicine* **38**, 29–35.

Uggla AH, Sundell-Bergman S. (1990) The induction and repair of DNA damage detected by the DNA precipitation assay in Chinese hamster ovary cells treated with acridine orange + visible light. *Mutation Research* **236**: 119–127.

VanDuijn C. (1961) Photodynamic effects of vital staining with diazine green (Janus green) on living bull spermatozoa. *Experimental Cell Research* **25**: 120–130.

VanRensburg CEJ, Anderson R, Myer MS, Joone GK, O'Sullivan JF. (1994) The riminophenazine agents clofazimine and B669 reverse acquired multidrug resistance in a human lung cancer cell line. *Cancer Letters* **85**: 59–63.

Vennerstrom JL, Makler MT, Angerhofer CK, Williams JA. (1995). Antimalarial dyes revisited: xanthenes, azines, oxazines, and thiazines. *Antimicrobial Agents and Chemotherapy* **39**: 2671–2677.

Wainwright M. (2000) The use of methylene blue derivatives in blood product disinfection. *International Journal of Antimicrobial Agents* **16**: 381–394.

Wainwright M. (2001) Acridine – a forgotten antimicrobial chromophore. *Journal of Antimicrobial Chemotherapy* **47**: 1–13.

Wainwright M. (2002) The emerging chemistry of blood disinfection. *Chemical Society Reviews* **31**: 126–136.

Wainwright M. (2003) The use of dyes in modern biomedicine. *Biotechnic and Histochemistry* **78**: 147–155.

Wainwright M, Giddens RM. (2003) Phenothiazinium photosensitisers: choices in synthesis and application. *Dyes and Pigments* **57**: 245–257.

Wainwright M, Grice NJ, Pye LEC. (1999) Phenothiazine photosensitisers. Part II. 3,7-Bis(arylamino) phenothiazines. *Dyes and Pigments* **42**: 45–51.

Wainwright M, Phoenix DA, Marland J, WareingDRA, Bolton FJ. (1997a) *In vitro* photobactericidal activity of aminoacridines. *Journal of Antimicrobial Chemotherapy* **40**: 587–589.

Wainwright M, Phoenix DA, Rice L, Burrow SM, Waring JJ. (1997b) Increased cytotoxicity and phototoxicity in the methylene blue series via chromophore methylation. *Journal of Photochemistry and Photobiology B: Biology* **40**: 233–239.

Wallis C, Melnick JL. (1963) Photodynamic inactivation of poliovirus. *Virology* **21**: 332–341.

Wallis C, Melnick JL. (1964) Irreversible photosensitization of viruses. *Virology* **23**: 520–527.

Wang YT, Zhao FL, Li KA, Tong SY. (1999) Molecular spectroscopic study of DNA binding with neutral red and application to assay of nucleic acids. *Analytica Chimica Acta* **396**: 75–81.

Williams J, Stamp J, Devonshire R, Fowler G. (1989) Methylene blue and the photodynamic therapy of superficial bladder cancer. *Journal of Photochemistry and Photobiology B: Biology* **4**: 229–232.

Wilson M, Mia N. (1994) Effect of environmental factors on the lethal photosensitization of *Candida albicans in vitro. Lasers in Medical Science* **9**: 105–109.

Woese CR, Naono S, Soffer R, Gross F. (1963) Breakdown of messenger ribonucleic acid (RNA). *Biochemical and Biophysical Research Communications* **11**: 435–440.

Zeina B, Greenman J, Purcell WM, Das B. (2001) Killing of cutaneous microbial species by photodynamic therapy. *British Journal of Dermatology* **144**: 274–278.

Zimmermann HW. (1986) Physicochemical and cytochemical investigations on the binding of ethidium and acridine dyes to DNA and to organelles in living cells. *Angewandte Chemie, International Edition* **25**: 115–130.

5

Triarylmethanes and xanthenes

The triarylmethanes and xanthene-type photosensitisers are included together in this chapter by virtue of several common structural factors: simple xanthyliums such as rosamine may equally be considered as oxygen-bridged crystal violet (Figure 5.1) and the fluoresceins as triarylmethanes having oxygen chromophores rather than amino. Certainly, there are significant differences in charge, λ_{max} and singlet oxygen generation.

5.1 Triarylmethanes

As mentioned in various parts of this book, crystal violet is the basis of the Gram stain in microbiology. Taken with the approach concerning positive staining of cells with photosensitisers, it is most surprising that crystal violet has not featured more widely as a lead compound for photobactericidal studies, although the logical lack of inherent (dark) toxicity to Gram-negative organisms has been appreciated since the early twentieth century (Crossley 1919). Crystal (gentian) violet has been used clinically to eradicate methicillin-resistant *Staphylococcus aureus* (MRSA) infection of decubitus ulcers (Saji *et al.* 1995). In addition, a range of derivatives have been reported to be active antibacterials *per se*. Wilson has shown that the activity of crystal violet against bacteria and yeasts is significantly increased on illumination (Dobson and Wilson 1992; Wilson and Mia 1993). Brilliant green and crystal violet have been employed in a standard antiseptic preparation for use in neonatal nurseries. The related dye, brilliant green, is so toxic to a range of common bacteria that it is used to keep agar slopes clear of competition in order to allow the slow growth of mycobacterial species in tuberculosis testing. Brilliant green was also used, along with proflavine, in Browning's pioneering wound therapy programme during World War II (see Chapter 11). Moreover, several triarylmethane examples were among those examined for antitumour activity in the 1940s (Lewis, Goland and Sloviter 1946), and this suggested the testing of other derivatives, such as the Victoria blue (VB) series for

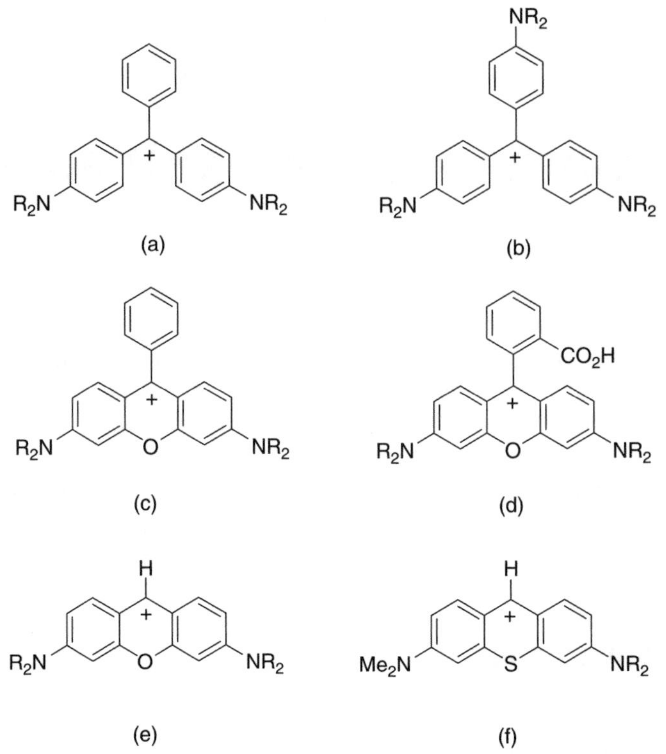

Figure 5.1 General structural-type relationships in triarylmethanes and xanthenes: (a) malachite green, (b) crystal violet, (c) rosamine, (d) rhodamine, (e) pyronin and (f) thiopyronin.

photodynamic therapy (PDT) potential. Several of the VB compounds appear to have some photodynamic utility in human tumour cell lines (Wainwright *et al.* 1999a,b). Plainly there is much scope for derivatisation here, and triarylmethanes must remain of considerable potential in photodynamic therapy.

The basic propeller structure of triphenylmethanes means that, in the simple derivatives at least, the energy accrued from photoexcitation can be dissipated easily via internal conversion. This makes it difficult to examine parameters such as singlet oxygen efficiencies. However, binding and rigidification of the structure in the biological milieu promote the photosensitisation route (Bartlett and Indig 1999).

As noted above, crystal violet, being the most investigated derivative here, has been screened against a wide range of microorganisms. It has also been shown to interact with both DNA and RNA, the former preferentially (Wakelin *et al.* 1981), thus illustrating the rigidification argument above. However, the interaction with nucleic acids is non-intercalative, since the propeller geometry, although altered, is not sufficiently flattened to allow intercalation in a fashion similar to that of the standard *in vitro* intercalator, methylene blue. Such structural conformational changes that occur on binding may be of use in biomolecular targeting, as has been shown by the work of Detty *et al.* on rosamine and pyrylium derivatives (see below).

5.2 Victoria blue series

The VBs are well-known, commercially available dyes and are similar in structure to crystal violet, one of the phenyls being replaced by a 1-naphthyl group with a secondary amino function at position 4 (Figure 5.2). The VBs were included in the original Lewis study (Lewis, Goland and Sloviter 1946), which demonstrated selective uptake into mammalian tumours. Given the increased long-wavelength absorption of dyes such as Victoria blue BO (VBBO) compared to that of crystal violet (612 nm *versus* 595 nm), such compounds were logical choices as anticancer photosensitiser candidates. Although no evidence of singlet oxygen production has been found in *in vitro* tests, the production of superoxide has been demonstrated by electron paramagnetic resonance (Viola *et al.* 1996). VBBO has been shown to be highly photoactive against several tumour cell lines, with a low level of dark toxicity, the combination of delocalised positive charge and high lipophilicity ensuring mitochondrial uptake.

Investigation of structure–activity relationships in the VB series shows that the naphthyl moiety is important for photoactivity, since similarly substituted triphenyl-methane analogues are often much less active in this respect. Compounds containing only tertiary amino functionality are also less active in terms of both uptake and photo-sensitising ability, indicating that the secondary amino group is involved in drug action, as is the case with Nile blue (benzo[*a*]phenoxazinium) derivatives (*q.v.*). The presence of the secondary amino group in the commercial VBs allows simple conversion of the dye cation to the neutral quinoneimine species, the *Homolka base* (Figure 5.3), and this has been used to explain the high activity of VBBO compared with analogues having all-tertiary amino character (Wainwright *et al.* 1999a).

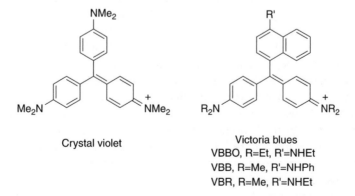

Crystal violet

Victoria blues
VBBO, R=Et, R'=NHEt
VBB, R=Me, R'=NHPh
VBR, R=Me, R'=NHEt

Figure 5.2 Crystal violet and the common Victoria blues.

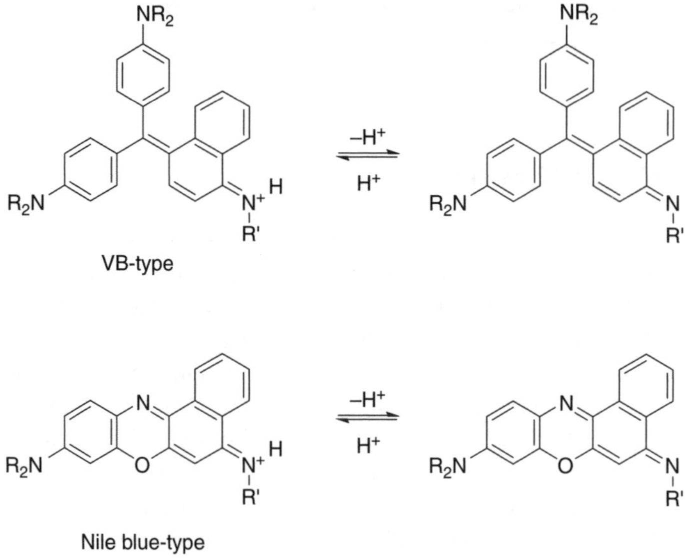

Figure 5.3   Homolka base/quinoneimine formation in Victoria blue and Nile blue derivatives.

5.3   Structure and photoproperties in triarylmethanes – extending the long-wavelength absorption

As has been mentioned, the number of triarylmethane derivatives tested for photodynamic activity is low in comparison to other chromophoric types. However, the enormous range of possible derivatives here demands some discussion in respect of potential. Consequently, the following section is a discussion on structure and function, which provides indications only as to prospective activity. This can be justified since several derivatives are known to bind to biomolecules, several have been shown to be photosensitisers insofar as Type I/Type II activity has been measured *in vitro* (Brezová *et al.* 2004), and several have been shown to be effective photosensitisers in cellular investigations (Wainwright *et al.* 1999a; Burrow *et al.* 2000). Close derivatives of such examples would thus be expected to show similar activities, so perhaps the question should be: Why have more derivatives not been investigated?

The commonly employed triphenylmethanes, such as crystal violet and malachite green, exhibit light absorption in the region 590–620 nm (usually measured in acetic acid). Clearly, this is at the low end of the useful therapeutic range for photodynamic activity where there is likely to be interference by endogenous absorbers. Consequently, considerable research has been carried out with the aim of producing longer-wavelength absorbers. However, it should be remembered that not all photodynamic applications have such a stringent wavelength requirement, for example the phototreatment of non-pigmented blood products, and this widens the range of useful structures. As a

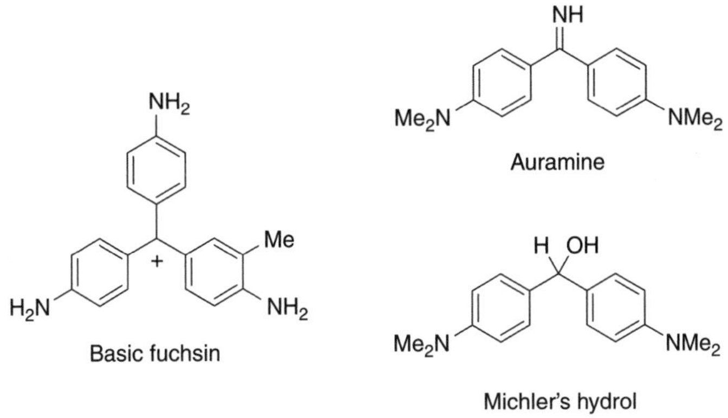

Figure 5.4 Other conventional derivatives of crystal violet.

comparison, the triphenylmethane basic fuchsin (Figure 5.4), due to its primary amino auxochromes, has a λ_{max} of 544 nm, significantly shorter than that of crystal violet. Auramine is a diphenylmethaneimine derivative and absorbs at 434 nm, whereas the hydroxyl analogue Michler's hydrol blue absorbs at 608 nm due to the simple diphenylmethane cation (Figure 5.4).

In line with other photosensitisers, simple substitutions of the auxochrome or chromophoric system usually lead to small wavelength shifts. Thus, the alkyl substituents in the crystal violet or malachite green structure may be lengthened or cyclised – to alter lipophilicity, for example – without causing too much variation in absorption (Figure 5.5).

Similarly, methylation of the chromophoric structure of malachite green causes little change in absorption but, due to steric interactions between pendant groups attached to the separate rings, can increase out-of-plane twisting. This is likely to affect binding to biomolecules, for example nucleic acids, and, in turn, to alter photodynamic efficacy.

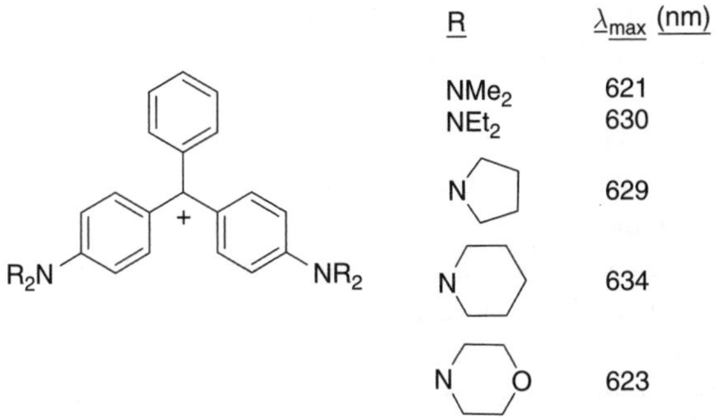

R	λ_{max} (nm)
NMe$_2$	621
NEt$_2$	630
N (pyrrolidine)	629
N (piperidine)	634
N O (morpholine)	623

Figure 5.5 Variation in λ_{max} of malachite green analogues with auxochrome character.

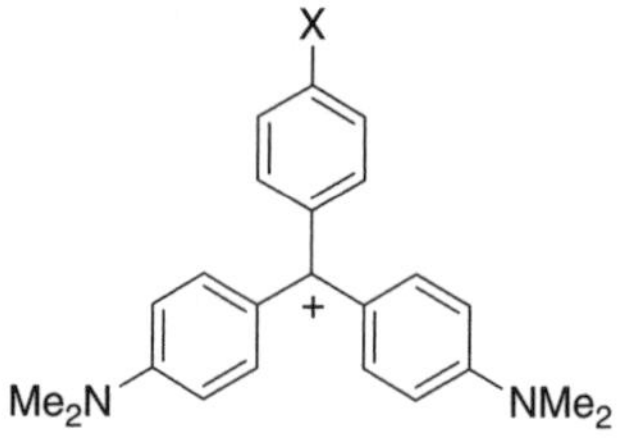

X	λ_{max} (nm)
NO$_2$	645
CF$_3$	634
Br	628
H	621
Me	616
OMe	608

Figure 5.6 Electronic effects in malachite green derivatives.

Conversely, the inclusion of atoms or groups with greater electronic influence (e.g. due to electronegativity) may cause a significant alteration in λ_{max}. For example, the decrease in wavelength may be correlated to increase in electron donation, depending on the substituent position (Figure 5.6).

In the VB series, extension of the π-system of crystal violet, etc. by the inclusion of a 1-naphthyl moiety produces a bathochromic shift of the long-wavelength absorption (to c. 610 nm). The resulting relatively small shifts observed with these compounds are due to the triarylmethane cation assuming a lower-energy, non-planar state by rotation about the bonds to the central carbon atom. Similar derivatives have been synthesised using the 2-naphthyl and various phenanthryl systems (Figure 5.7). While none of the reported singly substituted compounds departs significantly from the lead compound in terms of long-wavelength absorption (mostly around 630 nm), there is plainly a huge opportunity for variation/positioning of functional groups for drug design and structure–activity relationship investigation. In the search for near-infrared absorbers, the use of three naphthyl residues moves the long-wavelength absorption out to around 700 nm for 1-naphthyl and around 800 nm for 2-naphthyl (Figure 5.7, Sanguinet *et al.* 2005). However, there are little data available regarding the photodynamic potential of these candidates, although such activity is likely, given the performance of close congeners among the VB analogues already reported (Burrow *et al.* 2000).

The aromatic ring rotation in triarylmethane cations mentioned above may be restricted by bridging the 2′ and 2″ positions, i.e. allowing only the free aryl ring to rotate (Figure 5.8).

Bridging in this way with an electron-donating heteroatom leads to large hypsochromic shift of the long-wavelength absorption band, for example in the oxygen-bridged analogue of crystal violet. However, whereas no shift is observed for the carbon (isopropylene)-bridged analogue, direct joining across the 2′ and 2″ positions gives rise to considerable differences. The consequential presence of the 9-fluorenyl moiety in the triarylmethane unit leads to bathochromic shifts of the long-wavelength absorption of 60 nm for the crystal violet analogue and 220 nm for the corresponding malachite green compound. In addition, further, new bands are exhibited c. 100 nm into the red (Guinot *et al.* 1999).

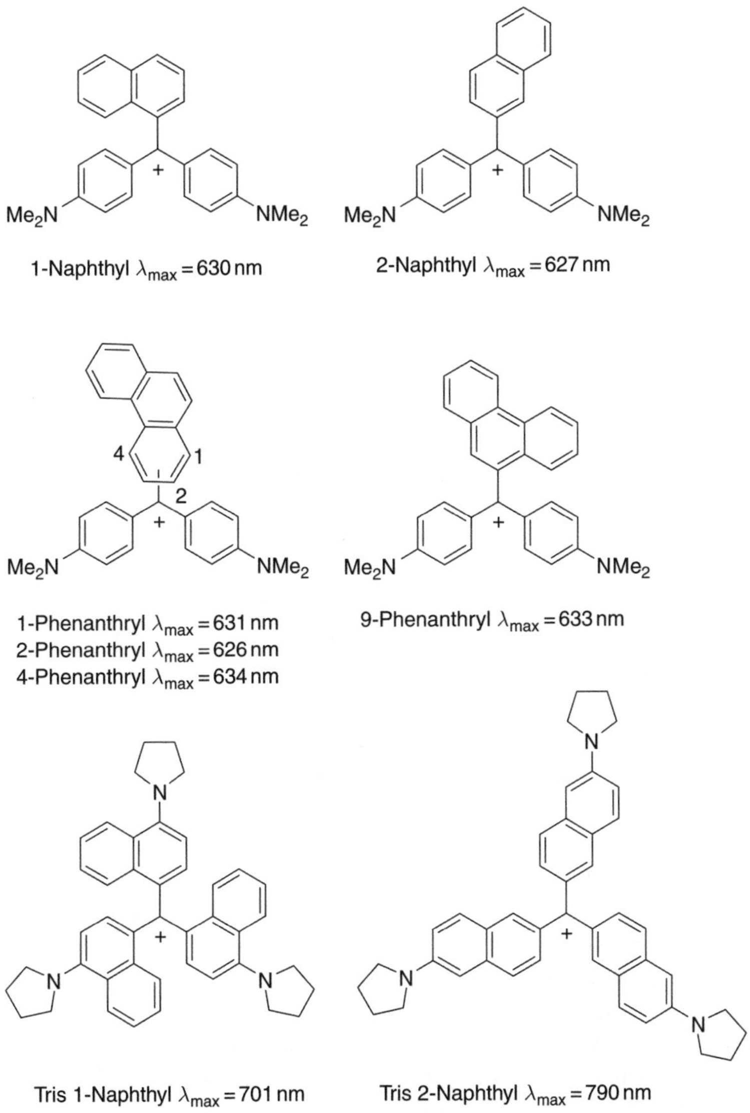

Figure 5.7 Alternative Victoria blue-type structures.

In terms of the long-wavelength absorption required for much of photosensitising drug design, shifts of this type are of considerable consequence, since longer wavelengths may be achieved without recourse to the large increases in molecular volume encountered with the inclusion of extra aromatic rings, benzannelation, etc.

Combination of the fluorenyl and naphthyl modifications have been shown by Guinot *et al.* (1999) to provide near-infrared absorption (Figure 5.9), although the photosensitising abilities of these compounds is unknown.

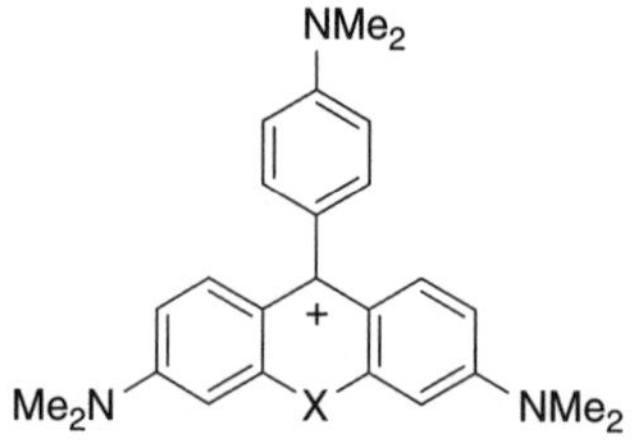

X	λ_{max} (nm)
2H	595
CMe$_2$	608
O	545

Figure 5.8 Effect of bridging on crystal violet long-wavelength absorption – see Figure 5.9 for fluorenyl analogue.

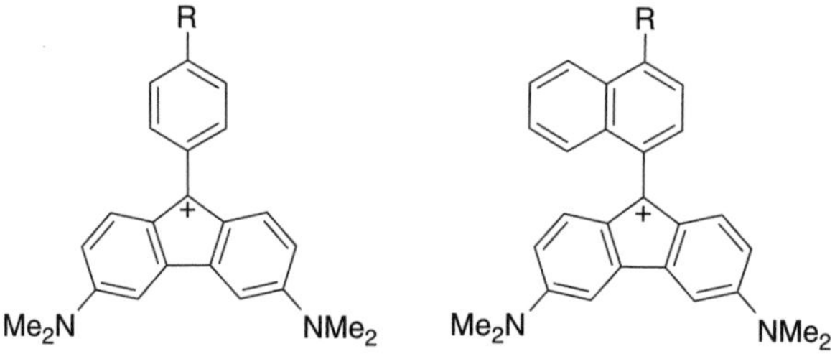

R=H, λ_{max}=955 nm
R=NMe$_2$, λ_{max}=850 nm

R=H, λ_{max}=>900 nm
R=NMe$_2$, λ_{max}=835 nm

Figure 5.9 Long-wavelength absorbers based on the fluorenyl cation.

Extension of the chromophoric system in triarylmethanes has also been achieved via the insertion of unsaturation in between the rings. Thus, both olefinic and acetylenic derivatives have been synthesised, leading to examples that absorb at longer wavelengths compared to the parent compound, e.g. closer to 700 nm for crystal violet and VB analogues (Figure 5.10) (Guinot *et al.* 2000).

Analogues having appended styryl moieties (Figure 5.11) have also been synthesised, the increase in delocalisation of the system leading to long-wavelength absorption in the region of 1000 nm (Sengupta and Sadhukhan 2000).

The activities of various triarylmethanes screened for use in PDT have been related to respirational processes in mitochondria. Thus, the usual VB series (BO, B and R) have been reported to cause uncoupling, whereas crystal violet, ethyl violet and malachite green may produce permeability transitions in the organelle (Kowaltowski *et al.* 1999). In this particular work, all of the triarylmethanes tested inhibited respiration in the presence or absence of photoexcitation, the uptake by mitochondria being in line with earlier findings for simple cationic photosensitisers (Modica-Napolitano *et al.* 1990).

In addition, triarylmethanes will react with nucleophilic species within the cell. This is logical, given the electrophilic centre offered by the photosensitiser that are routinely

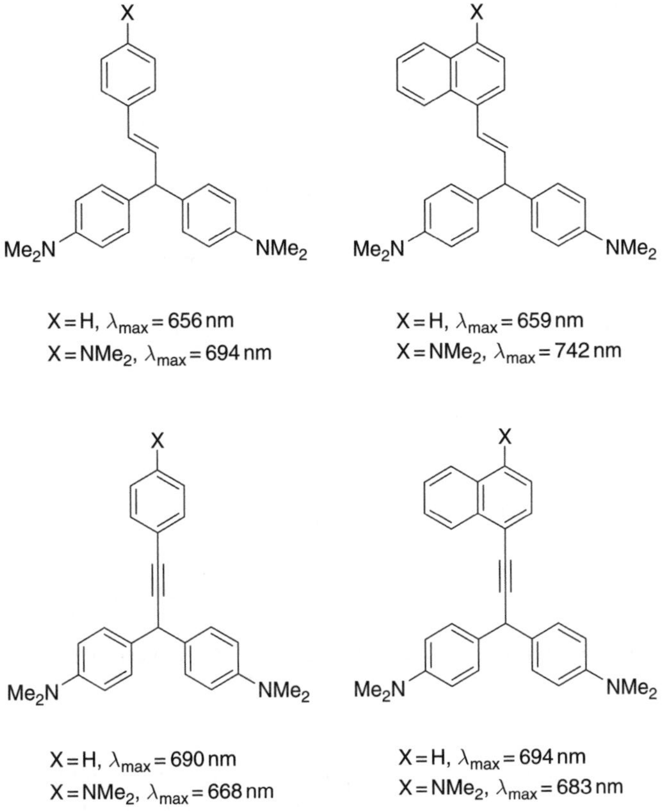

X = H, λ_{max} = 656 nm
X = NMe₂, λ_{max} = 694 nm

X = H, λ_{max} = 659 nm
X = NMe₂, λ_{max} = 742 nm

X = H, λ_{max} = 690 nm
X = NMe₂, λ_{max} = 668 nm

X = H, λ_{max} = 694 nm
X = NMe₂, λ_{max} = 683 nm

Figure 5.10 Crystal violet/Victoria blue derivatives having additional intra-chromophoric unsaturation.

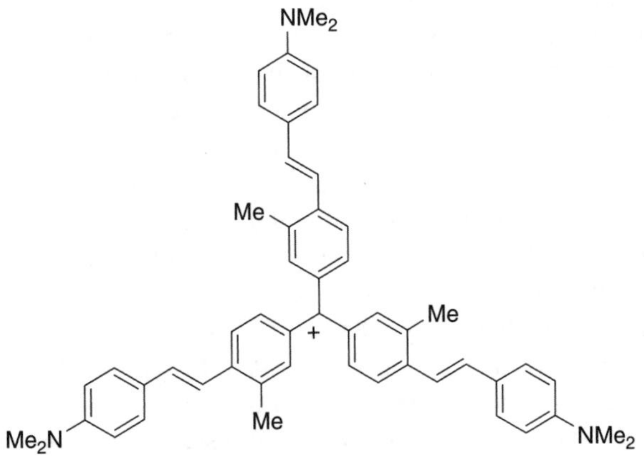

Figure 5.11 Trivinyl derivative of crystal violet, λ_{max} 1003 nm (dichloromethane).

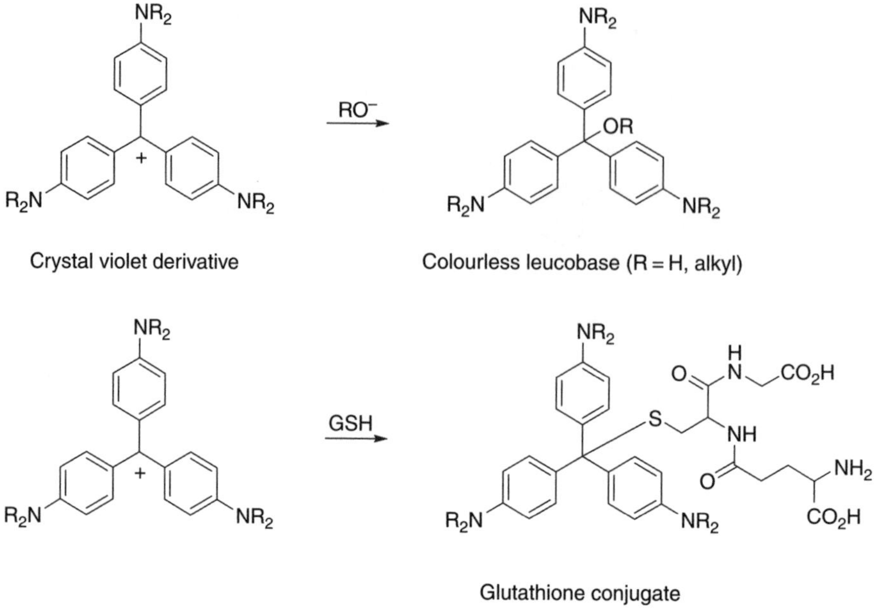

Figure 5.12   Reaction of crystal violet derivatives with nucleophilic agents.

neutralised in the laboratory as part of the purification process, simply by the addition of hydroxide or alkali metal alkoxides (Figure 5.12). Intracellular reaction with molecules such as glutathione (GSH) may lead to cytotoxicity, due to its depletion, but also offers a resistance mechanism in cells where such detoxifying molecules are over-expressed (Debnam, Glanville and Clark, 1993). Such effects may, of course, be seen with other thiol-containing proteins (e.g. human serum albumin) (Özer and Çağlar 2002). Orally administered triarylmethanes are usually excreted as the GSH conjugate.

Several triarylmethanes have also been shown to inhibit human plasma cholinesterase in similar ranges to conventional drugs such as donepezil and the phenothiazines (Küçükkilinç and Özer 2005). The activity of crystal violet against the protozoan *Trypanosoma cruzi* is well established and has led to the use of CV in blood decontamination procedures in Latin America (Temperton *et al.* 1998).

The bioconjugation of triarylmethanes should be a cause for concern to those involved in photosensitising drug discovery, especially those working in anticancer PDT. Here, there is far more likelihood of eventual intravenous administration for tumour delivery – far more so at the time of writing than for the same derivative employed in local photodynamic antimicrobial chemotherapy (PACT)-type disinfection. Intravenous administration inevitably requires the mixing of the photosensitiser with blood proteins, with a concomitant likelihood of reaction as discussed above, and removal from the system without hitting the proposed target. However, even given this scenario, it should be possible to design triarylmethanes that are less likely to bind to blood lipoproteins, e.g. by the use of hydrophilic moieties (e.g. morpholine, diethanolamine, etc. as aux-ochromes). In addition, molecules can be so designed that the reaction centre is crowded, either locally or by overall molecular conformational change, inhibiting the ingress of, in the present case, the protein thiol moiety. Also, standard chemical theory dictates that the rate of reaction for a given protein and different triarylmethanes will vary due to structural and electronic diversity in the derivatives (Eldem and Özer 2004).

Since polysaccharides are intimately involved in biofilm formation and structure (Izano *et al.* 2008), it is sensible to employ photosensitisers that exhibit binding to such biomolecules in order to attack biofilms. Since there are several lead photosensitisers that can perform in this way as well as demonstrating photobactericidal activity, it makes further sense to employ these agents in disinfection in patients, colonised individuals and the physical infrastructure of the healthcare environment. Both crystal violet and phe-nothiazinium methylene blue are premier examples of such cases.

Crystal violet may also be employed in the detection of Barrett's oesophagus and related carcinomas (Amano *et al.* 2004). The lower efficacy of crystal violet in tumour cell lines has been remedied by the co-application of the anionic photosensitiser, mer-ocyanine 540 (Miyagi *et al.* 2003).

5.4 Synthesis

Given the long-established use of triarylmethanes (Figure 5.13), it is not surprising that the synthetic routes entailed are well understood and that 'new' routes are usually, in fact, elaborations of previous approaches.

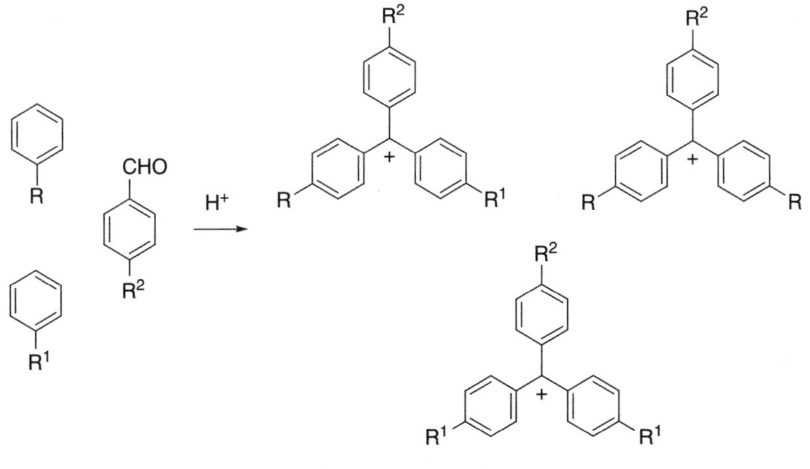

Figure 5.13 Triphenylmethane synthesis via acid condensation of anilines and a benzaldehyde derivative. R, R^1 are usually *N*-alkyl or *N,N*-dialkyl, and R^2 may additionally be H, *N*-alkyl or *N,N*-dialkyl.

Michler's ketone   Leucobase   Triarylmethane cation

MK derivative   Leucobase   Triarylmethane cation

Figure 5.14 Triarylmethane synthesis from Michler's ketone.

The former method as described above may be employed in the production of symmetrical or asymmetrical derivatives, as shown in Figure 5.14, depending on the choice of R, R^1 and R^2, although it should be noted that product purification in the asymmetrical derivatives is often troublesome and might benefit from a stepwise synthesis of the target molecule, i.e. via the asymmetrical Michler's ketone equivalent (Figure 5.14).

The two general approaches to triarylmethanes involve the reaction of an aromatic molecule bearing a reactive single carbon (typically an aldehyde) with two other electron-rich aromatics or the reaction of a suitably substituted benzophenone derivative (e.g. Michler's ketone) with an electron-rich aromatic.

While modern workers might use aryllithiums rather than simple aniline derivatives as the electron-rich components, and much interesting chemistry is often entailed in the synthesis of precursors, the basic principles remain the same. Plainly, the progress made in heterocyclic chemistry is such that triarylmethanes having partially or wholly heteroaromatic ring systems are also available (Figure 5.15), again extending the

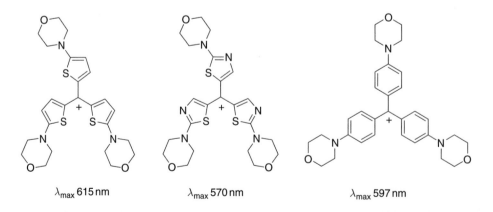

λ_{max} 615 nm λ_{max} 570 nm λ_{max} 597 nm

Figure 5.15 Heterocyclic triarylmethanes based on thiophene and thiazole derivatives.

range of possible structures for examination as photosensitisers (Noack, Schroder and Hartmann 2002).

5.5 Xanthene derivatives

The transition from the triarylmethanes discussed above to the xanthenes is achieved via two main operations: the inclusion of a carboxylic acid residue on one of the rings next to the central carbon (phthaleins) and the bridging of two of the carbons on separate rings adjacent to the central methane with an oxygen atom, i.e. providing the tricyclic xanthene system. In addition, the pendant auxochromic groups may be phenolic or amino in character. There are three main series emanating from these arrangements that are relevant to the current discussion: fluoresceins, rhodamines and rosamines (Figure 5.16).

Loss of the pendant aromatic group from the rosamine structure provides the pyronins, but these are dealt with elsewhere, due to their structural commonality with the azine-type photosensitisers (Chapter 4).

In terms of gross properties, it can be seen from Figure 5.16 that one of the canonical forms of each structure has a positive charge localised on the central carbon. This charge is, of course, delocalised around the xanthene ring system, but this also means that it may be neutralised by proton loss from a conjugated phenolic group, as in the fluorescein-type structures. This is not the case where the phenolic residue is replaced by an amino – in this

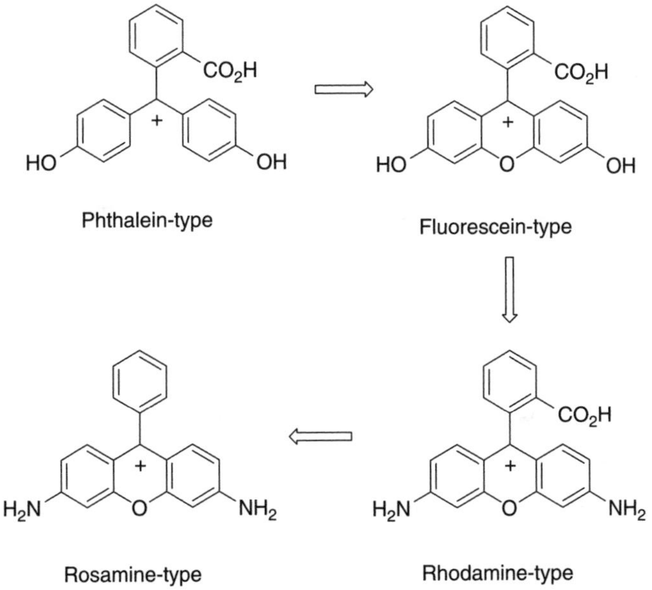

Figure 5.16 Structural evolution of xanthene photosensitisers.

case the positive charge is delocalised onto the amino nitrogen. A further consideration comes from the presence of the carboxylic residue on the pendant aromatic ring since, deprotonated, this may also neutralise the positive charge by addition of the resulting carboxylate anion to the central carbon. Thus, the fluorescein derivatives are considered to be anionic/neutral, the rhodamines cationic/neutral and the rosamines cationic. Clearly there is a sensitivity to pH to be considered here, which may be implicated in photo-activation and photosensitisation, as well as in pharmacological matters. The situation may, of course, be simplified by the esterification of the pendant carboxylic acid. The synthesis of some simple derivatives is given in Figure 5.17, while the range of ionic and neutral species possible is demonstrated in Figure 5.18.

As mentioned elsewhere, heavy atom substitution has other associated outcomes, perhaps the most obvious being that of raised lipophilicity (hydrophobicity). For this reason, most of the fluoresceins can be considered as lipophilic, anionic photosensitisers in the same way as porphyrins such as haematoporphyrin derivative. Indeed, the applica-tion targets of the two classes are similar: tumour PDT, Gram-positive bacteria, viruses

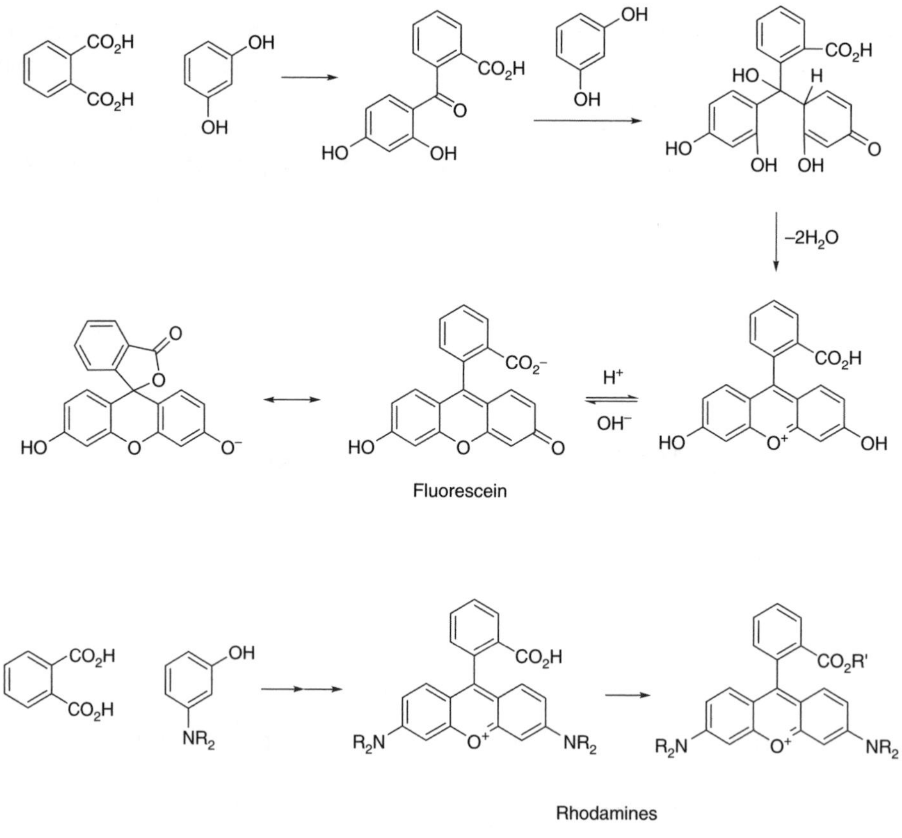

Figure 5.17   Fluorescein/rhodamine synthesis.

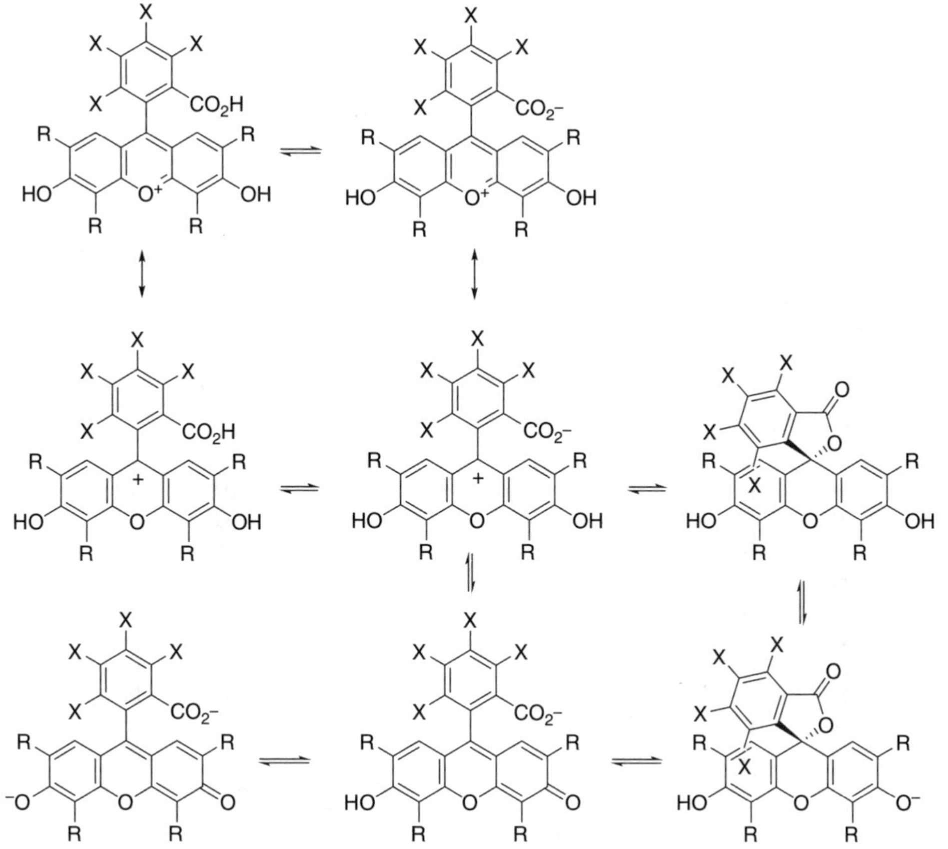

Figure 5.18 Dissociative and resonance forms of fluorescein-type photosensitisers.

via the envelope and PDT 'spin-offs' based around vascular occlusion. The rhodamines, having cationic character, have been more involved in intratumoural PDT, the positive ionisation allowing mitochondrial targeting – rhodamine 123 was one of the early synthetic dyes used as a lead compound in anticancer PDT for this reason, although their previous use as laser dyes was undoubtedly also helpful.

In terms of photosensitising action and photosensitiser structure, there can be few better examples than the heavy atom effect demonstrated by the xanthene series from fluorescein to rose Bengal (Table 5.1). Indeed, given the relatively simple synthesis of the fluoresceins and rhodamines, it is again strange that a greater range of derivatives has not been reported in the literature.

Thus, if fluorescein is considered to be the parent compound, the remaining congeners can be seen as having greater levels and/or varying distribution of halogenation. There is also variety in the halogen employed (Table 5.1). From the table it can be seen that the photosensitising efficacy and λ_{max} follow standard heavy atom substitution patterns (Cl < Br < I), but also that halogenation, logically, has greater effect in the chromophore

Table 5.1 Commonly used fluorescein derivatives

X	X'	R	R'	R''	λ_{max}(nm)	Φ_Δ	Common name
H	H	H	H	H	490	0.03	Fluorescein
H	Br	H	H	H	502	—	—
H	Cl	H	H	H	504	—	—
H	H	H	Cl	Cl	502	0.07	—
H	H	H	Br	Br	504	0.42	—
H	H	H	I	I	506	0.48	—
H	H	Br	Br	HgOH	480	0.23	Mercurochrome
H	H	NO_2	Br	Br	517	0.52	Eosin B
H	H	Cl	Cl	Cl	510	0.05	—
H	H	Br	Br	Br	516	0.50	Eosin Y
H	H	I	I	I	526	0.63	Erythrosine
Cl	H	Br	Br	Br	531	—	Phloxine
Cl	Cl	Br	Br	Br	538	0.65	Phloxine B
Cl	Cl	I	I	I	548	0.75	Rose Bengal

Sources: Pooler and Valenzo (1979) and Gandin, Lion and van de Vorst (1983).

itself, chlorination of the pendant phenyl having added little in terms of singlet oxygen production. Considering the chromophore halogenated derivatives, it is noticeable with bromine or iodine, again, that there is little difference in terms of photosensitising efficacy between di- and tetrasubstitution, for example the tetraiodinated erythrosine has a quantum efficiency of 0.63, compared to 0.48 for the 4,5-diiodo analogue. This may be important in the photosensitiser design, since lower levels of iodination for a similar ($^1O_2$) effect would both allow further functionality to be included in the molecule and also correspond to a lower molecular lipophilicity.

As seen in Table 5.1, the fundamental (9-phenylxanthylium) structure of the fluoresceins is slightly modified to furnish the various halogenated photosensitisers. However, beside halogenation, there has been very little attempted in terms of novel chemistry, for example to extend the active wavelength range. Clearly, both the carboxylic and the phenolic residues constitute reactive moieties, but their use is more likely to be effective in transport and binding.

As with other chromophoric types, it is possible to produce increased wavelength absorption by extending the aromatic (π) system of the chromophore. For xanthenes, this obviously suggests the use of benzannelation, as in Figure 5.19 (Hilderbrand and

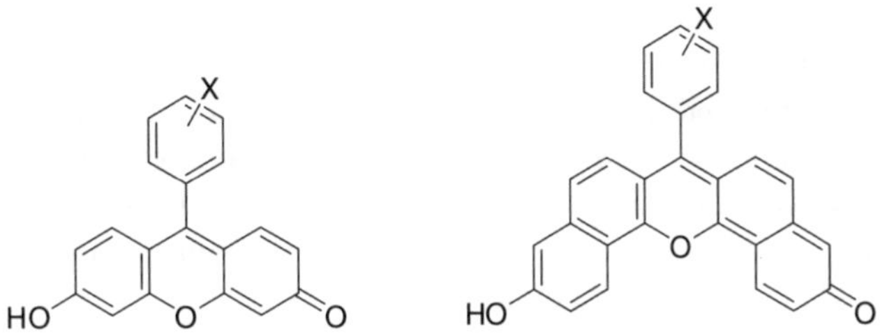

Fluoresceins ($\lambda_{max} \sim 490$ nm) Naphthofluoresceins($\lambda_{max} \sim 610$ nm)

Figure 5.19 Longer-wavelength absorbers via benzannelation.

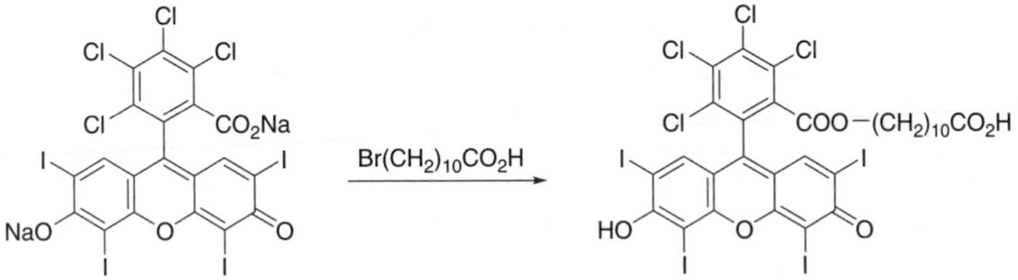

Rose Bengal (548 nm) (553 nm)

Figure 5.20 Formation of long-chain functional esters of rose Bengal.

Weissleder 2007). Although this does move the long-wavelength absorption into the red, the associated photosensitising efficacies of such compounds are likely to be low without halogenation.

Concerning side-chain alteration, since the carboxylate residue is more nucleophilic than the xanthenyl phenoxide, it is logical that the former group will be most easily functionalised. Substituted alkyl moieties have been added to, for example, rose Bengal without loss of photoproperties, the ester-linked groups being sufficiently removed from the photosensitiser chromophore (Figure 5.20) (Paczkowski 1987).

5.6 Biological uses

Rose Bengal is widely employed as a standard compound in photosensitiser studies, for example it is often found as a reference against which new photosensitising efficacies – singlet oxygen yields – are measured. The high efficiency of singlet oxygen production associated with rose Bengal itself (c. 0.75) has also led to its use in industrial photo-oxidation processes, often as an immobilised agent. Its use as a vital stain in ophthalmology has also suggested photodynamic application in this field. Whereas toxicity issues appear to have inhibited its use in anticancer PDT, there are applications in blood vessel occlusion and tissue welding, similarly to those of the lipophilic anionic porphyrin types, but being activated by green rather than red light (Tseng, Feenstra and Watson 1994). However, the anionic nature of the photosensitiser (and its close derivatives) means that it cannot be considered as a broad-spectrum photodisinfectant in the same way as methylene blue, for example.

The inherent Gram-positive/Gram-negative cut-off, mentioned at various points in this book, was well demonstrated in a survey carried out in the early 1970s (Fung and Miller 1973). A range of xanthene and related (unbridged) phthalein dyes were tested against 16 Gram-positive and 14 Gram-negative strains (including *Salmonella typhimurium*), giving the now accepted pattern of antibacterial activity.

As a photobactericide, rose Bengal has been shown to kill the Gram-positive *Streptococcus mutans* in culture media using a broad-band light source (400–500 nm). Bacteria grown in the presence of fibroblasts were destroyed at low concentrations of rose Bengal (*c.* 0.5 µM) and light doses around 300 mJ/cm^2 (Paulino *et al.* 2005).

Unusually, for an anionic photosensitiser, rose Bengal has also been shown to be photobactericidal towards the Gram-negative pathogen *Salmonella typhimurium*, reportedly binding slowly to a lipophilic region of the outer membrane (Dahl *et al.* 1989). As this is the converse situation to that expected, along with the stated slow uptake of the photosensitiser, it may be that the process is allowed via the uptake of a neutral, i.e. lactone, form of rose Bengal from the extracellular medium. Certainly, given the excellent singlet oxygen yield associated with the photosensitiser, the concentration required to cause fatal damage to the bacterial cell might also be expected to be low.

The Gram-negative activity of phloxine may be increased by the adjuvant use of ethylenediaminetetraacetic acid (EDTA), which causes increased membrane permeability, allowing greater access to the photosensitiser (Rasooly 2005).

As well as a considerable collection of research on its antibacterial applications, rose Bengal has also been shown to be an active antiviral. Given that the standard approach to targeting photoantivirals involves the nucleic acid intercalation paradigm, the xanthene derivative might not at first appear to be a strong candidate. However, there are other target sites available, particularly with enveloped viruses. Rose Bengal is known to inactivate fusion proteins, thus affecting viral infectivity (Lenard and Vanderoff 1993) and other external proteins affecting antigenicity (Turner and Kaplan 1968).

The mercury-containing compound mercurochrome (merbromin) is a direct derivative of eosin, a bromine being replaced by —HgOH. This compound has a lower singlet oxygen efficiency than its precursor (Table 5.1) (Martínez *et al.* 1993). However, the main use of mercurochrome was made in local antisepsis, prior to the general realisation of the toxic effects of mercury-containing preparations. Clearly, toxicity issues should be a major consideration in designing therapeutics, and it is unfortunate for the present argument that a compound employed as an antibacterial *per se*, and which is subsequently discovered to be a photosensitising agent, should pose a toxic threat. This does not, of course, disqualify mercurochrome as a lead compound in design, but it is very likely that the original antibacterial use came about due to the microbial susceptibility to mercury, i.e. the remainder of the mercurochrome molecule acts as a vehicle for the heavy metal. Little research has been published to clarify this, but it is evident that various dye types were mercurated with the aim of producing useful therapeutics, for example azoics (Proskouriakoff and Raiziss 1925) and triarylmethanes (Whitmore and Leuck 1929; Chalkley 1941), as well as other xanthenes – the fluorescein derivative was called *Flumerin* (White 1924). Notwithstanding the toxicity issue, mercurochrome is derived from Eosin Y (Table 5.1), which has a high singlet oxygen yield in its own right. The inefficacy of mercury as a heavy atom,- i.e. ^{202}Hg *versus* ^{79}Br, is not well understood, although it may be due to pseudo-ionic character in the aryl–mercury bonding.

In addition, in the area of photoantibacterials, the related compounds phloxine B and erythrosine (Table 5.1) are commonly used as food dyes (with regulatory approval). While it is again unfortunate that compounds such as these, approved for use in humans, are Gram-specific due to their anionic nature, they have been demonstrated to be effective photobactericidal agents, for example against typical oral pathogens.

Interestingly, the inherent antibacterial activities of a range of fluoresceins have been investigated using both standard and methicillin-resistant strains of *Staphylococcus aureus* (Rasooly and Weisz 2002). Derivatives having fully chlorinated pendant phenyl moieties were significantly more active.

Erythrosine, which is not chlorinated in the pendant phenyl moiety, is often used as a disclosing agent for dental plaque. Although, as mentioned above, not an effective antibacterial *per se*, erythrosine has been reported to be a highly effective photobactericide, for example against *Streptococcus mutans*. Indeed, using white light, the xanthene derivative was more effective in biofilms than either methylene blue or Photofrin at equal concentrations (Wood *et al.* 2006). Erythrosine, as noted, has the added advantage over many proposed photosensitisers of being clinically approved for human use.

5.7 Rhodamines

In comparison to the fluorescein derivatives, the rhodamines are generally simpler compounds – particularly the esterified forms – with fewer equilibrium structures (Figure 5.21). For the esters, the xanthene chromophore behaves similarly to the azines discussed elsewhere, being based on a tricyclic heteroaromatic cation.

The discovery that rhodamine 123 (Rh123, Figure 5.22) is taken up specifically by mitochondria, together with its known use as a fluorescent indicator, led to its use as a fluorescent probe in sub-cellular studies and was explained by the cationic, lipophilic nature of this class of agents. Subsequently, and with the modern evolution of PDT, this was extended to the investigation of Rh123 treatment of animal tumours. Rh123 has a fluorescence quantum yield ($\Phi_f = 0.9$), correlating with low photosensitising activity *in vitro* – rhodamine 123 has a singlet oxygen quantum yield of 0.01 and, not surprisingly, shows considerable dependence for cell killing on high levels of tumour cell oxygenation (Richmond and O'Hara 1993).

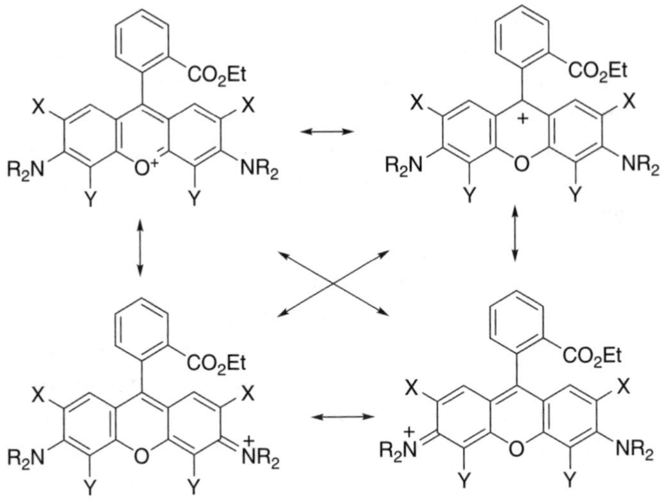

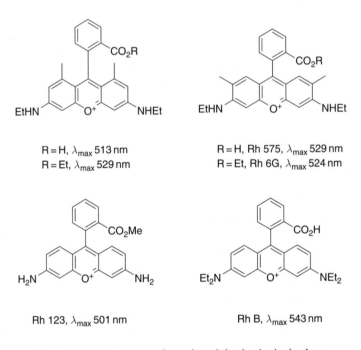

Figure 5.21 Simpler (cationic) delocalisation in rhodamines (compared to the fluoresceins, Figure 5.18). R = H, alkyl; X, Y = H, Br.

R = H, λ_{max} 513 nm
R = Et, λ_{max} 529 nm

R = H, Rh 575, λ_{max} 529 nm
R = Et, Rh 6G, λ_{max} 524 nm

Rh 123, λ_{max} 501 nm

Rh B, λ_{max} 543 nm

Figure 5.22 Structure and wavelength in simple rhodamines.

The associated cell photoinactivation is likely to be due to electron transfer interactions from the excited states of the molecule *in situ*, given the importance of redox activity within mitochondria. In addition, at least *in vitro*, there appears to be a selective inactivation (at 514 nm) between tumour and normal cell types (Saxton *et al.* 1994). Rhodamine 6G (Figure 5.22) offers some improvement in tumour uptake when administered systemically (Haghighat *et al.* 1992).

Analogue preparation has been carried out to produce improved localisation or singlet oxygen yields, normally via simple chromophore functionalisation (Figure 5.22). As with other chromophoric types, ring or auxochrome alkylation generally has only limited effects on photoproperties.

Similarly to the fluorescein derivatives, halogenation of the xanthene moiety increases photosensitising activity. Thus, the 4,5-dibrominated analogue of rhodamine123 (DBR123, Figure 5.23), simply prepared by direct bromination at room temperature, has a significantly higher quantum yield of singlet oxygen than the parent compound. The tetrabrominated derivative again exhibits an increased photosensitising ability, being almost equal to rose Bengal in terms of singlet oxygen production. Differing ester functionality has also been included in recent work to improve cellular photosensitising ability, although related increases in dark toxicity were reported for the longer-chain butyl ester, presumably due to the higher lipophilicity of this compound leading to a lower rate of efflux from cells (Pal *et al.* 1996).

Attachment of two more bromines to the rhodamine nucleus furnishes tetrabromorhodamine 123 (TBR123, Figure 5.23). The extra halogenation of the parent nucleus causes, not surprisingly, considerably increased lipophilicity, and TBR123 is reportedly found in hydrophobic regions of malignant cells (Kessel and Woodburn 1995). However, when used against multidrug-resistant (MDR) cell lines, photoinactivation was not observed, but rather inhibition of the cellular detoxification agent P-glycoprotein (PGP), allowing the uptake of other toxic compounds such as Rh123 or the conventional anticancer agent daunorubicin. Plainly, this suggests the use of such compounds as TBR123 in the

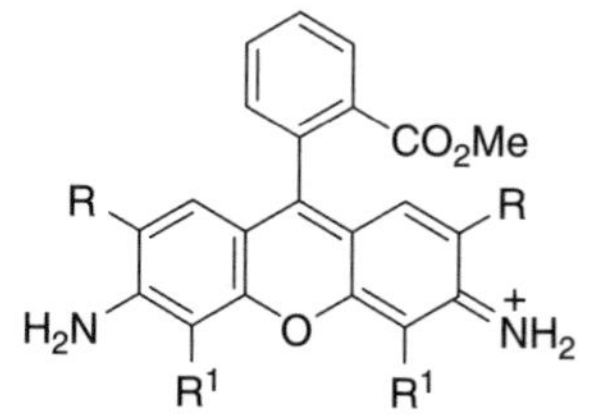

	$\underline{R}$	$\underline{R^1}$	$\underline{\Phi_\Delta}$
Rh123	H	H	0.01
DBR123	H	Br	0.47
TBR123	Br	Br	0.70

Figure 5.23 Rhodamine 123, brominated analogues and singlet oxygen yields.

adjuvant therapy of MDR disease, either with standard chemotherapeutic agents or with other photosensitisers.

While the poor photosensitising activity of Rh123 is a drawback, given its selectivity for mitochondria, this can be addressed by functionalisation as discussed above. However, such alteration does not properly address the other shortcoming, *viz.* the relatively short wavelength of absorption available. Rh123 absorbs light at just above 500 nm, alkyl substitution at the amino groups increasing this by *c.* 20 nm (rhodamine 6G, $\lambda_{max} = 528$ nm; rhodamine B, $\lambda_{max} = 543$ nm). As mentioned previously, this alteration of the parent compound decreases uptake or introduces unwanted dark toxicity. However, their proven *in vitro* activity against malignant cells strongly indicates the use of rhodamines in the eradication of leukaemic cells from bone marrow extracts in preparation for transplantation (Pal *et al.* 1996).

In addition to the halogenated and simple alkyl derivatives, there have been amino variants synthesised, which included cyclic and tethered amino derivatives, as shown in Figure 5.24 (Vogel *et al.* 1988).

The use of tethered amino functionality, as in the julolidino derivative (Figure 5.24), also endows longer-wavelength ranges to be achieved. For example, rhodamine 101, the free carboxylic acid form of the julolidino analogue, absorbs at 568 nm in methanol, a bathochromic shift of 25 nm relative to the rhodamine B parent. Given the potential use of rhodamines in antitumour PDT, the requirement for longer-wavelength absorption follows the usual, logical progression, as with other classes of photosensitisers, although much of the published work in this area pertains to the use of the rhodamines produced in fluorescence detection (David *et al.* 2008). The fusion of more complex heterocyclics such as benzothiophene or indole (Figure 5.25) produces true red absorbers. Again, as these compounds were deigned for fluorescence use, presumably recourse would be required to heavy atom substitution for effective photosensitising action. However, the indole derivative exhibited an associated fluorescence quantum yield of only 15% of that of rhodamine 101.

As with other chromophores in this class, the use of lower chalcogen atoms is associated with bathochromic shift of the λ_{max}, as shown in Figure 5.26 (Calitree *et al.* 2007).

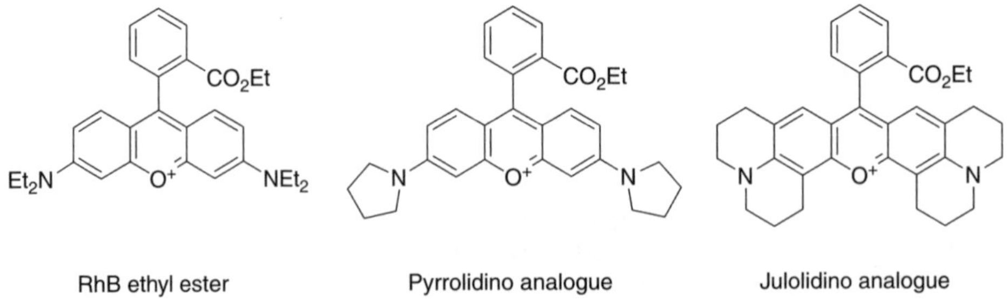

RhB ethyl ester   Pyrrolidino analogue   Julolidino analogue

Figure 5.24 Amino variants of rhodamine B.

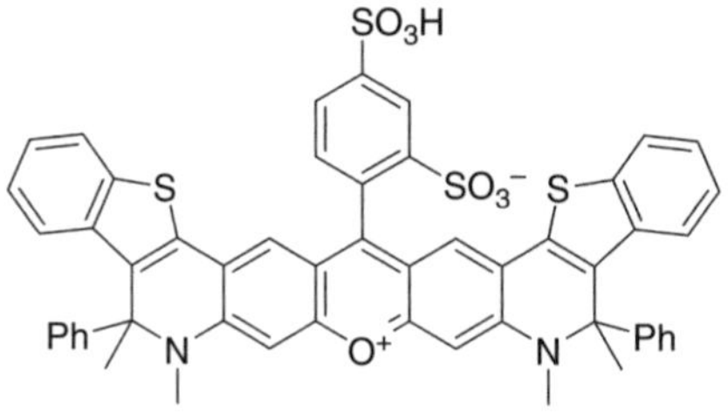

Benzo[*b*]thiophene analogue (613 nm) 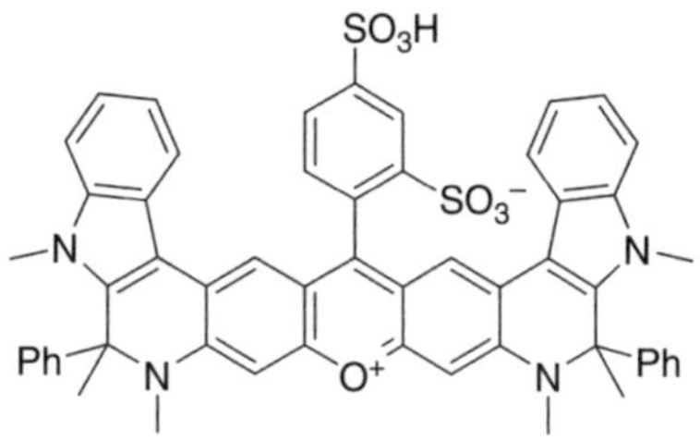   Indolean alogue (634 nm)

Figure 5.25   Red-absorbing rhodamine fluorophores.

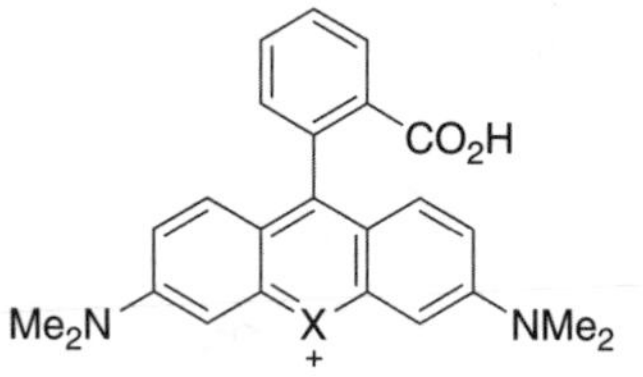

X	λ_{max} (nm, H_2O)
S	572
Se	580
Te	595

Figure 5.26   Rhodamine chalcogenologues.

5.8 Rosamines

From a structural point of view, rosamines have much in common with both rhodamines and azines. The overall, bridged triarylmethane structure is, of course, identical with that of the rhodamines but for the carboxylic acid residue on the pendant phenyl. However, the inclusion of bridging heteroatoms other than oxygen produces structures with associated singlet oxygen yields of the same order as the phenothiaziniums. Clearly this is a manifestation of the heavy atom effect (Figure 5.27).

Detty and co-workers have demonstrated the photodynamic efficacy of the rosamine series, initially as inhibitors of PGP, an often over-expressed transport protein in multi-drug resistance. Compounds such as the rhodamines and tetramethylrosamine (TMR, Figure 5.27) have been used for the fluorescence staining of PGP in resistant tumour cell lines, thus 'boosted' TMR derivatives, i.e. containing heavy atoms, are able both to bind to PGP and to cause photodamage (Detty *et al.* 2004).

The synthesis of simple rosamine analogues is relatively straightforward (three main steps), thus facilitating compound series preparation. The synthetic route employed for sulphur and seleno analogues is shown in Figure 5.28 (Detty *et al.* 2004).

The synthetic procedure may also be altered slightly to include heteroaromatic rings at position 9. For example, the use of 2-lithiothiophene in the final stage of Figure 5.28 leads to the production of the 9-thienoanalogue (Wagner *et al.* 2005), while alternative substituted aromatics produce functionalised rosamine derivatives (Figure 5.29; Holt *et al.* 2006).

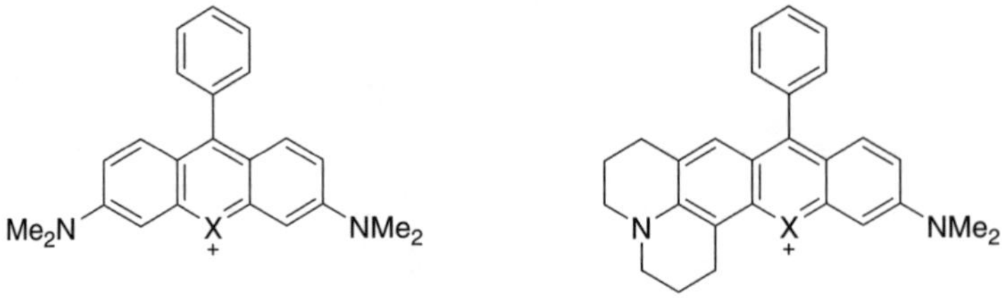

X	λ_{max} (nm)	Φ_Δ
O (TMR)	552	0.08
S	571	0.21
Se	581	0.87

X	λ_{max} (nm)	Φ_Δ
S	582	0.28
Se	588	0.85

Figure 5.27 Rosamine photosensitisers.

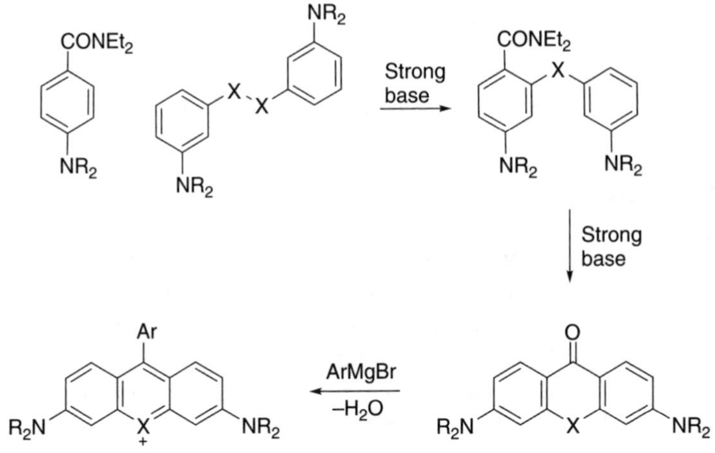

Figure 5.28 Synthesis of sulphur and selenium rosamine analogues (X = S and Se).

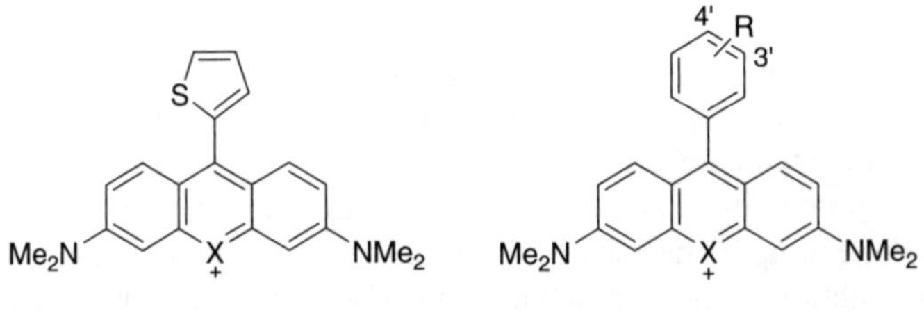

Figure 5.29 9-Thiopheno- and 9-(substituted)phenyl rosamine derivatives (X = S and Se; R = 3'/4'-NH$_2$, -NMe$_2$, -OMe, or -H).

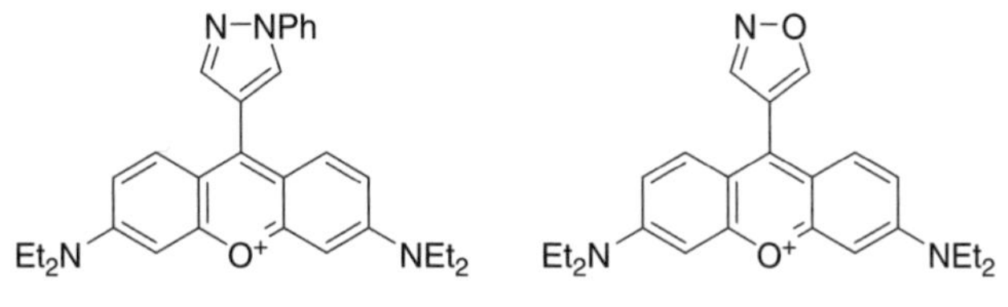

Pyrazole analogue (λ_{max} 559 nm)

Isoxazole analogue (λ_{max} 570 nm)

Figure 5.30 9-Heterocyclic rosamine derivatives.

Relative to the 9-phenyl thiotetramethylrosamine base compound (Figure 5.29, X = S, λ_{max} = 571 nm in methanol), synthetic attempts at shifting the maximum wavelength again into the red region have been based on increased electron release at C-9 (e.g. 9-thiopheno, λ_{max} = 590 nm) or on rigidification of one of the amino auxochromes (e.g. the julolidino derivative, Figure 5.27). As usual, the heavy atom effect also tends to longer wavelengths. The use of structural alterations other than that of heavy atom substitution does not appear to alter the singlet oxygen yield significantly, although the reader is once again reminded that this may not be of particular relevance since the structurally varied molecules may have quite different cellular localisations or interactions with biomolecules thereat.

The incorporation of heterocyclic nuclei into rhodamine molecules as the pendant aromatic (at C-9) is also possible via the active methylene chemistry of 9-methylpyronine, a different route from the traditional aldehyde/aminophenol condensation. Thus, various five-membered heterocycles may be furnished, including pyrazoles and isoxazoles (Figure 5.30). Interestingly, the heterocyclic derivatives produced demonstrated little variation in (low) fluorescence yields compared to the parent 9-methylpyronine (Shandura, Poronik and Kovtun 2007).

The cellular (*in vitro*) targets intended for the rosamine derivatives, as with many other examples in this book, emanated from anticancer PDT, several molecules being intended for use against PGP-mediated multidrug-efflux capability (Gibson *et al.* 2004).

Based on their structural commonalities with methylene blue, rosamine derivatives have also been tested as intracellular antivirals for red blood cell fractions, the long-wavelength band being increased to 601 nm via the use of a 2-thienyl (thiophene) moiety in place of the phenyl (Wagner *et al.* 2005).

5.9 Pyrylium compounds

As can be seen from Figure 5.31, there are obvious structural similarities in the rosamines and the pyryliums. Moreover, both types are selective fluorescent probes for mitochondria but are not phototoxic. Consequently, and also in common with the rosamines, the ring oxygen has been replaced by lower Group VI atoms in the search for mitochondria-specific

Figure 5.31 Structural comparison between pyrylium (upper) and rosamine (xanthylium, lower) photosensitisers.

photosensitisers. However, the range of pyrylium derivatives also includes examples having tellurium as the heteroatom.

Whereas heteroatom replacement in rosamines has not led to significant increases in λ_{max}, providing candidates still having sub-optimal absorption characteristics, this is not the case with the pyrylium system. Both selenium and tellurium analogues of the original pyrylium salt (Figure 5.32, $X = O$) have absorption profiles closer to that expected in photodynamic applications, and this is further aided by introducing asymmetry into the system (Figure 5.32).

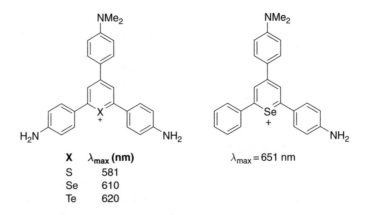

X	λ_{max} (nm)
S	581
Se	610
Te	620

$\lambda_{max} = 651$ nm

Figure 5.32 Pyrylium analogues from photosensitiser development.

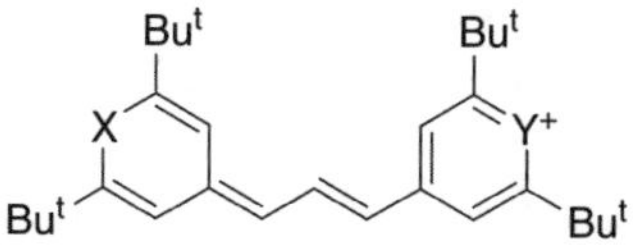

X	Y	λ_{max} (nm, H$_2$O)	Φ_Δ
O	O	593	0.0004
S	S	685	0.0006
Se	Se	730	0.014
Te	Te	810	0.12

Figure 5.33 Binuclear pyrylium derivatives.

It should be noted, however, that although it is possible to produce more suitable photoproperties in this fashion, the yield of singlet oxygen in solution is usually around 10% (Detty, Gibson and Wagner 2004).

Similarly, binuclear pyrylium derivatives (Figure 5.33) have been designed as mitochondrial photosensitisers by Detty *et al.* (1990). Again, the increase in heavy chalcogen content reportedly led to bathochromic shifts and increased singlet oxygen yields, although the latter were low for potential photosensitisers, being of a similar magnitude to those noted above.

The low yields of singlet oxygen encountered with the pyrylium derivatives are mainly due to the ease of ring rotation, which facilitates rapid decay of the excited state. However, this situation is altered on adsorption or binding to biomolecules, the pyrylium molecule becoming more rigid. This approach was taken by Wagner and co-workers in targeting microbial DNA in blood products (McKnight *et al.* 2007).

5.10 Pyronines

Pyronine Y (Figure 5.34, X = O) has an associated λ_{max} value of 546 nm and exhibits a very low singlet oxygen yield (0.05 relative to that of methylene blue). On both criteria this should make it an unsuitable candidate for PDT or PACT applications. However,

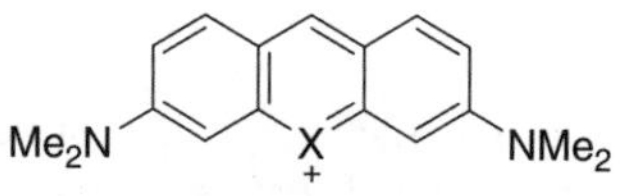

X = O, pyronine
X = S, thiopyronine

Figure 5.34 Pyronine and thiopyronine chromophores.

both pyronin and thiopyronin have been shown to cause photodamage to both bacteria and viruses or virus models, although this may be tempered by the fact that it is irreversibly reduced by bacteria (Jacob 1974).

Yeasts such as *Saccharomyces* spp. are similarly sensitive, and the photodamage in cell culture has not been shown to involve DNA (Roth, Takamori and Lochmann 1994). Indeed damage to DNA and RNA, although exhibited in cell-free systems, has not been reported in cell culture (either eukaryotic or prokaryotic) (Ehrlich *et al.* 1987).

The triarylmethane series dates back to the earliest days of dye use in biology and medicine, having a particular link to antimicrobial chemotherapy via the Gram stain and clinical use as both an antibacterial and an antifungal. While the closely related fluorescein, rhodamine and rosamine families are perhaps better known for fluorescence, several examples are also established photosensitisers. However, few groups have made the link between biological activity in one subset and photoactivity in the other. Consequently – for the most part – this group as a whole remains in something of a backwater. This should therefore be seen as a positive aspect, since there is much to discover in terms of new photosensitisers.

References

Amano Y, Komazawa Y, Ishimura N, *et al.* (2004) Two cases of superficial cancer in Barrett's esophagus detected by chromoendoscopy with crystal violet. *Gastrointestinal Endoscopy* **59**: 143–146.

Bartlett JA, Indig GL. (1999) Spectroscopic and photochemical properties of malachite green noncovalently bound to bovine serum albumin. *Dyes and Pigments* **43**: 219–226.

Brezová V, Pigošová J, Havlínová B, Dvoranová D, Ďurovič; M. (2004) EPR study of photochemical transformations of triarylmethane dyes. *Dyes and Pigments* **61**: 177–198.

Burrow SM, Phoenix DA, Wainwright M, Waring JJ. (2000) Reduced cellular glutathione levels do not affect the cytotoxicity or photocytotoxicity of the cationic photosensitiser VBBO. *Membrane and Cell Biology* **14**: 357–366.

Calitree B, Donnelly DJ, Holt JJ, *et al.* (2007) Tellurium analogues of rosamine and rhodamine dyes: synthesis, structure, 125Te NMR, and heteroatom contributions to excitation energies. *Organometallics* **26**: 6248–6257.

Chalkley L. (1941) Organic mercury derivatives of basic triphenylmethane dyes: dimercuri derivatives of malachite green. *Journal of the American Chemical Society* **63**: 981–987.

Crossley ML. (1919) Gentian violet – its selective bactericidal action. *Journal of the American Chemical Society* **41**: 2083–2090.

Dahl TA, Valdes-Aguilera O, Midden WR, Neckers DC. (1989) Partition of rose Bengal anion from aqueous medium into a lipophilic environment in the cell envelope of *Salmonella typhimurium*: implications for cell-type targeting in photodynamic therapy. *Journal of Photochemistry and Photobiology B: Biology* **4**: 171–184.

David E, Lejeune J, Pellet-Rostaing S, *et al.* (2008) Synthesis of fluorescent rhodamine dyes using an extension of the Heck reaction. *Tetrahedron Letters* **49**: 1860–1864.

Debnam P, Glanville S, Clark AG. (1993) Inhibition of glutathione *S*-transferases from rat liver by basic triphenylmethane dyes. *Biochemical Pharmacology* **45**: 1227–1233.

Detty MR, Gibson SL, Wagner SJ. (2004) Current clinical and preclinical photosensitizers for use in photodynamic therapy. *Journal of Medicinal Chemistry* **47**: 3897–3915.

Detty MR, Merkel PB, Hilf R, Gibson SL, Powers SK. (1990) Chalcogenopyrylium dyes as photochemotherapeutic agents. 2. Tumor uptake, mitochondrial targeting, and singlet oxygen-induced inhibition of cytochrome C oxidase. *Journal of Medicinal Chemistry* **33**: 1108–1116.

Detty MR, Prasad PN, Donnelly DJ, Ohulchanskyy T, Gibson SL, Hilf R. (2004) Synthesis, properties and photodynamic properties in vitro of heavy-chalcogen analogues of tetramethylrosamine. *Bioorganic & Medicinal Chemistry* **12**: 2537–2544.

Dobson J, Wilson M. (1992) Sensitization of oral bacteria in biofilms to killing by light from a low-power laser. *Archives of Oral Biology* **37**: 883–887.

Ehrlich W, Mangir M, Nothelfer R, Baumgärtel H, Lochmann ER. (1987) Thiopyronine-sensitized photodynamic effect on cell growth, RNA and DNA synthesis of Chinese hamster ovary cells. *International Journal of Radiation Biology* **52**: 207–212.

Eldem Y, Özer I. (2004) Electrophilic reactivity of cationic triarylmethane dyes towards proteins and protein-related nucleophiles. *Dyes and Pigments* **60**: 49–54.

Fung DYC, Miller RD. (1973) Effect of dyes on bacterial growth. *Applied Microbiology* **25**: 793–799.

Gandin E, Lion Y, vande Vorst A. (1983) Quantum yield of singlet oxygen production by xanthene derivatives. *Photochemistry and Photobiology* **37**: 271–278.

Gibson SL, Hilf R, Donnelly DJ, Detty MR. (2004) Analogues of tetramethylrosamine as transport molecules for and inhibitors of P-glycoprotein-mediated multidrug resistance. *Bioorganic & Medicinal Chemistry* **12**: 4625–4631.

Guinot SGR, Hepworth JD, Wainwright M. (1999) Extended conjugation in di- and tri-arylmethane dyes. Part 3. The effects of increased planarity in Victoria blue dyes. *Dyes and Pigments* **40**: 151–156.

Guinot SGR, Hepworth JD, Wainwright M. (2000) Extended conjugation in di- and tri-arylmethane dyes. Part 5. Vinylogues and ethynologues. *Dyes and Pigments* **47**: 129–142.

Haghighat S, Castro DJ, Lufkin RB, *et al.* (1992) Laser dyes for experimental phototherapy of human cancer: comparison of three rhodamines. *Laryngoscope* **102**: 81–87.

Hilderbrand SA, Weissleder R. (2007) One-pot synthesis of new symmetric and asymmetric xanthene dyes. *Tetrahedron Letters* **48**: 4383–4385.

Holt JJ, Gannon MK, Tombline G, *et al.* (2006) A cationic chalcogenoxanthylium photosensitizer effective in vitro in chemosensitive and multidrug-resistant cells. *Bioorganic & Medicinal Chemistry* **14**: 8635–8643.

Izano EA, Sadovskaya I, Wang H, *et al.* (2008) Poly-*N*-acetylglucosamine mediates biofilm formation and detergent resistance in *Aggregatibacter actinomycetemcomitans*. *Microbial Pathogenesis* **44**: 52–60.

Jacob HE. (1974) Photo-oxidation sensitized by methylene blue, thiopyronine, and pyronine-IV. The behaviour of thiopyronine in suspensions of bacteria. *Photochemistry and Photobiology* **19**: 133–137.

Kessel D, Woodburn K. (1995) Selective photodynamic inactivation of a multidrug transporter by a cationic photosensitising agent. *British Journal of Cancer* **71**: 306.

Kowaltowski AJ, Turin J, Indig GL, Vercesi AE. (1999) Mitochondrial effects of triarylmethane dyes. *Journal of Bioenergetics and Biomembranes* **31**: 581–590.

Küçükkilinç; T., Özer İ (2005) Inhibition of human plasma cholinesterase by malachite green and related triarylmethane dyes: mechanistic implications. *Archives of Biochemistry and Biophysics* **440**: 118–122.

Lenard J, Vanderoff R. (1993) Photoinactivation of influenza virus fusion and infectivity by rose Bengal. *Photochemistry and Photobiology* **58**: 527–531.

Lewis MR, Goland PP, Sloviter HA. (1946) Selective action of certain dyestuffs on sarcomata and carcinomata. *Anatomical Record* **96**: 201–206.

Martínez G, Bertolotti SG, Zimerman OE, Miártire DO, Braslavsky SE, García NA. (1993) A kinetic study of the photodynamic properties of the xanthene dye merbromin (mercurochrome) and its aggregates with amino acids in aqueous solutions. *Journal of Photochemistry and Photobiology B: Biology* **17**: 247–255.

McKnight RE, Ye M, Ohulchanskyy TY, *et al.* (2007) Synthesis of analogues of a flexible thiopyrylium photosensitizer for purging blood-borne pathogens and binding mode and affinity studies of their complexes with DNA. *Bioorganic & Medicinal Chemistry* **15**: 4406–4418.

Miyagi K, Sampson RW, Sieber-Blum M, Sieber F. (2003) Crystal violet combined with Merocyanine 540 for the *ex vivo* purging of hematopoietic stem cell grafts. *Journal of Photochemistry and Photobiology B: Biology* **70**: 133–144.

Modica-Napolitano JS, Joyal JL, Ara G, Oseroff AR, Aprille JR. (1990). Mitochondrial toxicity of cationic photosensitizers for photochemotherapy. *Cancer Research* **50**: 7876–7881.

Noack A, Schroder A, Hartmann H. (2002) Synthesis and spectral characterisation of a series of triphenylmethane analogues. *Dyes and Pigments* **57**: 131–147.

Özer I, Çağlar A. (2002) Protein-mediated nonphotochemical bleaching of malachite green in aqueous solution. *Dyes and Pigments* **54**: 11–16.

Paczkowski J. (1987) The water-soluble rose Bengal derivatives and their spectroscopic behaviours. *Tetrahedron* **43**: 4579–4589.

Pal P, Zeng H, Durocher G, *et al.* (1996) Phototoxicity of some bromine-substituted rhodamine dyes: synthesis, photophysical properties and application as photosensitizers. *Photochemistry and Photobiology* **63**: 161–168.

Paulino TP, Ribeiro KF, Thedei G, Tedesco AC, Ciancaglini P. (2005) Use of hand held photopolymerizer to photoinactivate *Streptococcus mutans*. *Archives of Oral Biology* **50**: 353–359.

Pooler JP, Valenzo DP. (1979) Physicochemical determinants of the sensitizing effectiveness for photooxidation of nerve membranes by fluorescein derivatives. *Photochemistry and Photobiology* **30**: 491–498.

Proskouriakoff A, Raiziss GW. (1925) Mercury derivatives of azo dyes. *Journal of the American Chemical Society* **47**: 1974–1979.

Rasooly R. (2005) Expanding the bactericidal action of the food color additive phloxine B to Gram-negative bacteria. *FEMS Immunology and Medical Microbiology* **45**: 239–244.

Rasooly R, Weisz A. (2002) In vitro antibacterial activities of phloxine B and other halogenated fluoresceins against methicillin-resistant *Staphylococcus aureus*. *Antimicrobial Agents and Chemotherapy* **46**: 3650–3653.

Richmond RC, O'Hara JA. (1993) Effective photodynamic action by rhodamine 123 leading to photosensitized killing of Chinese hamster ovary cells in tissue culture and a proposed mechanism. *Photochemistry and Photobiology* **57**: 291–297.

Roth RM, Takamori Y, Lochmann ER. (1994) Thiopyronine- and 8-methoxypsoralen-sensitized photodynamic effect on cell growth, colony forming ability and RNA synthesis in *Saccharomyces* mutants deficient in DNA repair. *Photochemistry and Photobiology* **59**: 627–630.

Saji M, Taguchi S, Uchiyama K, Osono E, Hayama N, Ohkuni H. (1995) Efficacy of gentian violet in the eradication of methicillin-resistant *Staphylococcus aureus* from skin lesions. *Journal of Hospital Infection* **31**: 225–228.

Sanguinet L, Twieg RJ, Wiggers G, Mao G, Singer KD, Petschek RG. (2005) Synthesis and spectral characterization of bisnaphthylmethyl and trinaphthylmethyl cations. *Tetrahedron Letters* **46**: 5121–5125.

Saxton RE, Haghighat S, Plant D, Lufkin R, Soudant J, Castro DJ. (1994) Dose response of human tumor cells to rhodamine 123 and laser phototherapy. *Laryngoscope* **104**: 1013–1018.

Sengupta S, Sadhukhan SK. (2000) Trivinylogs of crystal violet: synthesis and absorption properties of new near-IR dyes. *Journal of the Chemical Society, Perkin Transactions 1*: 4332–4334.

Shandura MP, Poronik YM, Kovtun YP. (2007) New heterocyclic analogues of rhodamines. *Dyes and Pigments* **73**: 25–30.

Temperton NJ, Wilkinson SR, Meyer DJ, Kelly JM. (1998) Overexpression of superoxide dismutase in *Trypanosoma cruzi* results in increased sensitivity to the trypanocidal agents gentian violet and benznidazole. *Molecular and Biochemical Parasitology* **96**: 167–176.

Tseng SCG, Feenstra RPG, Watson BD. (1994) Characterization of photodynamic actions of Rose Bengal on cultured cells. *Investigative Ophthalmology & Visual Science* **35**: 3295–3307.

Turner GS, Kaplan C. (1968) Photoinactivation of vaccinia virus with rose Bengal. *Journal of General Virology* **3**: 433–443.

Viola A, Hadjur C, Jeunet A, Julliard M. (1996) Electron paramagnetic resonance evidence of the generation of superoxide and hydroxyl radicals by irradiation of new photodynamic therapy photosensitizer, Victoria blue BO. *Journal of Photochemistry and Photobiology, B: Biology* **32**: 49–58.

Vogel M, Rettig W, Sens R, Drexhage KH. (1988) Structural relaxation of rhodamine dyes with different *N*-substitution patterns: a study of fluorescence decay times and quantum yields. *Chemical Physics Letters* **147**: 452–460.

Wagner SJ, Skripchenko A, Donnelly DJ, Ramaswamy K, Detty MR. (2005) Chalcogenoxanthylium photosensitizers for the photodynamic purging of blood-borne viral and bacterial pathogens. *Bioorganic & Medicinal Chemistry* **13**: 5927–5935.

Wainwright M, Phoenix DA, Burrow SM, Waring JJ. (1999a) Uptake and cell-killing activities of a series of Victoria blue derivatives in a mouse mammary tumour cell line. *Cytotechnology* **29**: 35–43.

Wainwright M, Phoenix DA, Burrow SM, Waring JJ. (1999b) Cytotoxicity and adjuvant activity of cationic photosensitisers in a multidrug-resistant cell line. *Journal of Chemotherapy* **11**: 61–68.

Wakelin LPG, Adams A, Hunter C, Waring MJ. (1981) Interaction of crystal violet with nucleic acids. *Biochemistry* **20**: 5779–5787.

White EC. (1924) New organic mercurials and their therapeutic applications. *Industrial & Engineering Chemistry* **16**: 1034–1038.

Whitmore FC, Leuck GJ. (1929) The mercuration of aurin and attempts to mercurate some other triphenylmethane dyes. *Journal of the American Chemical Society* **51**: 2782–2784.

Wilson M, Mia N. (1993) Sensitisation of *Candida albicans* to killing by low-power laser light. *Journal of Oral Pathology and Medicine* **22**: 354–357.

Wood S, Metcalf D, Devine D, Robinson R. (2006) Erythrosine is a potential photosensitizer for the photodynamic therapy of oral plaque biofilms. *Journal of Antimicrobial Chemotherapy* **57**: 680–684.

6
Porphyrins

The commonly reported discoveries in the early twentieth century by von Tappeiner and Jesionek and by Policard, and later by Figge (Figge, Weiland and Manganiello 1948), among others, led to the modern usage of photodynamic therapy. Obviously, the relatively rudimentary procedures involved in the original production of haematoporphyrin derivative (HpD) have left a clinical legacy that employs impure mixtures of porphyrins, although porfimer sodium is a purer product than HpD. Subsequent investigations involving such mixtures have underlined their non-ideal nature with respect to clinical side effects, such as significant skin photosensitisation. In addition, the absorption spectra of the mixtures are unsatisfactory with respect to endogenous light absorption, from haem and melanin pigments.

Along with the related phthalocyanine class (Chapter 7), the porphyrins constitute the largest chemically related group of chromophores employed in the photodynamic milieu. Some base structures are given in Figure 6.1.

Tetraazaporphyrins offer advantages in relatively straightforward synthetic routes and ease of inclusion of functional groups, coupled again with improved photoproperties – λ_{max} 700–800 nm, depending on substitution pattern – compared to both haemato- and *meso*-porphyrin types (Vesper *et al.* 2006).

The texaphyrins have a great advantage over conventional porphyrins in having long-wavelength absorption band above 700 nm, for example that of lutetium texaphyrin occurs at 732 nm (Sessler and Miller 2000). Photodynamic therapy (PDT) drugs based on texaphyrins have been clinically trialled for recurrent breast cancer and atherosclerosis (see below).

Porphycenes are also seen as future-generation porphyrin-type PDT agents, particularly given the long-wavelength absorption wavelengths associated with sulphonated derivatives, the trisulphonic acid having a λ_{max} in water at just below 700 nm (Baba, Shimakoshi and Hisaeda 2004).

Sapphyrins may also be considered as extended porphyrins, although when they are photoexcited at long wavelengths, e.g. in viral inactivation research, it is the mono-protonated form that is responsible (Judy *et al.* 1991).

The extended porphyrin/'stretched porphycene' chromophore, having a much larger π-system, also exhibits longer-wavelength absorptions than are usually associated with standard porphyrins, being in the region of 800–900 nm (Berman *et al.* 1993).

Photosensitisers in Biomedicine   Mark Wainwright
© 2009 John Wiley & Sons, Ltd

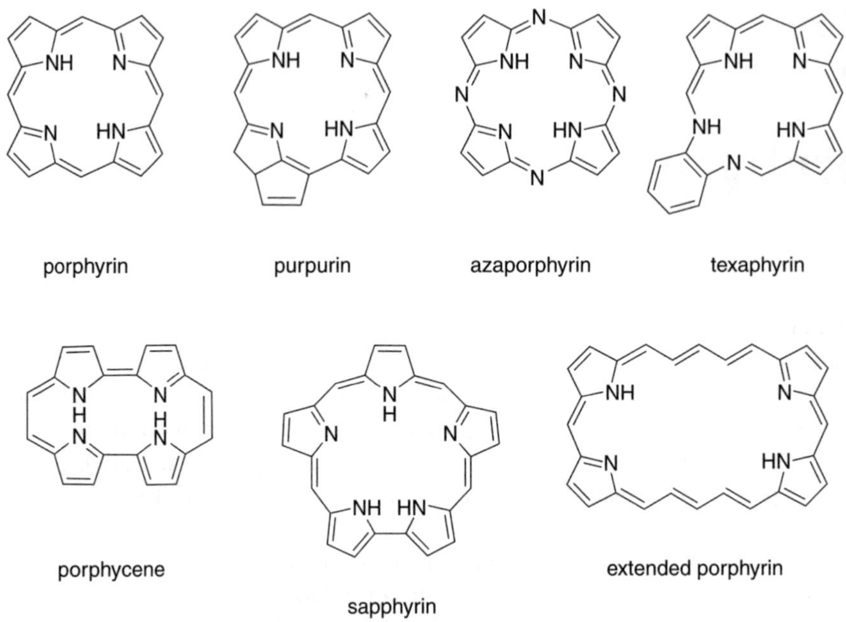

porphyrin   purpurin   azaporphyrin   texaphyrin

porphycene   extended porphyrin

sapphyrin

Figure 6.1 Porphyrynoid chromophores used in photosensitiser discovery.

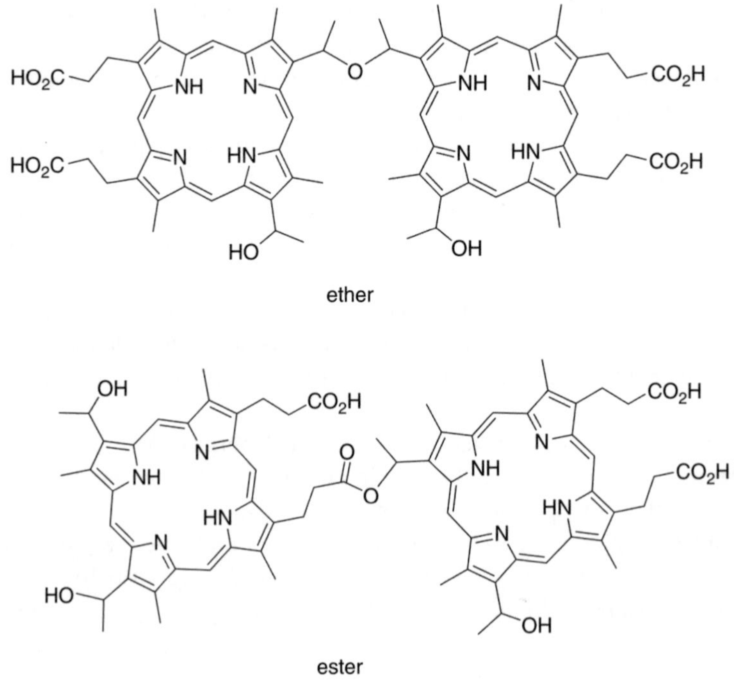

ether

ester

Figure 6.2 Possible dihaematoporphyin ether and ester structures.

As noted, the original photosensitiser preparation used in clinical PDT was HpD, the activity of which was due to porphyrin oligomers, including dihaematoporphyrin ethers and esters (Figure 6.2). The formation of these oligomers results from the use of sodium hydroxide as part of the treatment of the original haematoporphyrin. Thus, the clinical photosensitiser (mixture) here was not designed but was rather a result of the chemical process employed to produce a more soluble photosensitiser than haematoporphyrin itself (Lipson and Baldes 1960).

Photofrin (QLT, Vancouver) gained clinical approval in Canada in 1993 and was launched in the USA in 1996. However, it seems unlikely that regulatory authorities such as the US Food & Drugs Administration would grant permissions to such a chemical mixture again. Certainly, the watchword in photodynamic drug development in the 21st century is 'purity'! It is, of course, very rare for a conventional drug to contain more than one chemical entity, outside enantiomeric forms.

Consequently, different research lines have been followed towards novel porphyrin development (some examples appear in Figure 6.3). Since the asymmetric nature of

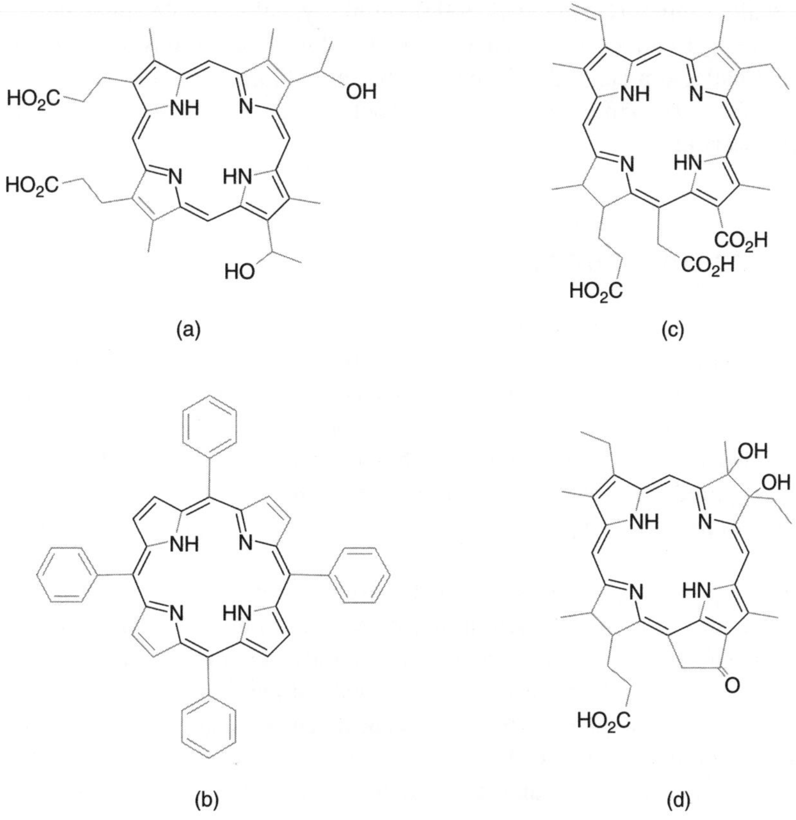

Figure 6.3 Porphyrin photosensitisers showing differences in ring structure: (a) haem, (b) TPP (porphyrin), (c) chlorin e$_6$ (chlorin) and (d) bacteriochlorin.

natural product porphyrins demands complex, difficult synthesis, provision of derivatives such as the chlorins normally relies on semi-synthetic methods, i.e. with pre-formed, natural-product starting materials. Conversely, more easily functionalised materials based on *meso*-tetraarylporphyrin derivatives are fully synthetic.

Nevertheless, there is a relatively wide structural range available (Wainwright 2008), and these will be dealt with in general structural sections based on ring character (reduced, extended, etc.) or functionalisation.

6.1   Central metals

While the majority of *meso*-porphyrin derivatives employed in PDT research are metal free, i.e. having two pyrrole residues with free NH moieties, this is by no means generally applicable. Concerning metal-substituted porphyrins, zinc complexes provide efficient singlet oxygen producers (see Table 6.2 towards the end of this chapter), whereas those containing copper(II) do not; both these situations are also found with metal phthalocyanines (Huang *et al.* 2006). Similarly, gallium-substituted porphyrin and pheophorbide derivatives have high associated yields of singlet oxygen (Litwinski *et al.* 2006), and palladium-centred chlorins and bacteriochlorins provide several promising candidates for PDT. Tin etiopurpurin and Tookad (palladium) are clinically approved photosensitisers (*q.v.*).

6.2   *Meso* compounds

Relative to the chlorins, bacteriochlorins and other chlorophyll-type derivatives, the far greater structural simplicity and ease of preparation of the *meso* compounds (Figure 6.4) makes them popular alternatives in photosensitiser design, while maintaining porphyrin character. It is also a relatively straightforward proposition with this type to synthesise series of closely related compounds with the small variations in functional group character required for structure–activity studies to be conducted. The main drawback from a photosensitiser viewpoint lies in the poor long-wavelength characteristics, as with HpD itself.

Parent compounds in this area are the tetrasubstituted derivatives based on phenyl-sulphonic acid and *N*-methylpyridinium, being tetranionic and tetracationic species respectively (Figure 6.5). Certainly, both compounds were in the vanguard of candidate photosensitisers for PDT in the 1970s (Diamond, Granelli and McDonagh 1977), and the sulphonic acid analogue was reported to show selective uptake in animal tumours as early as 1962 (Winkelman 1962).

Thus, in many cases, compound series are based on the tetraaryl (typically tetraphenyl or tetrapyridyl) nucleus, with group variation included in the pendant aromatic rings. The simplicity of this approach may be understood on inspection of the synthetic route involved (Figure 6.6).

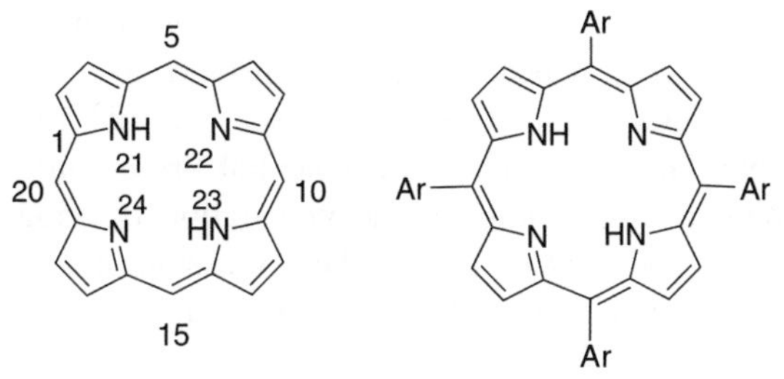

Figure 6.4 Numbering in the porphyrin system and aryl *meso*-substitution.

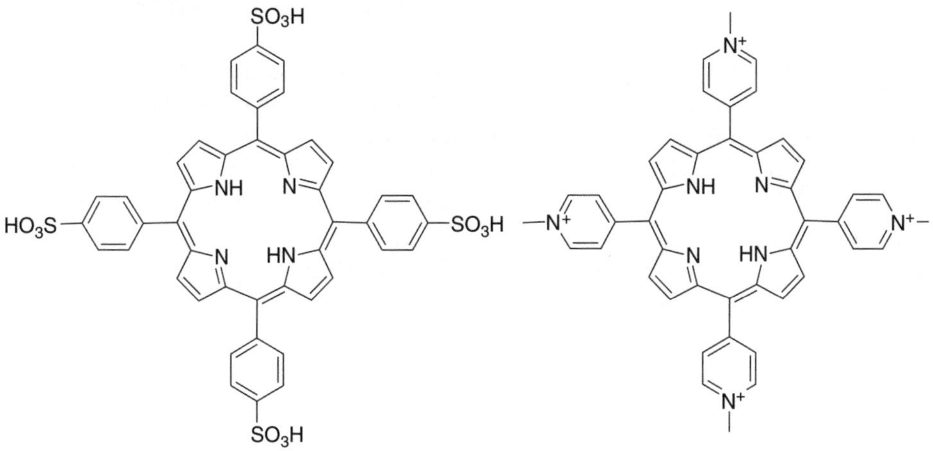

Figure 6.5 *meso*-Tetra(4-sulphonatophenyl)- and tetra(*N*-methyl-4-pyridyl)porphyrins.

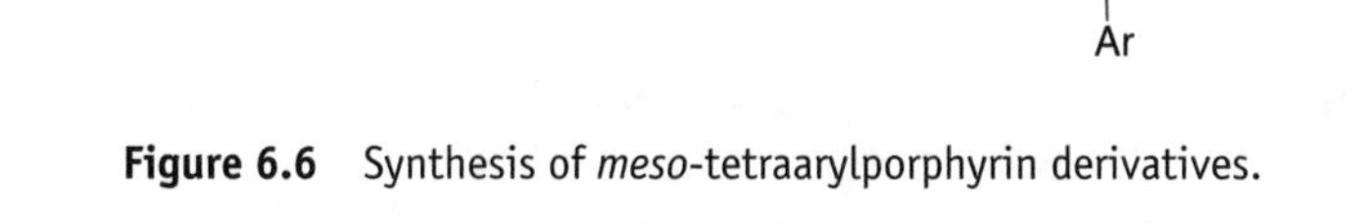

Figure 6.6 Synthesis of *meso*-tetraarylporphyrin derivatives.

Thus, the basic *meso*-compound 5,10,15,20-tetraphenylporphyrin can be synthesised from benzaldehyde and pyrrole with an acid catalyst, and the facility of this approach can be employed to produce a wide range of tetrasubstituted derivatives, i.e. having four identical aryl units.

The provision of derivatives where not all of the *meso*-positions are substituted is somewhat more complex, depending on the pattern required. For example, the tetrasulphonic acid discussed above is synthesised via the direct sulphonation of *meso*-tetraphenylporphyrin, thus there is the potential for production of a mixture of porphyrinphenylsulphonic acids, although the preparation is straightforward if carried out correctly (Srivastava and Tsutsui 1973). The discussion of partial sulphonation and other functionalisations is more relevant to the phthalocyanines and is therefore covered in Chapter 7.

For derivatives where the asymmetry depends on the reaction of different starting benzaldehydes, advances have been made in tetraaryl-substituted syntheses, so that routes to asymmetric derivatives are relatively straightforward, with A_2B-, A_3-, A_3B-, and A_2BC-porphyrins being obtainable (Figure 6.7, where *meso*-tetraphenylporphyrin = A_4) (Wiehe *et al.* 2005).

The availability of such substitution patterns allows some control of the hydrophilic/lipophilic balance of the resulting molecules, which is important in tailoring the physicochemical profile of candidate structures. Also, the inclusion of reactive groups in one or more of the aryl moieties allows the construction of more complex molecules.

The use of dipyrrolemethane allows the synthesis of 5,15-disubstituted derivatives (Figure 6.8), again allowing some control over the physicochemical properties of the resulting compounds. This can be extended to 5,15-A_2-10,20-B_2-type compounds via the use of substituted dipyrrolemethanes.

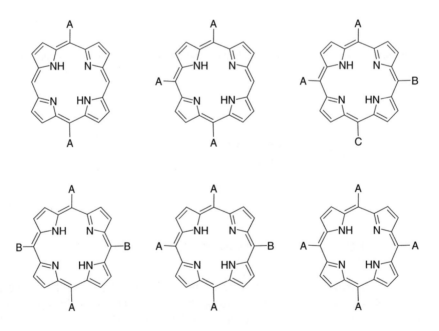

Figure 6.7 Substitution patterns in *meso*-porphyrins.

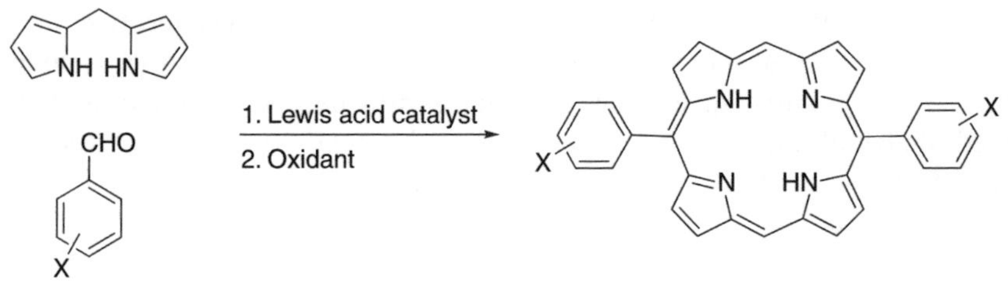

Figure 6.8 Synthesis of 5,15-disubstituted porphyrins.

Synthesis of *meso*-porphyrins having adjacent aryl pendants is slightly more involved, relying on monoformylation of the aryldipyrrolmethane followed by aroylation of the remaining pyrrole residue and subsequent cyclisation with dipyrrolemethane (e.g. Figure 6.9).

Functional groups may be included via the above substitution patterns, not only to dictate the hydrophilic/hydrophobic balance but also to endow positive or negative charge, which can exert a considerable effect on cellular uptake and distribution. Neutral groups may also, of course, be included.

While the most common of the anionic *meso*-compounds is the tetrasulphonic acid, its cationic analogue is undoubtedly the tetrapyridinium salt, derived from pyridine-4-carboxaldehyde, the tetrasubstituted material often being converted to the tetraquaternary salt with iodomethane or dimethylsulphate.

Perhaps unsurprisingly, the tetracationic derivative is known to interact with DNA, although there has been considerable debate as to the nature of this interaction. Although intercalation has been proposed, as with other cationic photosensitisers

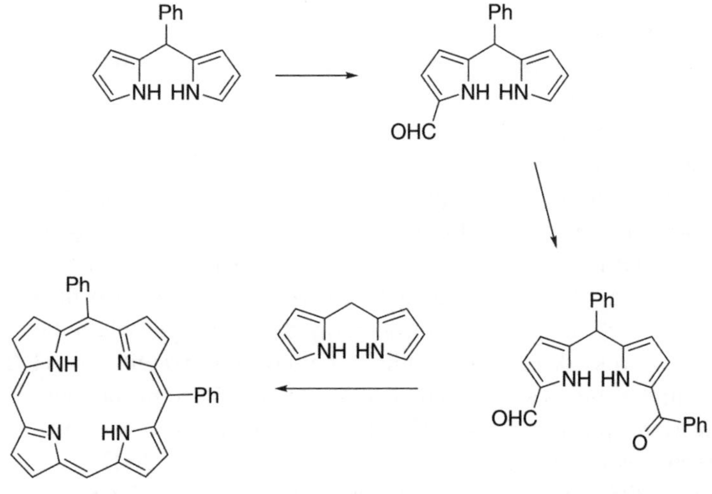

Figure 6.9 Synthesis of 5,10-diphenylporphyrin (Briñas and Brückner 2002).

(Lipscomb *et al.* 1996), external binding is also a possibility (Fedoroff *et al.* 2000). However, it may be that molecules such as *meso*-tetra(*N*-methyl-4-pyridyl)porphyrin are specific for guanine quadruplex structures prevalent in telomers and that the expanded shape of these quadruplexes allows intercalation (Lubitz, Borovok and Kotlyar 2007).

6.3   Amino derivatives

The efficacy of cationic *meso*-porphyrins, such as the pyridyl derivatives, in PDT has encouraged the synthesis of primary amino ($-NH_2$) analogues. As with the sulphonic acid derivatives, this is not a straightforward proposition, in this case due to cross-reaction. However, it is possible to nitrate *meso*-tetraphenylporphyrin using sodium nitrite and trifluoroacetic acid (Luguya *et al.* 2004). The reaction is reportedly selective for the 4-phenyl position and the resulting nitro derivatives are easily reduced to the corresponding amines using tin(II) chloride and hydrochloric acid. As mentioned elsewhere, the amino function allows linking to other molecules, targeting moieties and the like.

Similarly, *meso*-porphyrins bearing aryl substituents may be further functionalised by nitration with fuming nitric acid. The strong electron withdrawing effect of the resulting nitro groups also allows further reaction of the functionalised rings with carbanions, e.g. of the methanesulphonyl type (Ostrowski, Mikus and Łopuszyńska 2004). *Meso*-tetraphenyl derivatives having up to 10 substituents in total (e.g. Figure 6.10) have been reported, with the potential, again, for further reaction via the nitro groups (≤ 3) as above. Clearly, in such situations, the greater the number of reactions carried out, the greater the need for purification in order to ensure optimal substitution at each stage.

The hydrophilic effect of the quaternised pyridine is a standard method of aiding aqueous solubility in drug molecule development and is a reasonably common phenomenon in porphyrin photosensitisers. This is illustrated by the series of *N*-methyl-4-pyridyl derivatives, with log*P* values decreasing with consecutive substitutions, although the tetraphenylsulphonic acid is considerably more hydrophilic (Table 6.1) (Meng *et al.* 1994). The slightly greater hydrophilicity of the 10,20-disubstituted derivative ('*trans*') relative to the 15,20- ('*cis*') analogue may be rationalised by the greater distribution of the dipositive charge in the former allowing more efficient salvation of the molecule.

As with other photosensitisers, differences in charge and lipophilicity have a profound effect on the cellular localisation of the *N*-pyridyl and phenylsulphonic acid derivatives. This can be seen with the tetrasubstituted derivatives, which obviously differ in charge by eight units, but which also exhibit considerably different hydrophilicities (Table 6.1). Differences are also stark in tumour cells in culture. For example in HT-29 cells, the cationic derivative exhibits a 10-fold higher uptake than the tetrasulphonic acid, with the *cis-/trans*-dicationic porphyrins 5- to 10-fold higher again. However, cell killing appears far more effective with the tetrasulphonic acid (Carvalho *et al.*2002).

There are also differences in organellar uptake for the two derivatives: nuclear for the cationic species, supported by its interactions with nucleic acids, while the tetrasulphonic

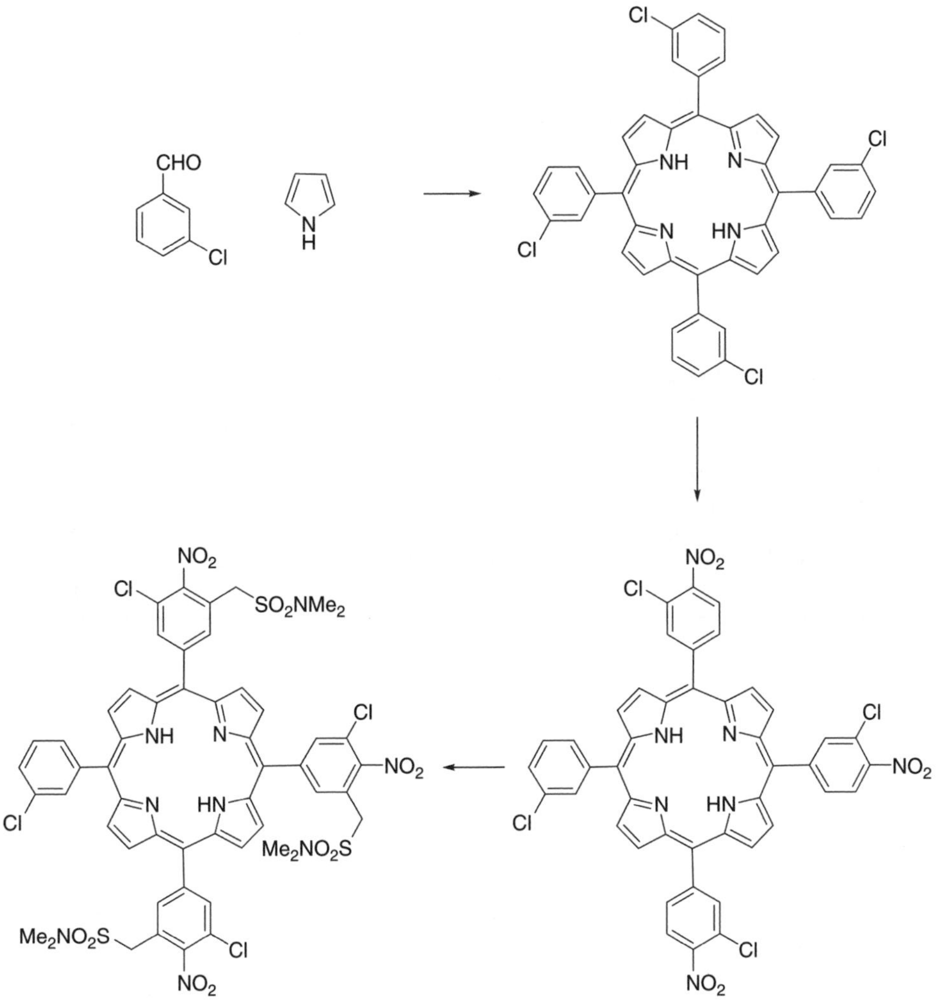

Figure 6.10 Synthesis of polysubstituted *meso*-tetraarylporphyrins.

acid derivative is known to be taken up by the lysosomes (Jiménez-Banzo *et al.* 2008). However, lysosomal targeting has also been reported for tetracationic *meso*-porphyrins having three pendant 4-*N*-methylpyridinium groups, and a fourth where the *N*-alkyl chain was varied in order to facilitate a range of lipophilicities (Figure 6.11). In cell culture, photoinactivation was observed to correlate well with accumulation and increasing lipophilic character, although the derivatives were not found in mitochondria (Ricchelli *et al.* 2005). It is also possible to target the mitochondria via increasing the lipophilic character of symmetrical tetracationic derivatives. For example, *meso*-tetra(*N*-decyl-4-pyridyl)porphyrin is highly selective for and photodamaging to mitochondria in HeLa cells (Cernay and Zimmerman 1996).

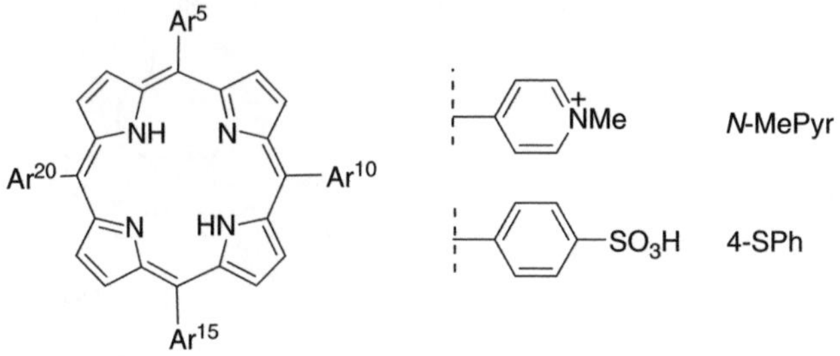

Table 6.1 Structure and hydrophilic/lipophilic balance in ionic *meso*-arylporphyrins

Ar^5	Ar^{10}	Ar^{15}	Ar^{20}	Charge	log$P$
Ph	*N*-MePyr	Ph	*N*-MePyr	+2	+1.11
Ph	Ph	*N*-MePyr	*N*-MePyr	+2	+1.23
Ph	*N*-MePyr	*N*-MePyr	*N*-MePyr	+3	−0.06
N-MePyr	*N*-MePyr	*N*-MePyr	*N*-MePyr	+4	−0.36
4-SPh	4-SPh	4-SPh	4-SPh	−4	< −2.0

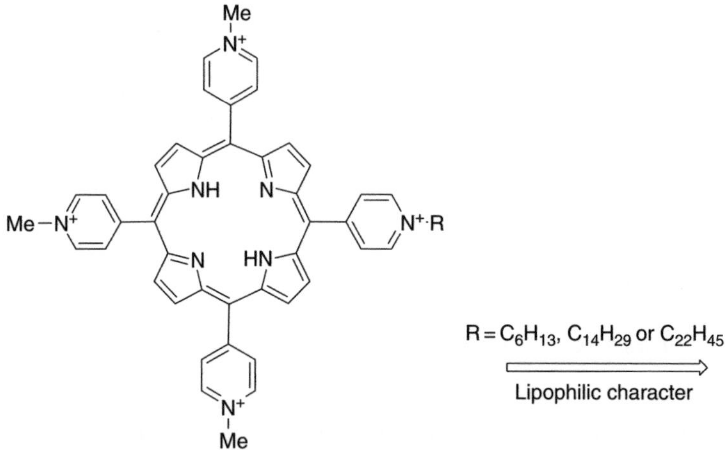

Figure 6.11 Asymmetrically substituted tetracationic *meso*-porphyrin allowing variation in lipophilicity.

Although the ionic *meso*-porphyrins were originally intended for anticancer application, the re-emergence of photoantimicrobial research in the 1990s has also seen various series screened for activity in this field also. Not surprisingly, given the appreciation of charge involvement in other chromophoric types, it is apparent that while Gram-positive bacteria are susceptible to anionic, neutral and cationic *meso*-porphyrins, only the latter group is regularly successful against Gram-negative bacteria (Banfi *et al.* 2006). Indeed, dicationic *meso*-porphyrins have been shown to be effective DNA photocleaving agents (Wu *et al.* 2006). The argument regarding the presumptive use of proven anticancer therapeutics in antibacterial applications remains illogical to the author.

Related derivatives to the quaternary nitrogen species are those based on phosphorus. Although this substitution represents a less active area of research, it is nevertheless relevant, particularly in view of the niche exploited using such compounds. While much is made of the PDT of cancer and photoantimicrobial approaches to bacteria, viruses and fungi, the field of tropical disease is rather less investigated. Work, particularly by Boyle in the UK, has involved the use of both quaternary nitrogen and phosphonium salt derivatives of *meso*-porphyrins for intended application in cutaneous leishmaniasis, a protozoal disease caused by sandfly bites, which had 12 million sufferers worldwide in 2005 (Bristow *et al.* 2006).

Highly lipophilic *meso*-derivatives may be synthesised from various patterns, i.e. via increasing the hydrocarbon content in side chains, etc. However, this can be approached symmetrically or asymmetrically, providing molecules that are evenly or 'regionally' lipophilic. For example, *meso*-compounds of the A_3B- pattern have been reported, with a single aromatic unit of the four having a long alkyl chain attached via an ether linkage.

Compounds of this sort having a single, linear C_{16} pendant chain (e.g. Figure 6.12) have been reported to exhibit effective uptake into mammalian cell cytoplasm *in vitro*, coupled with comparable yields of singlet oxygen to precursor porphyrins, and high photocytotoxicity/low dark toxicity, again *in vitro*. A likely problem with this approach lies in the low aqueous solubility of such derivatives, which requires the use of co-solvents, liposomes, etc. for drug delivery.

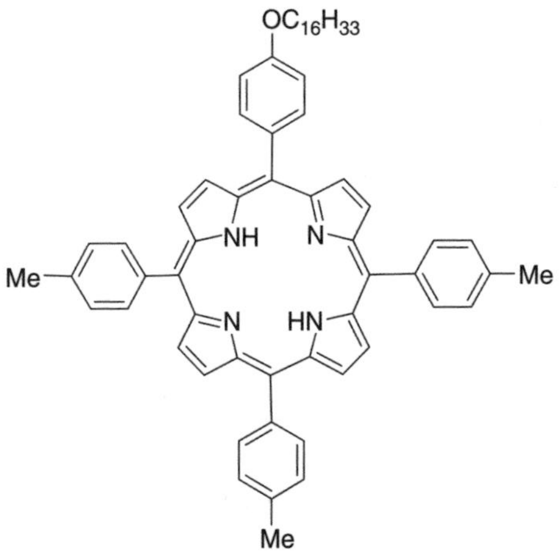

Figure 6.12 5-Hexadecyloxy-10,15,20-trimethylporphyrin.

The use of aryl pendants having reactive groups provides an ease of functionalisation of the resulting tetraarylporphyrins and allows the tethering of biologically significant molecules to aid in photosensitiser targeting. For example, KB nasopharyngeal tumour cells are known to overexpress receptors for folate, thus the attachment of folic acid, via diamino-tether molecules, to 5(4-carboxyphenyl)-10,15,20-triphenylporphyrin (Figure 6.13) reportedly furnishes conjugate compounds exhibiting large increases (more than seven-fold) in tumour uptake and concomitant photocytotoxicities compared to tetraphenylporphyrin in this cell line. The uptake was antagonised by folic acid (Schneider *et al.* 2005). A similar approach has been reported for oestrogen-dependant tumours, as in early-stage breast cancer, the porphyrin nucleus being attached to oestradiol (Figure 6.13) (James *et al.* 1999).

In terms of the effects of substitution in *meso*-compounds, clearly maintenance of photosensitisation ability is an essential criterion, but this may be compromised by the structural requirements of a target molecule. Thus, it is important to build up structure—function relationships involving the various 'building blocks' used. For example, amino and carboxylic acid residues are often employed as linkers between the porphyrin chromophore and a biological targeting residue, as above. However, work concerning *meso*-tetraphenylporphyrins has shown that a free amino ($-NH_2$) function at position 4 in a single phenyl ring (Figure 6.14) leads to higher triplet and singlet oxygen yields than either the weaker electron-releasing acetamido ($-NHCOMe$) group or the electron-withdrawing carboxylic acid ($-CO_2H$) derivative (Dudkowiak, Teślak and

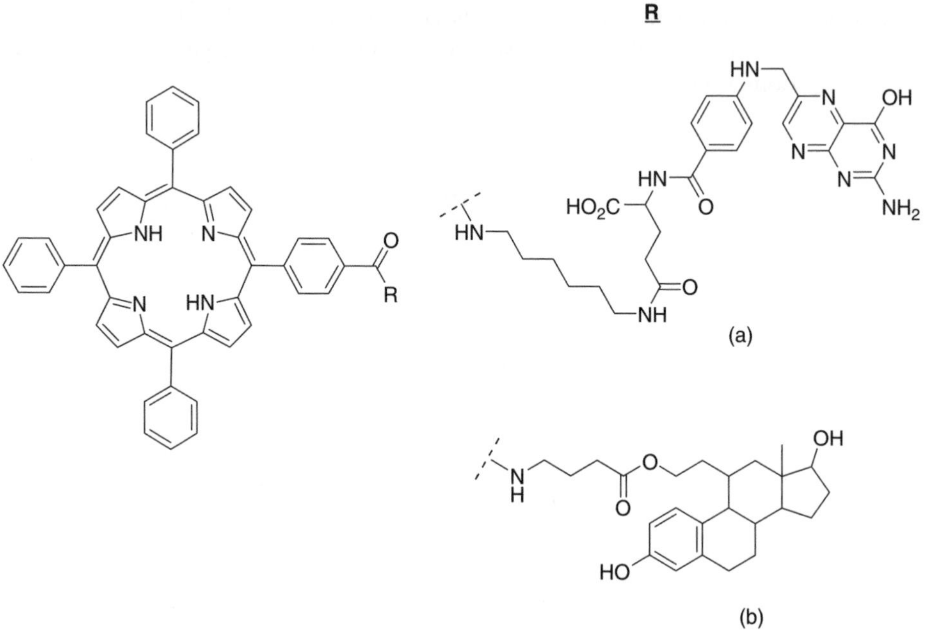

Figure 6.13 (a) *meso*-Porphyrin–folic acid conjugate and (b) oestradiol conjugate.

Figure 6.14 *meso*-5-(4-Aminophenyl)-10,15,20-(4-methylphenyl)porphyrin.

Habdas 2006). Thus, in order to maintain optimal photosensitising activity, as with other classes of photosensitisers, it is perhaps sensible to link via a substituted amine residue.

The activity of porphyrin oligomers in preparations such as HpD (e.g. dihaematoporphyrin ether) has led to the investigation of model dimers as photosensitisers (Pandey, Smith and Dougherty 1990). In addition, the use of sugar moieties as pendant property modifiers ($\log P$, $^1O_2$) and recognition units in linked porphyrin photosensitisers (Figure 6.15) has been employed as a method of improving antitumour profiles, although such glycosylated bis-porphyrin derivatives appear to be less efficient gross cell killers than Photofrin.

Other derivatives having lower glycosylation levels, e.g. single aminoglucose moieties, reportedly exhibit improved tumour recognition/uptake profiles (Di Stasio *et al.* 2005). In addition, the inclusion of glycosyl units in quaternised monopyridyl-triphenylporphyrin analogues has been associated with decreased dark toxicity, compared with cationic porphyrins with no glycosyl moieties (Ahmed *et al.* 2004), underlining the suggestion of increased essential targeting due to the inclusion of the sugar residues.

In contrast to the *meso*-tetraaryl systems, octasubstitution of the porphyrin nucleus has also been reported, i.e. each of the β-pyrrole positions of the ring system bearing a substituted phenyl residue (Figure 6.16) (Gan *et al.* 2005). However, in order to achieve singlet oxygen yields suitable for use in PDT, the inclusion of halogen atoms was required.

Analogous *meso*-substituted compounds have also been produced with cationic moieties to facilitate DNA interaction (Figure 6.16) (Chen *et al.* 2006). Unfortunately in each case, the long-wavelength band around 650 nm was not significantly more intense than normal *meso*-tetraarylporphyrins ($\log\varepsilon_{max} \sim 3$).

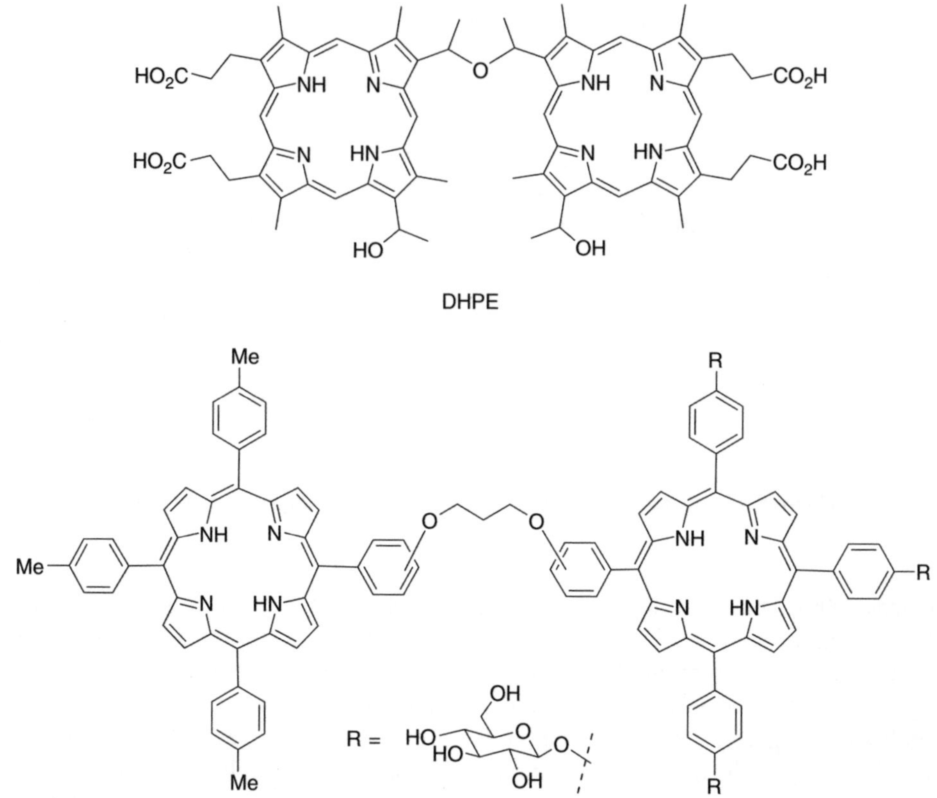

DHPE

Figure 6.15 Comparison of dihaematoprphyrin ether (DHPE) and an ether-linked glycosylated di-*meso*-porphyrin.

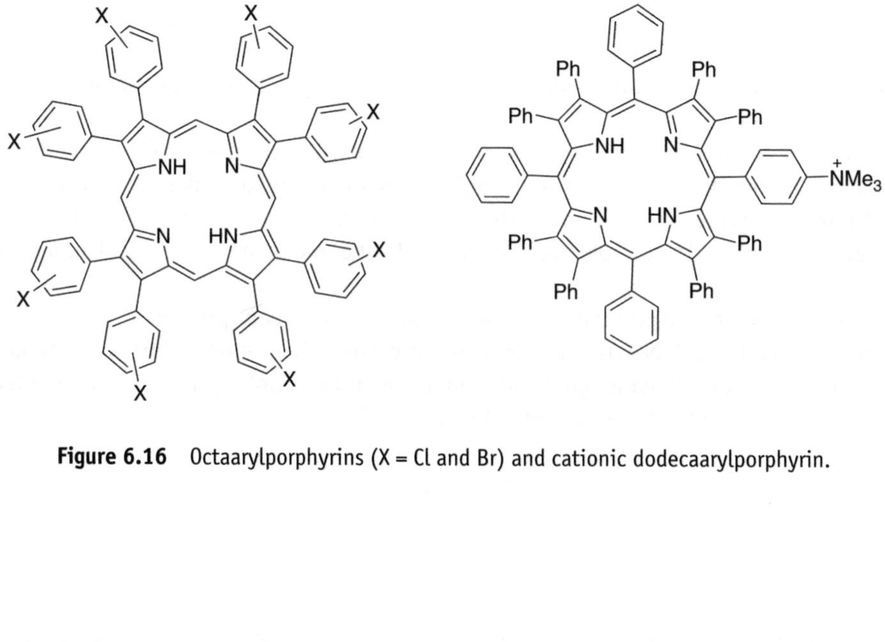

Figure 6.16 Octaarylporphyrins (X = Cl and Br) and cationic dodecaarylporphyrin.

6.4 Hetero-porphyrins

As with other areas of photosensitiser development, heteroatom replacement provides altered chromophore electron density/distribution and thus can enhance singlet oxygen production, in line with heavy atom effects discussed elsewhere.

For example, the replacement of nitrogen for sulphur or selenium as the central ligand atoms in the porphyrin system furnishes the 21,23-dithia- or diselenoporphyrin (Figure 6.17). In culture, tetrasulphonic acid derivatives are reportedly more effective photosensitisers against tumour cell lines than the corresponding porphyrin derivative. In addition, the thio- or seleno-based photosensitisers are activated at significantly longer wavelengths (*c.* 690 nm) (Stilts *et al.* 2000).

The synthesis of sulphur/selenium *meso*-analogues starts from reaction of the dilithiated thiophene or selenaphene with a benzaldehyde or a derivative, with subsequent reaction of the produced diol with pyrrole (Figure 6.17). As with conventional *meso*-porphyrins, the heterocyclic product may be functionalised further to provide water-solubilising groups (Hilmey *et al.* 2002).

Peripheral groups used in these heteroporphyrins are similar to those used in the parent *meso*-porphyrins, including sulphonic/carboxylic acids and quaternary nitrogen

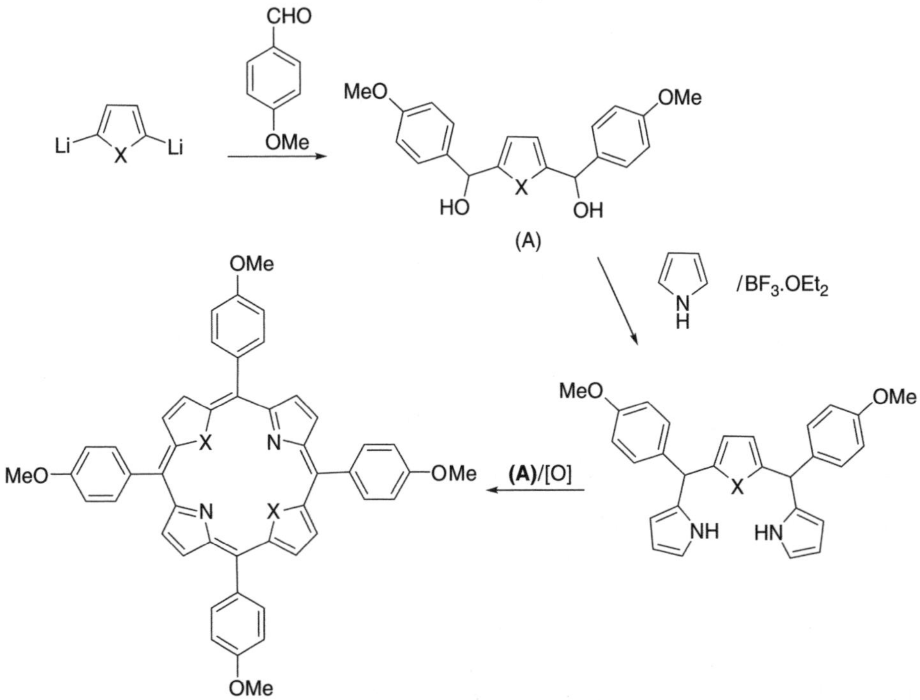

Figure 6.17 Synthesis of thia-/selenaporphyrins (X = S and Se).

moieties (Gupta and Ravikanth 2006). Reduction of the β-olefinic moieties is also possible, yielding chlorin-type derivatives.

6.5 Chlorins

Compared to the standard porphyrin ring system, chlorins (e.g. chlorin e_6, Figure 6.18) usually have high triplet yields and concomitant photoproperties, offering advantages over standard HpD-type photosensitisers (Shan *et al.* 2006).

The weak red light absorption associated with the porphyrin nucleus is much more intense for the chlorins. This change is achieved by the reduction of one of the peripheral (β-)olefinic

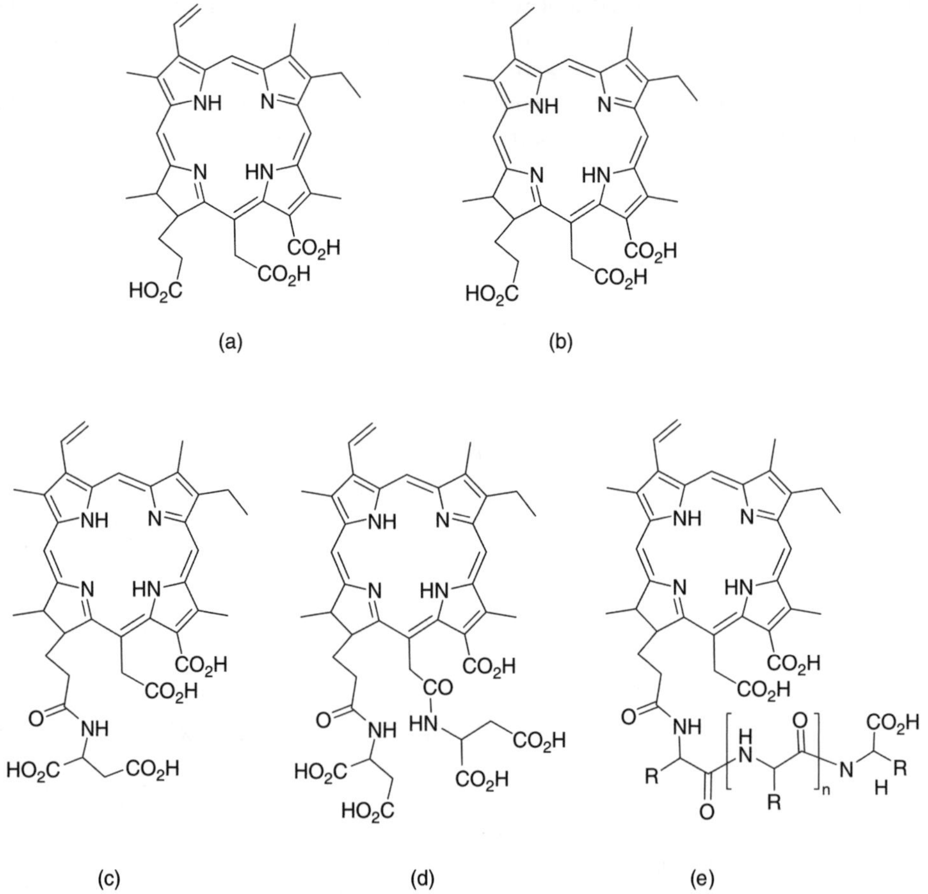

Figure 6.18 Chlorin e_6 derivatives: (a) chlorin e_6, (b) mesochlorin e_6, (c) mono-L-aspartyl chlorin e_6 (Talaporfin), (d) di- L-aspartyl chlorin e_6 and (e) chlorin e_6-poly(lysine) conjugate (R = —CH$_2$CH$_2$CH$_2$CH$_2$NH$_2$).

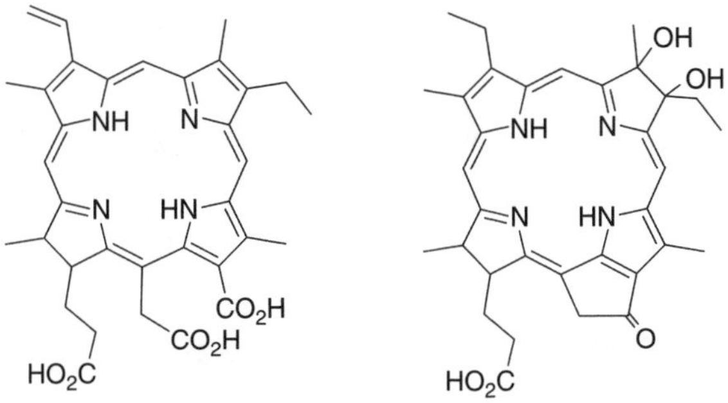

Figure 6.19 Differences in ring structure between chlorins and bacteriochlorins.

double bonds of the porphyrin system (Figure 6.19), which also moves the electronic absorption bands to considerably longer wavelengths. Similar changes are apparent with bacteriochlorins.

While the chlorins and bacteriochlorins (e.g. Figure 6.19) owe their greater λ_{max} values to the reduction of β-olefinic regions in the porphyrin chromophore, both chlorins and bacteriochlorins are often more structurally complex than the *meso*-porphyrins discussed above, e.g. due in many cases to natural product derivation and also where there has been peripheral extension following Diels–Alder reactions of simpler vinylporphyrins.

There are many different structures available of these types, which are derived from porphyrin type structures that occur naturally, for example those derived from chlorophylls, as shown in Figure 6.20.

The much improved photoproperties of the reduced-chromophore photosensitisers coupled with their relatively difficult syntheses have led to semi-synthetic approaches being adopted, based on both the chlorins themselves and also the basic porphyrin structures, where reaction generates new compounds using – and thus reducing – the β-olefinic groups, e.g. by Diels–Alder addition.

Although, as mentioned, the majority of chlorin and bacteriochlorin photosensitisers are natural products or their semi-synthetic derivatives, structurally simpler bacteriochlorophyll derivatives have also been investigated, for example based on reduced *meso*-porphyrins (Figure 6.21). The reduced tetra(*N*-methyl-3-pyridyl) tetrahydro *meso*-substituted porphyrin shown has an associated long-wavelength absorption (760 nm) and both good tumour selectivity and tumour kill on illumination (Schastak *et al.* 2005).

Similarly to the conversion of vinyl porphyrins to benzannelated derivatives via Diels–Alder reactions (e.g. benzoporphyrin derivative (BPD)), the extension of the ring system via reaction of the peripheral pyrrole olefinic moiety (β,β-double bonds, suitably activated by an electron-withdrawing group) of *meso*-tetraarylporphyrins with *ortho*-xylylene (from 1,3-dihydrobenzo[*c*]thiophene 2,2-dioxide) has been reported, furnishing derivatives such as that shown in Figure 6.22 (Ostrowski and Wyrębek 2006). Large bathochromic shifts reported resulted (to *c.* 710 nm).

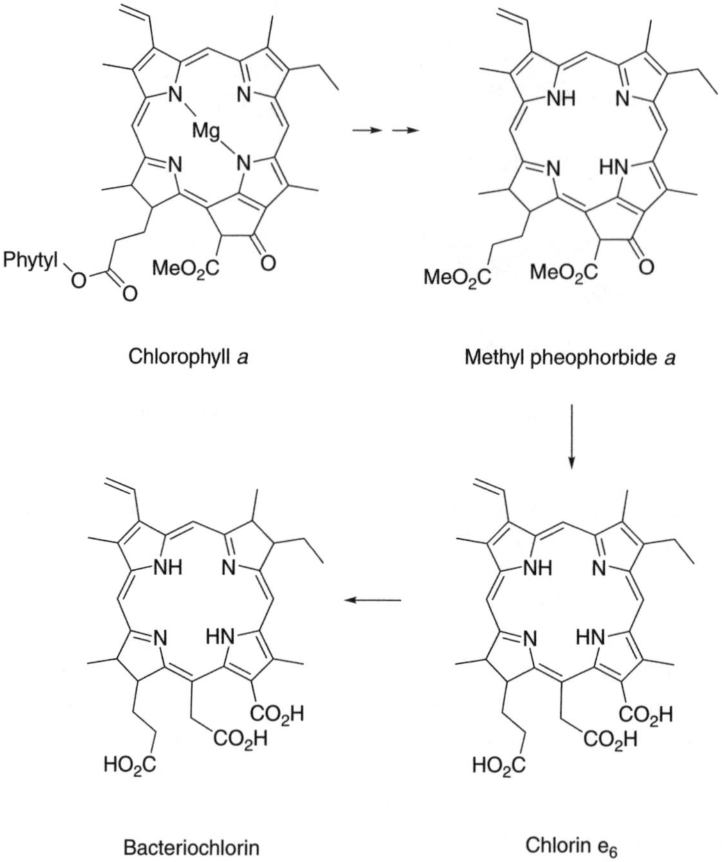

Figure 6.20 Photosensitisers derived from chlorophyll *a*.

Other porphyrin types offer different opportunities for functionalisation. For example, several bacteriochlorins, e.g. bacteriochlorin *p*, have been produced containing a six-membered anhydride ring formed from two of the carboxylic acid residues of a precursor molecule such as chlorin e_6. (Figure 6.23, X = O). This anhydride may be converted to a cyclic imide via reaction with a suitable primary amine, the amine side chain allowing access to a range of derivatives for structure–activity study. In addition, the cycloimide derivatives (Figure 6.23, X = N-alkyl) have improved light absorption characteristics (*c.* 800 nm) and good singlet oxygen yields (0.54–0.57) and were considerably more active against tumour cells *in vitro* than a standard photosensitiser (Photofrin) (Sharonov *et al.* 2006).

Similarly improved performance has been demonstrated by chlorin cycloimides. For example, [*N*-(3-hydroxypropyl)]cycloimide chlorin p_6 exhibited excellent short drug–light interval antitumour toxicity in mice, having a high tumour:skin ratio (1:8–10) at 1.5 hours post injection. Again, the cycloimide derivative also exhibited a suitable long-wavelength absorption for superficial illumination of the tumour site (711 nm) (Karmakova *et al.* 2006).

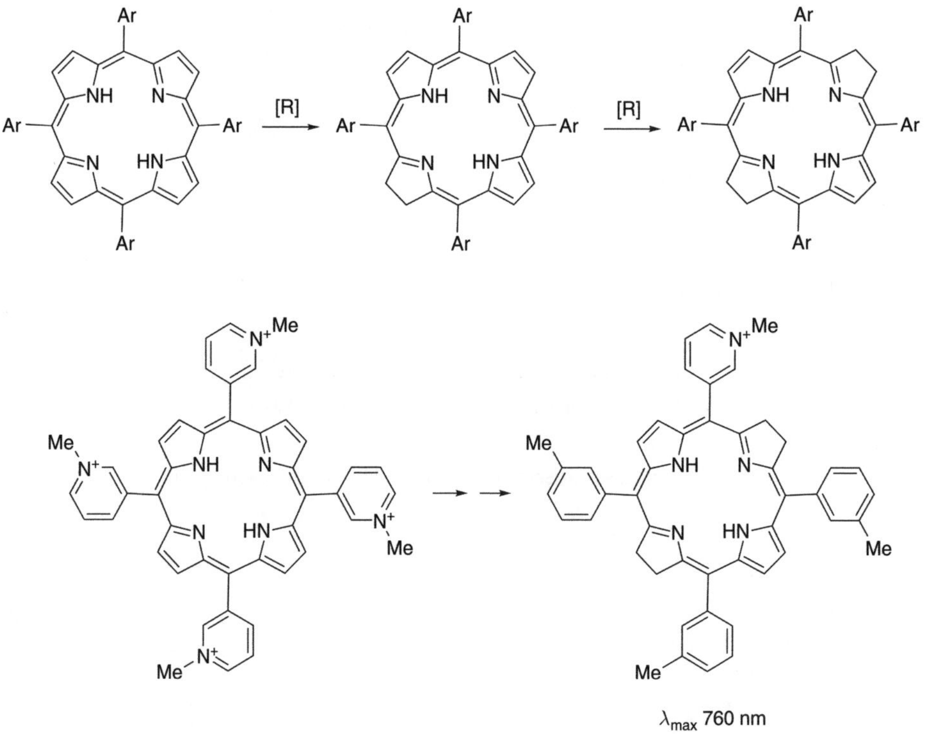

Figure 6.21 Reduced *meso*-porphyrins.

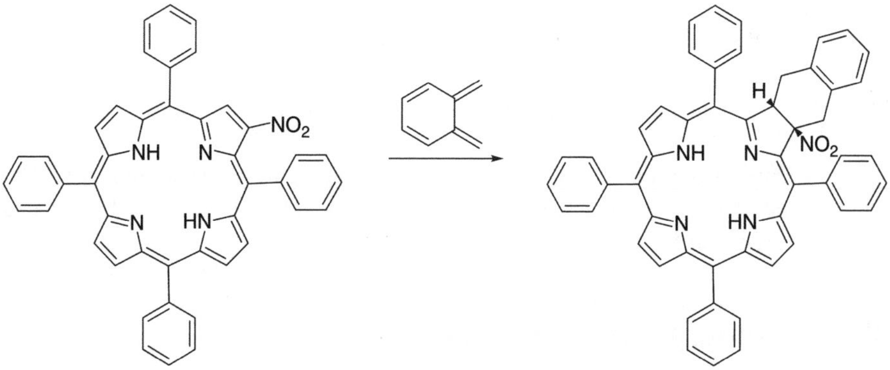

Figure 6.22 Diels–Alder reaction of an activated *meso*-porphyrin.

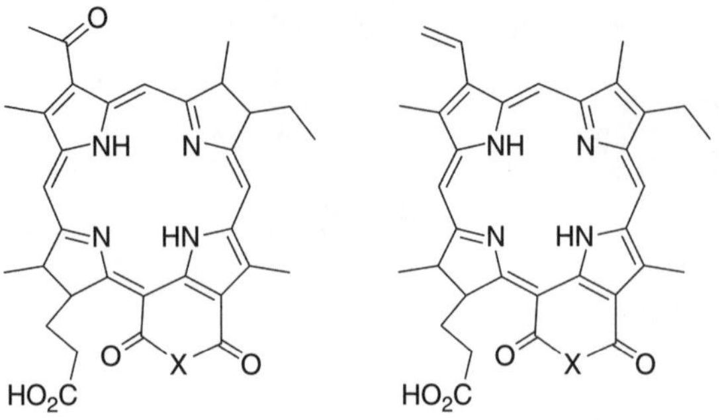

Bacteriochlorin *p*   Chlorin p_6

X = O → N-Hexyl   X = O → $NCH_2CH_2CH_2OH$

Figure 6.23 Photosensitisers derived from chlorins and bacteriochlorins.

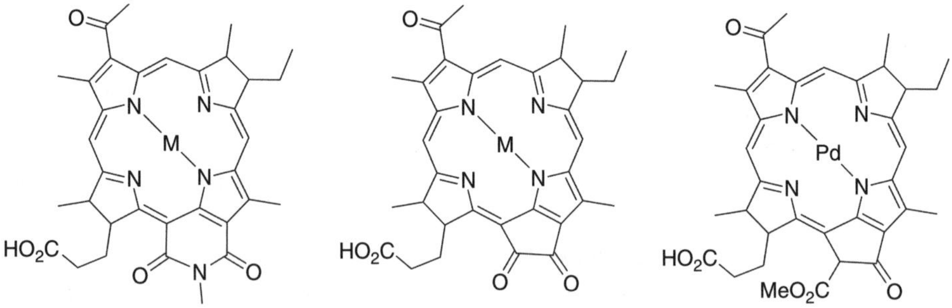

M = Zn, 838 nm   M = Zn, 788 nm   Tookad 762 nm
M = Pd, 819 nm   M = Pd, 771 nm

Figure 6.24 Long-wavelength-absorbing metal bacteriochlorins.

Bacteriochlorins having central metal ion inclusion (Figure 6.24) are red shifted from the non-metallated parent and have associated high singlet oxygen yields (Fukuzumi *et al.* 2008). See also Section 6.8.

6.6 Benzoporphyrin derivative

Reaction of vinylogous porphyrin with activated olefins via the Diels–Alder route can lead to several extended chromophore products. Of these, the premier is BPD, from which is derived the highly successful mono-acid ring A derivative (BPD-MA, verteporfin, Figure 6.25).

The initial BPD adduct from the cycloaddition may be altered at either pyrrole ring A or pyrrole ring B. However, after rearrangement of the new diene and partial hydrolysis to

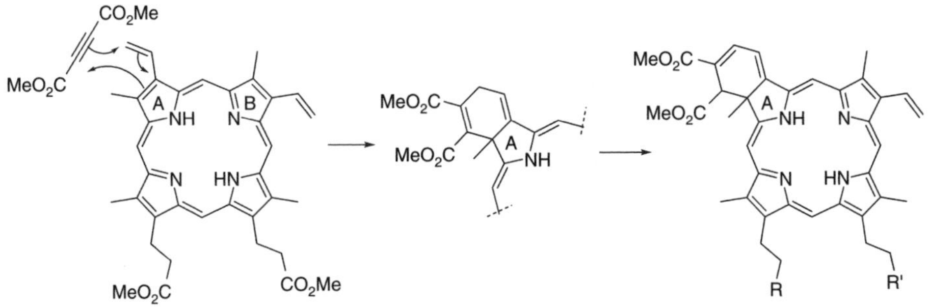

Figure 6.25 BPD-MA formation from PPIX. Either R or R' = CO_2H, the other is $-CO_2Me$.

convert one of the propionic esters, the greater photodynamic activity lies with the ring A derivative. Plainly, there are many different dienophiles that would lead to similar products, but no overall improvements on BPD-MA have been reported via this route.

Improvements have been made to the periphery of the BPD macrocycle to provide photosensitisers that are, in cell culture at least, significantly less dark toxic, while maintaining the same or similar phototoxicity to the lead compound (Yao *et al.* 2008). The successful variants were synthesised from chlorophyll-*a* derivatives, thus providing a single diene moiety in the correct region (Figure 6.26).

Interestingly, there are known naturally occurring benzoporphyrins (Figure 6.27), isolated from crude oil, these being formed via sedimentary Diels–Alder reactions (Lash 1993). It has not yet been established whether this may be a useful source of new porphyrinoid photosensitisers.

BPD-MA has been introduced (as Visudyne) for the treatment of age-related macular degeneration, rather than for conventional PDT of tumours, once again exhibiting the vascular localisation of lipophilic anionic porphyrins. The short drug-to-light interval and lack of skin photosensitivity reactions were key factors in the selection of this derivative, which has become an important drug in ophthalmology.

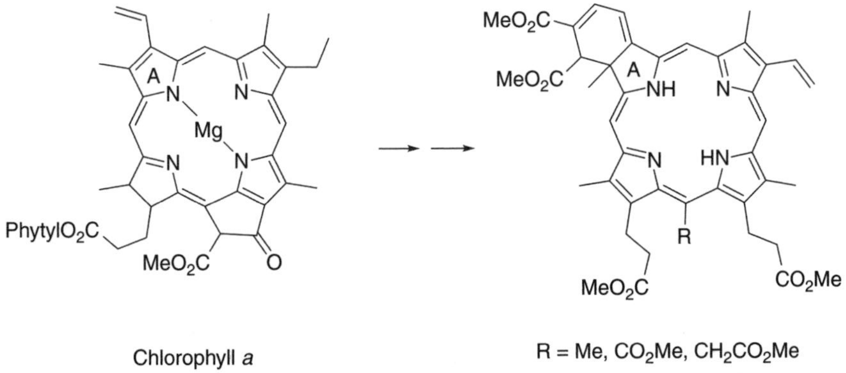

Chlorophyll *a*

R = Me, CO_2Me, CH_2CO_2Me

Figure 6.26 Synthesis of improved BPDs from chlorophyll *a*.

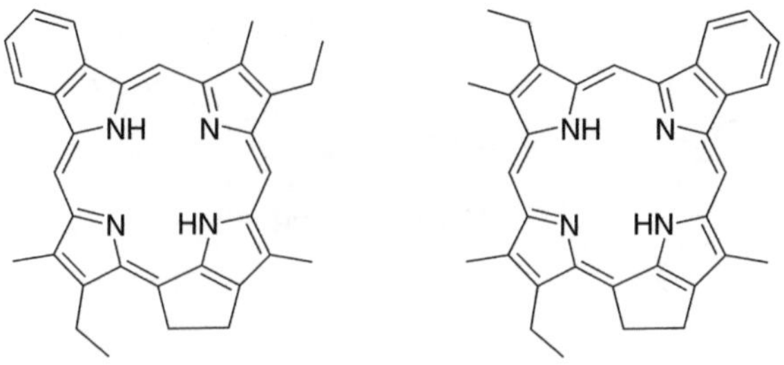

Figure 6.27   Naturally occurring BPDs.

6.7 Temoporfin

Along with BPD-MA, Temoporfin ((*m*THPC); Figure 6.28) has been the most successful second-generation porphyrin-derived photosensitiser. It has a range of proposed PDT indications: malignant mesothelioma (Ris *et al.* 1997), skin cancer (Kubler *et al.* 1999), prostate (Zhu and Finlay 2006), malignant glioma (Molinari *et al.* 2007), Barrett's oesophagus (Lovat *et al.* 2005) and head and neck cancer (Dilkes and Bapat 2004).

The bacteriochlorin *m*THBPC (i.e. further reduced form of Temoporfin) has a significantly greater λ_{max} (740 nm *versus* 652 nm in water), but poorer singlet oxygen efficiency compared to the chlorin and significant oxidation to the chlorin have been reported. The obvious gain in tissue penetration with *m*THBPC, which is also much more strongly light absorbing, may thus be balanced by poorer photosensitising performance (Grahn *et al.* 1997).

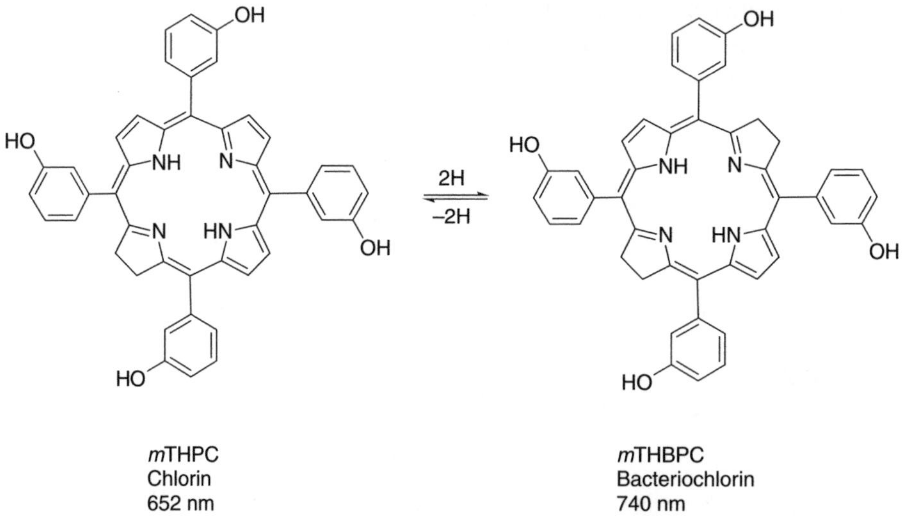

*m*THPC
Chlorin
652 nm

*m*THBPC
Bacteriochlorin
740 nm

Figure 6.28   *m*THPC and its reduced form, *m*THBPC.

The success of Temoporfin obviously suggests its use as a lead compound in the search for new, improved PDT photosensitisers. Given the level of sophistication exhibited in modern porphyrin synthesis, there are various routes available to this end, allowing both symmetrical and asymmetrical derivative production with a range of lipophilicities. Such work has generated a significant collection of Temoporfin analogues for biological testing (Wiehe *et al.* 2005). Increased aqueous solubility has been achieved by the addition of hydrophilic groups to the phenolic residues, e.g. to give four $-OCH_2CO_2H$ groups (Griffiths *et al.* 1998).

6.8   Tookad

Tookad is a palladium bacteriochlorin derivative (Figure 6.24). Given the lipophilic, anionic nature of the molecule (in common with most of the porphyrin derivatives employed in PDT), it is not surprising that its mode of action against tumours occurs via peritumoural vasculature shutdown. Although Tookad was originally intended for the treatment of bulky tumours, it has since been employed in the clinical treatment of prostate cancer, e.g. as a salvage treatment (Trachtenberg *et al.* 2007), and macular degeneration (Ruck, Bohmler and Steiner 2005). Usually after intravenous administration, Tookad is activated at 763 nm, and, as it is rapidly cleared from the circulation, the systemic photosensitisation problems associated with, for example, Photofrin or Temoporfin, are absent.

6.9   Purpurins

Tin ethyl etiopurpurin (SnET2, Figure 6.29) has been trialled for use in the ablation of prostate cancer (Selman 2007), in the treatment of choroidal neovascularisation

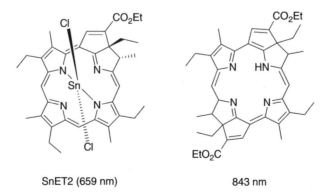

SnET2 (659 nm)   843 nm

Figure 6.29   Tin ethyl etiopurpurin (SnET2) and a long-wavelength-absorbing bacteriopurpurin.

(Renno and Miller 2001) and was used to treat the HIV-associated skin cancer, Kaposi's sarcoma (Allison *et al.* 1998). SnET2 is also known as Purlytin and Rostaporfin. Bacteriopurpurins (e.g. Figure 6.29) have also been designed for PDT purposes, notably due to their very-long-wavelength absorption at *c.* 850 nm (Robinson 2000).

6.10 Texaphyrins

Being an extended porphyrinoid, by virtue of the larger central space caused by the absence of one of the conventional nitrogen atoms of the tetrapyrrole structure, texaphyrins are designed to be able to hold larger metal ions, typically of the lanthanide type (Sessler *et al.* 1988). Consequently, various texaphyrin complexes have been investigated for anticancer activity, both conventional and photodynamic, such as those containing lutetium, gadolinium, europium, dysprosium and manganese (Donnelly *et al.* 2004).

The lutetium derivative (Figure 6.30) has entered clinical trials for the PDT of recurrent breast cancer, for the phototreatment of atherosclerosis in peripheral arteries and for the phototreatment of macular degeneration. Conversely, the gadolinium analogue may be employed as a radiation enhancer in metastatic brain cancer (Sessler and Miller 2000).

From a photosensitising point of view, lutetium texaphyrin exhibits much longer wavelength absorption than traditional porphyrins (732 nm) but retains associated similar yields of singlet oxygen (≤ 0.70). Other derivatives have much longer wavelength absorption, for example one of the cadmium texaphyrins reported had a λ_{max} value of 864 nm (Maiya *et al.* 1990). The magnesium-centred analogue (Figure 6.30) has also been suggested for use in PDT (Lanzo, Russo and Sicillia 2008), although the inclusion of smaller ions is associated with less impressive long-wavelength characteristics (Domenico Quartarolo *et al.* 2007). A range of other figures for singlet oxygen yield is given in Table 6.2.

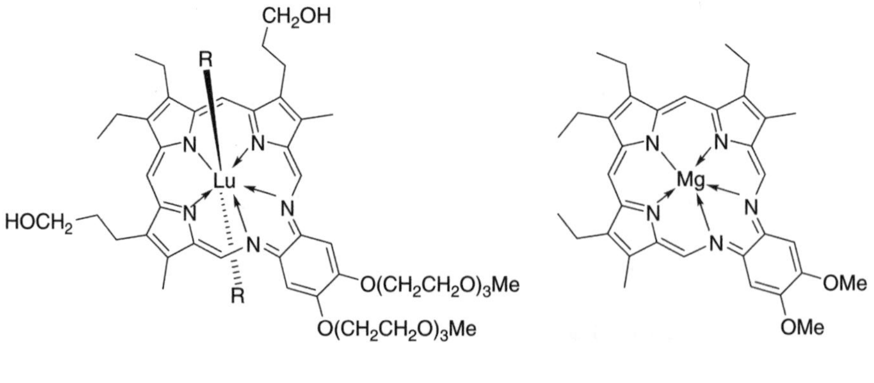

Figure 6.30 Lutetium and magnesium texaphyrins.

Table 6.2 Singlet oxygen yields for various porphyrin derivatives measured in pH 7.4 buffer/Triton X-100

Derivative	$^1O_2$ quantum yield at λ_{max}
Haematoporphyrin	0.74
Protoporphyrin IX	0.60
Zinc protoporphyrin IX	0.82
meso-Tetra(4-sulphonatophenyl)porphyrin	0.62
Photofrin	*c.* 1
Chlorin e$_6$	0.75
Benzoporphyrin derivative monoacid ring A	0.79
Pheophorbide-*a*	0.71
Methylene blue	0.49

Source: Fernandez *et al.* (1997).

6.11 Porphycenes

The porphycene chromophore is another porphyrin-related molecule – the porphyrin and porphycene structures are isomeric (Figure 6.31) – investigated as an alternative to haematoporphyrin photosensitisers, due to its increased red region absorption (due to lower molecular symmetry).

In vitro PDT results, as with other porphyrinoids, have been encouraging, with both peripheral alteration of the structure and metal inclusion, e.g. palladium(II) (Stockert *et al.* 2007). Tetraaza analogues (Figure 6.31) have been found to have more intense near-infrared absorption and higher internal conversion quantum yields (Rubio *et al.* 2006). Unsurprisingly, anionic porphycenes are known to exert a broad-spectrum photoantibacterial effect when bound to polylysine chains (Lauro *et al.* 2002).

6.12 5-Aminolaevulinic acid

Being a small simple amino acid, 5-aminolaevulinic acid (ALA) is a hydrophilic, low molecular weight molecule. It is absorbed easily through abnormal, but not through normal, keratin. ALA is absorbed by epidermal cells and is converted to the established photo-sensitiser protoporphyrin IX (PPIX), but not to any significant concentrations of haem, due

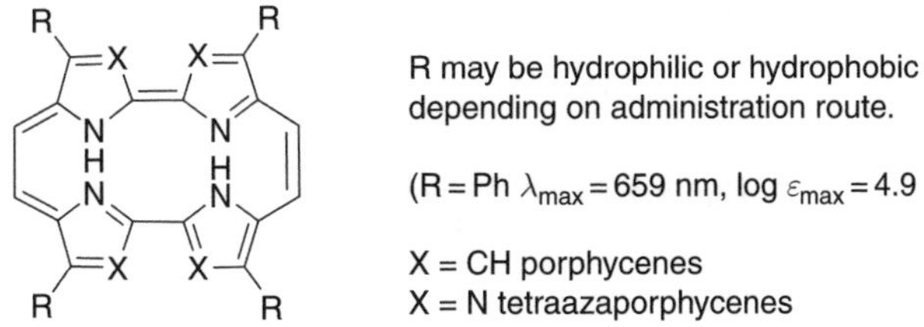

R may be hydrophilic or hydrophobic depending on administration route.

(R = Ph λ_{max} = 659 nm, log ε_{max} = 4.9

X = CH porphycenes
X = N tetraazaporphycenes

Figure 6.31 Basic porphycene photosensitiser geometry.

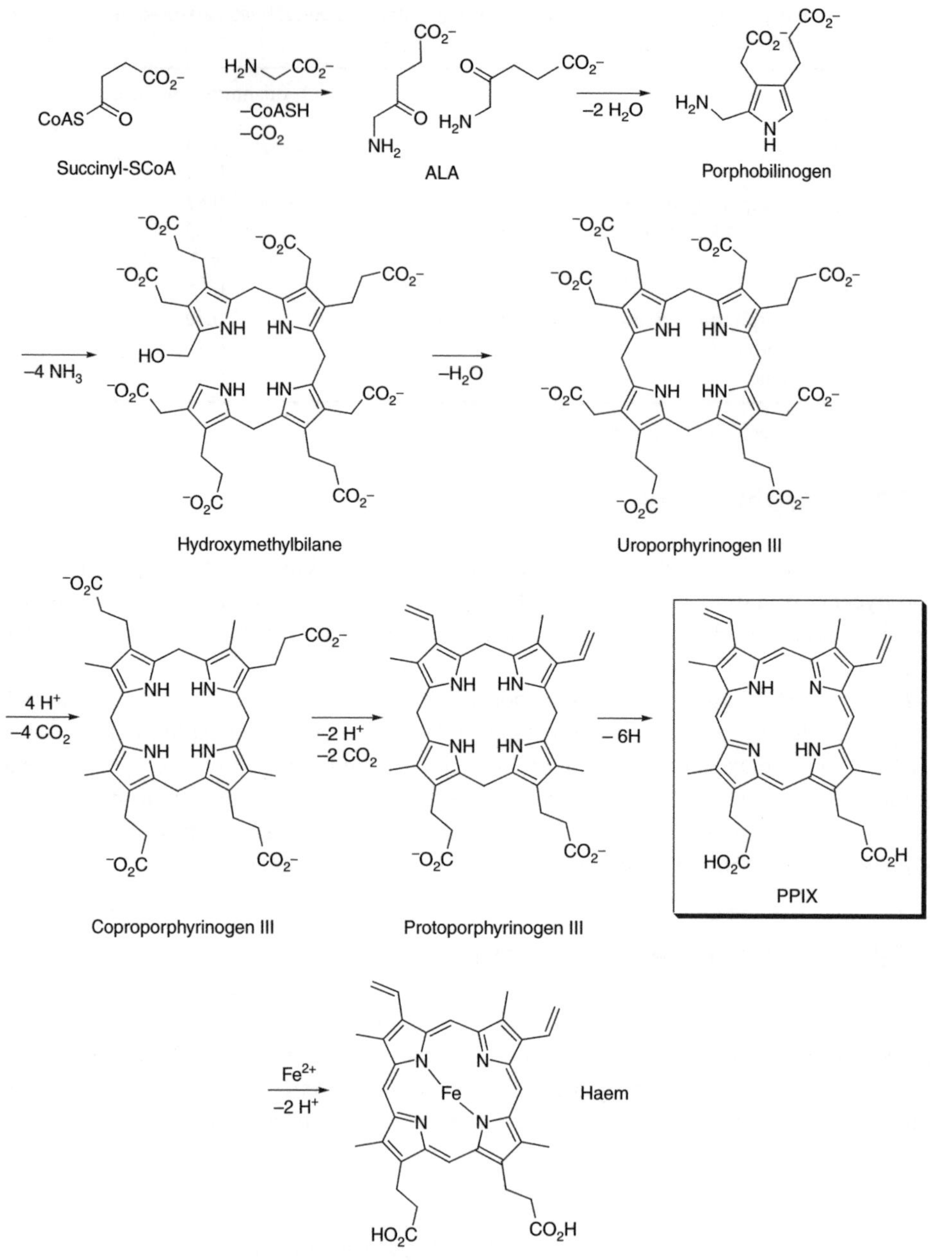

Figure 6.32 ALA and haem synthesis.

to limited supplies of iron, a necessary catalyst for ferrochelatase (Figure 6.32). Thus PPIX is accumulated in the target cells over a short period of time (≤ 4 hours), although a larger area of reaction may develop subsequently.

The utilisation of haem metabolism (i.e. the biosynthesis of PPIX) should be considered as one of the more important routes to selective tumour destruction using PDT. This is undoubtedly aided by increased rates of haem synthesis in tumour cells compared to normal cells, thus allowing rapid increase of the endogenous photosensitiser and thus PDT potentiation *in situ*. The fluorescence associated with protoporphyrin also allows useful tumour delineation via this route (see Chapter 12). Esters of ALA (e.g. methyl and hexyl) have been investigated over an extended period in order to increase uptake and ease of formulation. Logically, this is not an area of PDT research that produces many new derivatives, mainly due to the enzyme-fit requirement for PPIX biosynthesis, and derivatives are normally simple hydrolysable esters (see below).

ALA is approved for the treatment of actinic keratoses, it being the sole topical (pro-)photosensitiser available for dermatologic use in the USA (MacCormack 2008).

6.13 Esters

Several simple derivatives of ALA have been investigated, all of which are aimed at increasing the bioavailability of the porphyrin precursor, theoretically using the increased lipophilicity of the esters to facilitate membrane passage and thus cellular penetration (Juzeniene *et al.* 2006). In fact, ALA is taken up via BETA transporters, whereas the esters have different mechanisms, the more lipophilic examples exhibiting uptake by simple diffusion (Rodriguez *et al.* 2006). Higher esters, such as the hexyl (C_6) and undecyl (C_{11}) analogues, lend themselves to liposomal delivery (Di Venosa *et al.* 2008).

The topical application of ALA gives greater (i.e. directed) target selectivity, thus the logical use of the precursor in dermatological and gynaecological cancers, for example the use of methyl aminolaevulinate in Paget's disease of the vulva (Raspagliesi *et al.* 2006). The hexyl derivative is used in the photodynamic detection of bladder cancer. Conversely, selectivity for malignant over normal cells may be achieved, as in other areas, by glycosydation. Several glycoside esters of ALA (Figure 6.33) have exhibited effective conversion to ALA in malignant cells but not in normal fibroblasts (Vallinayagam *et al.* 2008).

ALA has also been shown, in some cases, to be useful in killing bacteria. Given that the porphyrin produced from ALA, PPIX, is non-photosensitising against Gram-negative bacteria when applied directly, in terms of putative photoantimicrobial action, the efficacy of the preparations here depends on the ability of the microbial target to metabolise ALA to PPIX, and this is by no means common to all. However, potential therapeutic utility has been demonstrated against a range of human pathogens, including both *Escherichia coli* and *Pseudomonas aeruginosa*, which are Gram-negative

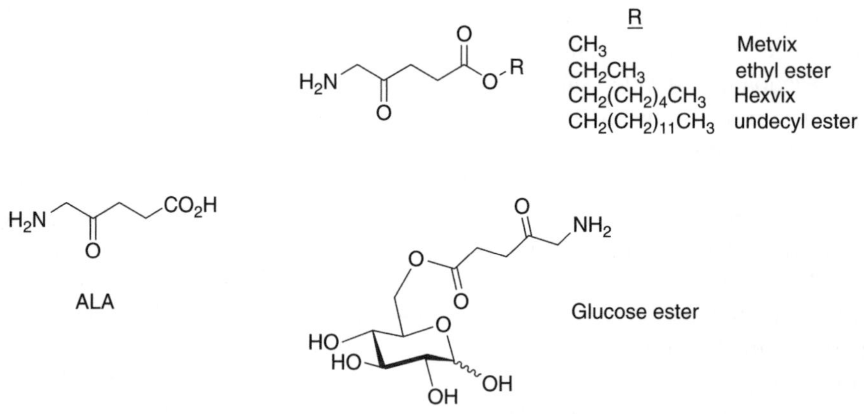

Figure 6.33 ALA and esters.

organisms. Incubation with ALA reportedly leads to an increase in PPIX production by a factor of 5–6, the resulting levels of protoporphyrin ensuring high photokilling (approximately 6 log kill) with white light (Szocs *et al.* 1999).

Conversely, *Staphylococcus aureus* and *Streptococcus sanguinis* have been reported to produce water-soluble porphyrins, after incubation with ALA, but none of the porphyrins encountered was PPIX (Dietel *et al.* 2006). It should be emphasised that, even with Gram-positive pathogens, PPIX alone applied directly is a far less effective photobactericide than PPIX produced *in situ* by the bacterial cell from ALA (O'Neill, Wilson and Wainwright 2003) (see PACT, Chapter 11).

Effective use of ALA against bacteria depends on the effective metabolism of the pro-photosensitiser by the organism. For example, strains of the Gram-negative microbe *Haemophilus parainfluenzae* capable of synthesising PPIX, when grown in the presence of ALA, may be inactivated by red light, while red light without ALA incubation has no effect on microbial longevity. *H. parainfluenzae* incapable of PPIX synthesis is unaffected by ALA incubation/illumination (van der Meulen *et al.* 1997). Facile uptake of ALA compared to the anionic porphyrins also potentiates greater antifungal activity, sufficient to support clinical trialling of ALA in the treatment of mycoses of the foot (Calzavara-Pinton *et al.* 2004).

While there is little doubt that without porphyrin-based photosensitisers PDT would have been much more slowly accepted, there is a danger in that new researchers in the field have it introduced to them as a porphyrins-only area. This is a result of the enormous amount of research carried out with HpD from the initial phase and its clinical acceptance, and this is currently being reinforced with ALA. From the point of view of the medicinal chemist/drug discoverer this can be unfortunate. However, it should always be remembered that the research must be patient-centred, and, as such, the descendants of HpD and various ALA derivatives are being used to great effect.

References

Ahmed A, Davoust E, Savoie H, Boa AN, Boyle RW. (2004) Thioglycosylated cationic porphyrins – convenient synthesis and photodynamic activity in vitro. *Tetrahedron Letters* **45**: 6045–6047.

Allison RR, Mang TS, Wilson BD, Vongtama V. (1998) Tin ethyl etiopurpurin-induced photodynamic therapy for the treatment of human immunodeficiency virus-associated Kaposi's sarcoma. *Current Therapeutic Research* **59**: 23–27.

Baba T, Shimakoshi H, Hisaeda Y. (2004) Synthesis and simple separation of β-pyrrole sulfonated porphycenes. *Tetrahedron Letters* **45**: 5973–5975.

Banfi S, Caruso E, Buccafurni L, *et al.* (2006) Antibacterial activity of tetraaryl-porphyrin photosensitizers: An in vitro study on Gram negative and Gram positive bacteria. *Journal of Photochemistry and Photobiology B: Biology* **85**: 28–38.

Berman A, Levanon H, Vogel E, Jux J. (1993) Triplet spin alignment of 'stretched porphycenes'. *Chemical Physics Letters* **211**: 549–554.

Briñas RP, Brückner C. (2002) Synthesis of 5,10-diphenylporphyrin. *Tetrahedron* **58**: 4375–4381.

Bristow CA, Hudson R, Paget TA, Boyle RW. (2006) Potential of cationic porphyrins for photodynamic treatment of cutaneous leishmaniasis. *Photodiagnosis and Photodynamic Therapy* **3**: 162–167.

Calzavara-Pinton PG, Venturini M, Capezzera R, Sala R, Zane C. (2004) Photodynamic therapy of interdigital mycoses of the feet with topical application of 5-aminolevulinic acid. *Photodermatology, Photoimmunology & Photomedicine* **20**: 144–147.

Carvalho VCM, Melo CAS, Bagnato VS, Perussi JR. (2002) Comparison of the effects of cationic and anionic porphyrins in tumor cells under illumination of argon ion laser. *Laser Physics* **12**: 1314–1319.

Cernay T, Zimmerman HW. (1996) Selective photosensitization of mitochondria by the lipophilic cationic porphyrin POR10. *Journal of Photochemistry and Photobiology B: Biology* **34**: 191–196.

Chen B, Wu S, Li A, *et al.* (2006) Synthesis of some multi-β-substituted cationic porphyrins and studies on their interaction with DNA. *Tetrahedron* **62**: 5487–5497.

Diamond I, Granelli SG, McDonagh AF. (1977) Photochemotherapy and photodynamic toxicity: simple methods for identifying potentially active agents. *Biochemical Medicine* **17**: 121–127.

Dietel W, Pottier R, Pfister W, Schleier P, Zinner K. (2006) 5-Aminolaevulinic acid (ALA) induced formation of different fluorescent porphyrins: a study of the biosynthesis of porphyrins by bacteria of the human digestive tract. *Journal of Photochemistry and Photobiology B: Biology* **86**: 77–86.

Dilkes M, Bapat U. (2004) Foscan photodynamic therapy for early head and neck cancer. *Otolaryngology – Head and Neck Surgery* **131**: 74–75.

Di Stasio B, Frochot C, Dumas D, *et al.* (2005) The 2-aminoglucosamide motif improves cellular uptake and photodynamic activity of tetraphenylporphyrin. *European Journal of Medicinal Chemistry* **40**: 1111–1122.

DiVenosa G, Hermida L, Batlle A, *et al.* (2008) Characterisation of liposomes containing aminolevulinic acid and derived esters. *Journal of Photochemistry and Photobiology B: Biology* **92**: 1–9.

Domenico Quartarolo A, Russo N, Sicilia E, Lelj F. (2007) Absorption spectra of potential photodynamic therapy photosensitizers texaphyrins complexes: a theoretical analysis. *Journal of Chemical Theory and Computation* **3**: 860–869.

Donnelly ET, Liu Y, Fatunmbi YO, Lee I, Magda D, Rockwell S. (2004) Effects of texaphyrins on the oxygenation of EMT6 mouse mammary tumors. *International Journal of Radiation Oncology Biology and Physics* **58**: 1570–1576.

Dudkowiak A, Teślak E, Habdas J. (2006) Photophysical studies of tetratolylporphyrin photosensitizers for potential medical applications. *Journal of Molecular Structure* **792–793**: 93–98.

Fedoroff OY, Rangan A, Chemeris VV, Hurley LH. (2000) Cationic porphyrins promote the formation of i-motif DNA and bind peripherally by a non-intercalative mechanism. *Biochemistry* **39**: 15083–15090.

Fernandez JM, Bilgin MD, Grossweiner LI. (1997). Singlet oxygen generation by photodynamic agents. *Journal of Photochemistry and Photobiology B: Biology* **37**: 131–140.

Figge FHJ, Weiland GS, Manganiello LOJ. (1948) Cancer detection and therapy. Affinity of neoplastic and traumatized regenerating tissues for porphyrins and metalloporphyrins. *Proceedings of the Society of Experimental Biology and Medicine* **68**: 540–641.

Fukuzumi S, Ohkubo K, Zheng X, *et al.* (2008). Metal bacteriochlorins which act as dual singlet oxygen and superoxide generators. *Journal of Physical Chemistry B* **112**: 2738–2746.

Gan Q, Xiong F, Li S, *et al.* (2005) Synthesis and photophysical properties of a series of octaphenyl-porphyrazine-magnesium. *Inorganic Chemistry Communications* **8**: 285–288.

Grahn MF, McGuinness A, Benzie R, *et al.* (1997) Intracellular uptake, absorption spectrum and stability of the bacteriochlorin photosensitizer 5,10,15,20-tetrakis(*m*-hydroxyphenyl)bacteriochlorin (*m*THPBC). Comparison with 5,10,15,20-tetrakis(*m*-hydroxyphenyl)chlorin (*m*THPC). *Journal of Photochemistry and Photobiology B: Biology* **37**: 261–266.

Griffiths SJ, Heelis PF, Haylett AK, Moore JV. (1998) Photodynamic therapy of ovarian tumours and normal cells using 5,10,15,20-tetra(3-carboxymethoxyphenyl)chlorin. *Cancer Letters* **125**: 177–184.

Gupta I, Ravikanth M. (2006) Recent developments in heteroporphyrins and their analogues. *Coordination Chemistry Reviews* **250**: 468–518.

Hilmey DG, Abe M, Nelen MI, *et al.* (2002) Water-soluble, core-modified porphyrins as novel, longer-wavelength-absorbing sensitizers for photodynamic therapy. II. Effects of core heteroatoms and *meso*-substituents on biological activity. *Journal of Medicinal Chemistry* **45**: 449–461.

Huang Q, Pan Z, Wang P, Chen Z, Zhang X, Xu H. (2006) Zinc(II) and copper(II) complexes of β-substituted hydroxylporphyrins as tumor photosensitizers. *Bioorganic & Medicinal Chemistry Letters* **16**: 3030–3033.

James DA, Swamy N, Paz N, Hanson RN, Ray R. (1999) Synthesis and estrogen receptor binding affinity of a porphyrin-estradiol conjugate for targeted photodynamic therapy of cancer. *Bioorganic & Medicinal Chemistry Letters* **9**: 2379–2384.

Jiménez-Banzo A, Sagristà ML, Mora M, Nonell S. (2008). Kinetics of singlet oxygen photosensitization in human skin fibroblasts. *Free Radical Biology & Medicine* **44**: 1926–1934.

Judy MM, Maithews JL, Newman JT, *et al.* (1991) *In vitro* photodynamic inactivation of *herpes simplex* virus with sapphyrins: 22 π-electron porphyrin-like macrocycles. *Photochemistry and Photobiology* **53**: 101–107.

Juzeniene A, Juzenas P, Ma LW, Iani V, Moan J. (2006) Topical application of 5-aminolaevulinic acid, methyl 5-aminolaevulinate and hexyl 5-aminolaevulinate on normal skin. *British Journal of Dermatology* **155**: 791–799.

Karmakova T, Feofanov A, Pankratov A, *et al.* (2006) Tissue distribution and in vivo photosensitizing activity of 13,15-[*N*-(3-hydroxypropyl)]cycloimide chlorin p6 and 13,15-(*N*-methoxy)cycloimide chlorin p6 methyl ester. *Journal of Photochemistry and Photobiology B: Biology* **82**: 28–36.

Kubler AC, Haase T, Staff C, Kahle B, Rheinwald M, Mühling J. (1999) Photodynamic therapy of nonmelanomatous skin tumours of the head and neck. *Lasers in Surgery and Medicine* **25**: 60–68.

Lanzo I, Russo N, Sicillia E. (2008). First-principle time-dependent study of magnesium-containing porphyrin-like compounds potentially useful for their application in photodynamic therapy. *Journal of Physical Chemistry B* **112**: 4123–4130.

Lash TD. (1993). Geochemical origins of sedimentary benzoporphyrins and tetrahydrobenzoporphyrins. *Energy & Fuels* **7**: 166–171.

Lauro FM, Pretto P, Covolo L, Jori G, Bertoloni G. (2002) Photoinactivation of bacterial strains involved in periodontal diseases sensitized by porphycene-polylysine conjugates. *Photochemical and Photobiological Sciences* **1**: 468–470.

Lipscomb LA, Zhou FX, Presnell SR, *et al.* (1996). Structure of a DNA-porphyrin complex. *Biochemistry* **35**: 2818–2823.

Lipson RL, Baldes EJ. (1960) The photodynamic properties of a particular hematoporphyrin derivative. *Archives of Dermatology* **82**: 508–516.

Litwinski C, Tannert S, Jesorka A, Katterle M, Röder B. (2006) Photophysical properties of gallium hydroxyl tetratolylporphyrin and 13^2-demethoxycarbonyl-(gallium hydroxyl)-methyl-pheophorbide a. *Chemical Physics Letters* **418**: 355–358.

Lovat LB, Jamieson NF, Novelli MR, *et al.* (2005). Photodynamic therapy with *m*-tetrahydroxyphenyl chlorin for high-grade dysplasia and early cancer in Barrett's columnar lined esophagus. *Gastrointestinal Endoscopy* **62**: 617–623.

Lubitz I, Borovok N, Kotlyar A. (2007) Interaction of monomolecular G4-DNA nanowires with TMPyP: evidence for intercalation. *Biochemistry* **46**: 12925–12929.

Luguya R, Jaquinod L, Fronczek FR, Vicente MGH, Smith KM. (2004) Synthesis and reactions of *meso*-(*p*-nitrophenyl)porphyrins. *Tetrahedron* **60**: 2757–2763.

MacCormack MA. (2008) Photodynamic therapy in dermatology: an update on applications and outcomes. *Seminars in Cutaneous Medicine and Surgery* **27**: 52–62.

Maiya BG, Mallouk TE, Hemmi G, Sessler JL. (1990) Effects of substituents on the spectral and redox properties of cadmium(II) texaphyrins. *Inorganic Chemistry* **29**: 3738–3745.

Meng GG, James BR, Skov KA, Korbelik M. (1994) Porphyrin chemistry pertaining to the design of anti-cancer drugs; part 2, the synthesis and in vitro tests of water-soluble porphyrins containing, in the meso positions, the functional groups: 4-methylpyridinium, or 4-sulfonatophenyl, in combination with phenyl, 4-pyridyl, 4-nitrophenyl, or 4-aminophenyl. *Canadian Journal of Chemistry* **72**: 2447–2457.

Molinari A, Colone M, Calcabrini A, *et al.* (2007) Cationic liposomes, loaded with m-THPC, in photodynamic therapy for malignant glioma. *Toxicology In Vitro* **21**: 230–234.

O'Neill J, Wilson M, Wainwright M. (2003) Comparative antistreptococcal activity of a range of photobactericidal agents. *Journal of Chemotherapy* **15**: 329–334.

Ostrowski S, Mikus A, Łopuszyńska B. (2004) Synthesis of highly substituted *meso*-tetraarylporphyrins. *Tetrahedron* **60**: 11951–11957.

Ostrowski S, Wyrębek P. (2006) The first example of a Diels-Alder cycloaddition of *ortho*-xylylenes to *meso*-tetraarylporphyrins containing electron-deficient β,β-double bonds. *Tetrahedron Letters* **47**: 8437–8440.

Pandey RK, Smith KM, Dougherty TJ. (1990) Porphyrin dimers as photosensitizers in photodynamic therapy. *Journal of Medicinal Chemistry* **33**: 2032–2038.

Raspagliesi F, Fontanelli R, Rossi G, *et al.* (2006) Photodynamic therapy using a methyl ester of 5-aminolevulinic acid in recurrent Paget's disease of the vulva: a pilot study. *Gynecologic Oncology* **103**: 581–586.

Renno RZ, Miller JW. (2001) Photosensitizer delivery for photodynamic therapy of choroidal neovascularization. *Advanced Drug Delivery Reviews* **52**: 63–78.

Ricchelli F, Franchi L, Miotto G, *et al.* (2005) Meso-substituted tetra-cationic porphyrins photosensitize the death of human fibrosarcoma cells via lysosomal targeting. *International Journal of Biochemistry & Cell Biology* **37**: 306–319.

Ris HB, Giger A, Im Hof V, *et al.* (1997) Experimental assessment of photodynamic therapy with chlorins for malignant mesothelioma. *European Journal of Cardio-Thoracic Surgery* **12**: 542–548.

Robinson BC. (2000) Bacteriopurpurins: synthesis from *meso*-diacrylate substituted porphyrins. *Tetrahedron* **56**: 6005–6014.

Rodriguez L, Batlle A, Di Venosa G, *et al.* (2006) Mechanisms of 5-aminolevulic acid ester uptake in mammalian cells. *British Journal of Pharmacology* **147**: 825–833.

Rubio N, Sánchez-Garcìa D, Jiménez-Banzo A, *et al.* (2006) Effect of aza substitution on the photophysical and electrochemical properties of porphycenes: characterization of the near-IR-absorbing photosensitizers 2,7,12,17-tetrakis(*p*-substituted phenyl)-3,6,13,16-tetraazaporphycenes. *Journal of Physical Chemistry A* **110**: 3480–3487.

Ruck A, Bohmler A, Steiner R. (2005) PDT with TOOKAD studied in the chorioallantoic membrane of fertilized eggs. *Photodiagnosis and Photodynamic Therapy* **2**: 79–90.

Schastak S, Jean B, Handzel R, *et al.* (2005) Improved pharmacokinetics, biodistribution and necrosis in vivo using a new near infra-red photosensitizer: tetrahydroporphyrin tetratosylat. *Journal of Photochemistry and Photobiology, B: Biology* **78**: 203–213.

Schneider R, Schmitt F, Frochot C, *et al.* (2005) Design, synthesis and biological evaluation of folic acid targeted tetraphenylporphyrin as novel photosensitizers for selective photodynamic therapy. *Bioorganic & Medicinal Chemistry* **13**: 2799–2808.

Selman SH. (2007) Photodynamic therapy for prostate cancer: one urologist's perspective. *Photodiagnosis and Photodynamic Therapy* **4**: 26–30.

Sessler JL, Miller RA. (2000) Texaphyrins: New drugs with diverse clinical applications in radiation and photodynamic therapy. *Biochemical Pharmacology* **59**: 733–739.

Sessler JL, Murai T, Lynch V, Cyr M. (1988) An 'expanded porphyrin': the synthesis and structure of a new aromatic pentadentate ligand. *Journal of the American Chemical Society* **110**: 5586–5588.

Shan X, Wang T, Li S, Yang L, Fu L, Yang G. (2006) Photophysical properties of diphenyl-2,3-dihydroxychlorin and diphenylchlorin. *Journal of Photochemistry and Photobiology, B: Biology* **82**: 140–145.

Sharonov GV, Karmakova TA, Kassies R, *et al.* (2006) Cycloimide bacteriochlorin *p* derivatives: photodynamic properties and cellular and tissue distribution. *Free Radical Biology and Medicine* **40**: 407–419.

Srivastava TS, Tsutsui M. (1973) Preparation and purification of tetrasodium meso-tetra(*p*-sulfophenyl)porphine. An easy procedure. *Journal of Organic Chemistry* **38**: 2103.

Stilts CE, Nelen MI, Hilmey DG, *et al.* (2000) Water-soluble, core-modified porphyrins as novel, longer-wavelength-absorbing sensitizers for photodynamic therapy. *Journal of Medicinal Chemistry* **43**: 2403–2410.

Stockert JC, Cañete M, Juarranz A, *et al.* (2007) Porphycenes: facts and prospects in photodynamic therapy of cancer. *Current Medicinal Chemistry* **14**: 997–1026.

Szocs K, Gabor F, Csik G, Fidy J. (1999) δ-Aminolaevulinic acid-induced porphyrin synthesis and photodynamic inactivation of *Escherichia coli*. *Journal of Photochemistry and Photobiology B: Biology* **50**: 8–17.

Trachtenberg J, Bogaards A, Weersink RA, *et al.* (2007) Vascular targeted photodynamic therapy with palladium-bacteriopheophorbide photosensitizer for recurrent prostate cancer following definitive radiation therapy: assessment of safety and treatment response. *Journal of Urology* **178**: 1974–1979.

Vallinayagam R, Schmitt F, Barge J, *et al.* (2008) Glycoside esters of 5-aminolevulinic acid for photodynamic therapy of cancer. *Bioconjugate Chemistry* **19**: 821–839.

van der Meulen FW, Ibrahim K, Sterenborg HJCM, Alphen LV, Maikoe A, Dankert J. (1997) Photodynamic destruction of *Haemophilus parainfluenzae* by endogenously produced porphyrin. *Journal of Photochemistry and Photobiology B: Biology* **40**: 204–208

Vesper BJ, Lee S, Hammer ND, *et al.* (2006) Developing a structure-function relationship for anionic porphyrazines exhibiting selective anti-tumour activity. *Journal of Photochemistry and Photobiology B: Biology* **82**: 180–186.

Wainwright M. (2008) Photodynamic therapy: the development of new photosensitisers. *Anti-Cancer Agents – Medicinal Chemistry* **8**: 280–289.

Wiehe A, Shaker YM, Brandt JC, Mebs S, Senge MO. (2005) Lead structures for applications in photodynamic therapy. Part 1: synthesis and variation of *m*-THPC (Temoporfin) related amphiphilic A₂BC-type porphyrins. *Tetrahedron* **61**: 5535–5564.

Winkelman J. (1962) The distribution of tetraphenylporphinesulfonate in the tumour-bearing rat. *Cancer Research* **22**: 589–593.

Wu S, Li Z, Ren L, *et al.* (2006) Dicationic pyridium porphyrins appending different peripheral substituents: synthesis and studies for their interactions with DNA. *Bioorganic & Medicinal Chemistry* **14**: 2956–2965.

Yao Z, Zhang W, Sheng C, *et al.* (2008) Design, synthesis and in vitro photodynamic activities of benzochloroporphyrin derivatives as tumor photosensitizers. *Bioorganic & Medicinal Chemistry Letters* **18**: 293–297.

Zhu TC, Finlay JC. (2006) Prostate PDT dosimetry. *Photodiagnosis and Photodynamic Therapy* **3**: 234–246.

7
Phthalocyanines

In the early years of haematoporphyrin-based photodynamic therapy (PDT), the realisation that compound mixtures such as those constituting haematoporphyrin derivative (HpD) itself were perhaps less than acceptable led to the search for alternative, more controllable, reproducible products. In the 21st century, from a regulatory viewpoint, the purity of potential drug candidates is highly important, since it is unlikely that clinical approval will be granted for anything other than a chemically and isomerically pure photosensitiser. Following HpD, plainly the requirements for intense, long-wavelength absorption and singlet oxygen production were (and remain) of great importance in alternative photosensitisers. The fact that phthalocyanine structures *look* very similar to those of porphyrins (Figure 7.1) probably also helped in their selection as the major 'second generation' class of photosensitiser. In addition, phthalocyanine chemistry was established in the 1930s (Byrne, Linstead and Lowe 1934; Linstead 1934; Linstead and Lowe 1934), although based on the inert, non-photosensitising copper(II)-centred chromophore. Indeed, such was the industry of the Linstead group in the 1930s that many alterations to the phthalocyanine chromophore were already established, particularly in the manner of hetero-atom inclusion (Bilton, Linstead and Wright, 1937; Cook and Linstead 1937; Linstead, Noble and Wright, 1937), although peripheral functionalisation was not investigated to any significant degree. However, during this period, the sulphonation of copper phthalocyanine using oleum was an established method of water solubilisation, and some cationic derivatives utilising quaternary nitrogen were also available for textile dyeing. At the outset of modern photosensitiser research, the replacement of copper(II) with diamagnetic metal ions such as zinc(II) or chloroaluminium was known to yield photosensitisers, a pattern was readily available for novel, synthetic photosensitisers, which, in theory at least, would be preferable to the oligomeric porphyrin mixtures available in the mid–late 1980s.

Formally, the phthalocyanines are tetrabenzo-fused 5,10,15,20-tetraazaporphyrins (Figure 7.1). Increased aromatic character due to the four fused benzene rings explains the more intense near-infrared absorption of these compounds compared to that of the parent porphyrin nucleus. In addition, relatively straightforward synthetic routes to the phthalocyanines have meant that a wide range of compounds are available in terms

Photosensitisers in Biomedicine   Mark Wainwright
© 2009 John Wiley & Sons, Ltd

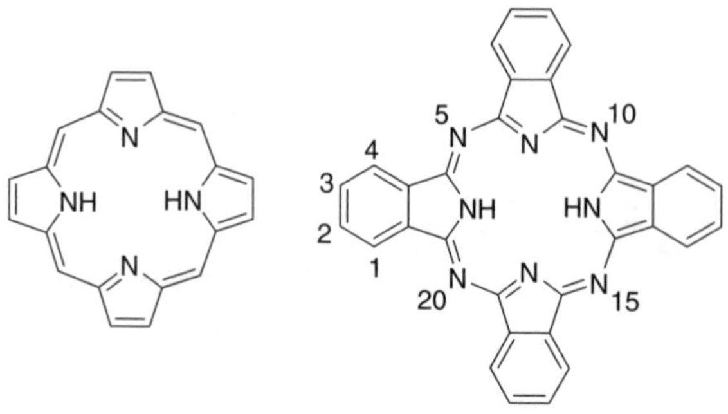

Figure 7.1   Porphyrin and phthalocyanine chromophores.

of the central metal/semi-metal atom and side-chain functionality. Concerning their photosensitising potential, phthalocyanines having a suitable central metal atom normally give high yields of singlet oxygen – greater than that of standard photosensitisers such as methylene blue (Griffiths *et al.* 1997). Suitable central atoms include aluminium, zinc, germanium, gallium and silicon (Darwent *et al.* 1982), whereas phthalocyanines based on open 'd' shell or paramagnetic metal ions such as V(II), Fe(II), Cr(II), Co(II), Ni(II) and Cu(II) have short triplet lifetimes (due to much more rapid reversion to the ground state) and are thus unsuitable as photosensitisers (Chan *et al.* 1987).

As with the porphyrin class, there is little evidence of a historic use of phthalocyanines either in the treatment or in the staining of microbes, and once again the considerable interest in this class as second-generation photosensitisers must be considered to be a development of the (PDT of cancer. In terms of photoproperties, the phthalocyanines normally exhibit intense absorption bands in the red region ($\varepsilon \geq 120\,000\,\mathrm{M}^{-1}\,\mathrm{cm}^{-1}$), usually at longer wavelengths than those of HpD, protoporphyrin IX, etc. These properties, along with (often) relatively straightforward synthesis, if not separation, have made the phthalocyanines an attractive class in terms of their screening as potential photosensitisers for PDT. Not surprisingly, they are second only in popularity to the porphyrin class in this respect.

In terms of clinical application, only one phthalocyanine example exists at the time of writing: Photosens is an isomeric mixture produced via the sulphonation of chloroaluminium phthalocyanine (see below). This mixture has been approved for use in various clinical protocols – cancer in particular – in the former Soviet Union.

7.1   General features

Phthalocyanines, like the porphyrinoid derivatives discussed in Chapter 6, may be considered as insoluble chromophores delivered to their site(s) of action by virtue of the solubilising groups in the periphery of the molecule. This differs from photosensitisers such as the azine, triarylmethane or cyanine classes where the peripheral groups also function as auxochromes, i.e. their presence is essential for the associated photosensitising action. For example, the chromophore zinc(II) phthalocyanine is insoluble in water but may be delivered to tumour cells using liposomes and activated using red light. This process would be unsuccessful using the phenothiazinium chromophore alone; this requires correctly positioned amino functionality for photosensitising ability *and* solubility.

As a consequence of this, the solubilisation of phthalocyanines for aqueous media has been attained using various polar groups, beginning with sulphonic acids ($-SO_3H$). As has been mentioned, this has been a standard route for this class in the dye industry almost since the discovery of phthalocyanines. In early PDT research (1980s), carboxylic acid derivatives (*q.v.*) were also examined. However, the rationale for the use of both of these groups was merely solubilisation, and to this end the nature of the groups was therefore considered as secondary. Given the difference in ionisation of aromatic sulphonic and carboxylic acids, this might be seen as somewhat naïve, particularly from the viewpoint of pharmacokinetics, in that the former group is usually completely ionised and very hydrophilic, whereas the properties of the latter are far more dependent on the electronic nature of the attached chromophore and the pH of the local environment.

To a certain extent, the hydrophilic–lipophilic balance has been addressed by partial functionalisation and the isomeric distribution of solubilising groups in order to furnish molecules with hydrophilic and lipophilic regions (Figure 7.2), thus allowing effective dissolution along with efficient uptake into cells.

The link between effective cellular uptake and amphiphilic character has also been addressed by the use of combinations of starting materials having lipophilic and hydrophilic substituents, for example tertiary-butyl and sulphonic acid, respectively. However, whether utilising partial sulphonation or mixed hydrophilic/hydrophobic groups, the resulting product will be a mixture of isomers.

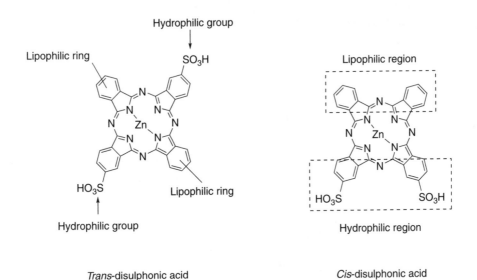

Figure 7.2 Effects of polar group substitution on molecular amphiphilicity in zinc phthalocyaninedisulphonic acids.

For relatively large, overall planar molecules, the photosensitising efficacy of the phthalocyanines may be compromised due to their ease of aggregation. Plainly, such behaviour might inhibit light transmission or provide facile relaxation of photoexcited molecules. Aggregation can be decreased by the use of bulky substituents.

Functionalisation of the phthalocyanine periphery can thus be utilised to aid in the improvement of photosensitising behaviour, although, as mentioned above, functional groups are not essentially auxochromic. This is not to say that peripheral groups in phthalocyanines do not affect photosensitising properties. Being directly attached to the phthalocyanine ring, a group having p or d orbitals, which can overlap with the chromophore, could alter the overall electron distribution of the ring. Depending on the extent and direction of electron flow thus engendered, electron promotion and thus access to Type I/II pathways might then be altered.

7.2   Phthalocyanine synthesis

There are three main routes to the phthalocyanine chromophore: the urea process, which requires the condensation of phthalic acids with urea at high temperatures, the resulting further condensation of imide species around a metal centre, and the similar template reaction, involving either phthalonitrile or diiminoisoindoline derivatives and a metal salt, at lower temperatures (Figure 7.3). Clearly there is commonality in the various approaches.

As one of the major drawbacks with the original porphyrin photosensitisers was purity, it is ironic that this is probably as much a problem with phthalocyanines, at least so far as functionalised derivatives are concerned.

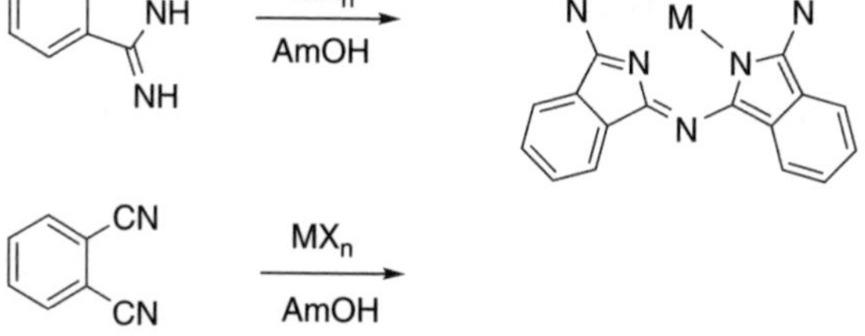

Figure 7.3 Synthetic routes to unsubstituted metal phthalocyanines.

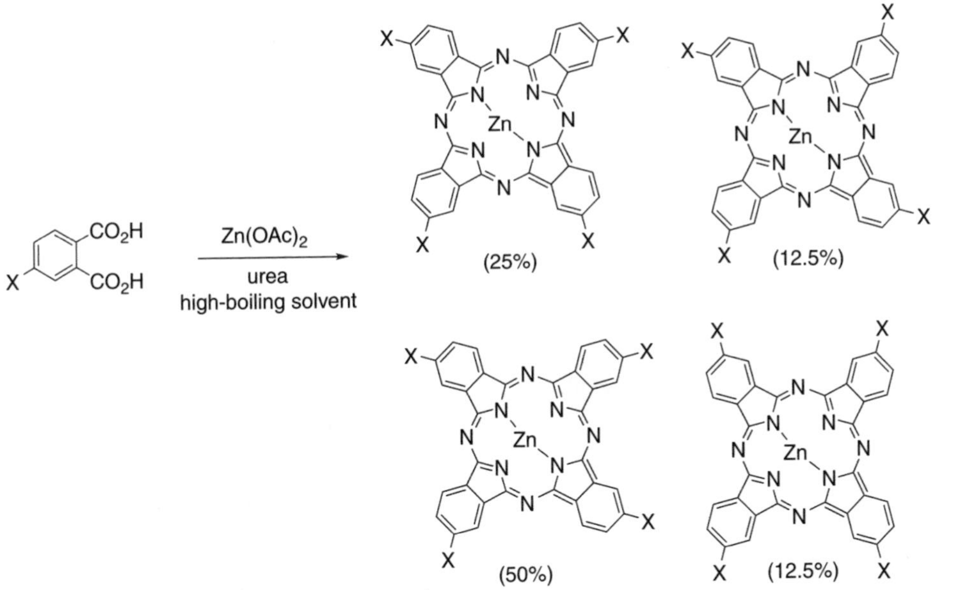

Figure 7.4 Tetrasubstituted zinc phthalocyanine isomer formation from a 4-substituted phthalic acid, and normal product distribution.

Typically, copper phthalocyanines have been applied as pigments, i.e. the raw chemical ground to a fine particle size and then dispersed within an ink formulation or a bulk plastic melt, etc. Phthalocyanines required for a medical end use, as in PDT, could not be applied as dispersions and so were initially required to be water soluble. In the early period of photosensitiser development, the main approach entailed the synthesis of sulphonic acids, either via sulphonation of the parent phthalocyanine (as with the copper dyes) or via direct synthesis from functionalised starting materials. In fact, neither method was satisfactory in terms of product purity. The sulphonation of metal phthalocyanines requires high temperatures and long reaction times. However, it is very difficult to achieve pure tetrasulphonation, e.g. of zinc phthalocyanine – there are always associated varying amounts of tri- and disulphonated material. In addition, longer reaction times and higher reaction temperatures lead to chromophore breakdown, the non-copper-containing phthalocyanines being somewhat less stable to such treatment. Approaching the problem from the synthetic angle, it was initially accepted that, for example, zinc phthalocyaninetetrasulphonic acid, produced from 4-sulphophthalic acid via the urea process, was a pure compound rather than a mixture. However, this requires a specific reaction orientation, which cannot be guaranteed in reality (Figure 7.4; $X = SO_3H$). In addition, 4-sulphophthalic acid is normally contaminated with a small amount of the 3-isomer, which leads to an even greater range of products.

One approach to the tetasubstituted isomer problem is to utilise steric hindrance between large groups on adjacent rings, for example the use of bulky tertiary butyl

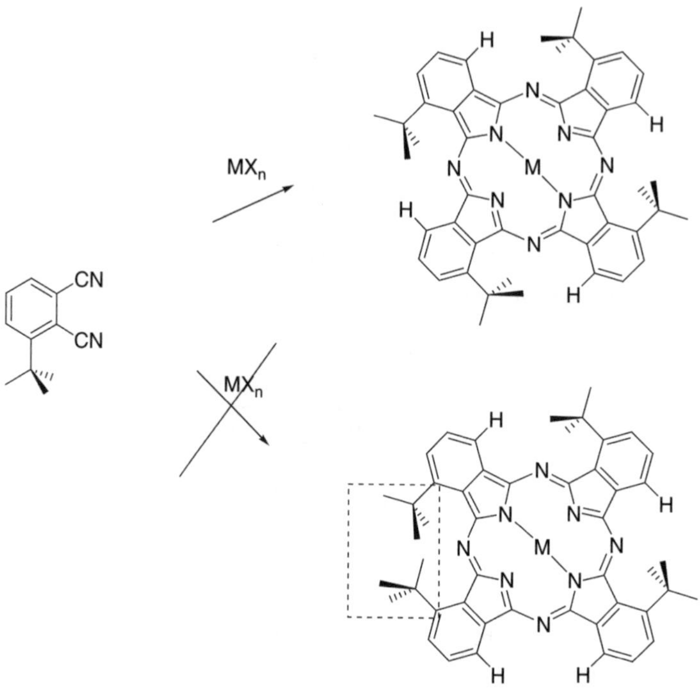

Figure 7.5 Steric hindrance of the formation of phthalocyanines bearing bulky groups.

groups at position 3 of the precursor phthalonitrile or phthalic acid. Given sufficient purity in the starting material, one isomer should result, as shown in Figure 7.5.

As a water-soluble variation, the tetracarboxylic acid derivatives mentioned above (Figure 7.4; X = CO_2H) are produced from the readily available trimellitic aid/anhydride (benzene-1,2,4-tricarboxylic acid/anhydride). Also synthesised via the urea method, the same isomer problem pertains with the resulting material. In addition, the high-temperature reaction with urea produces not a tetracarboxylic acid mixture but the corresponding tetracarboxamide derivatives (Griffiths *et al.* 1997). These, in turn, require hydrolysis to produce water-soluble material. Although this is achievable, it is doubtful if complete hydrolysis occurs, chromophore stability again constituting a problem. Consequently, many published tetracarboxylic acids are even more complex mixtures than the tetrasulphonic acids already discussed. To a certain extent, compound purity has been addressed by the use of *o*-phthalonitriles (1,2-dicyanobenzenes), which require less forcing conditions and produce fewer side reactions, but the synthesis of precursors is often difficult. Use of benzenetetracarboxylic acid (pyromellitic acid) in the urea process leads – theoretically at least, due to the problems related above – to phthalocyanineoctacarboxylic acids.

Something of the complexity of substituted phthalocyanine isomer formation may be appreciated from the mixture of disubstituted isomers produced by stoichiometric reaction (1:1) of phthalic acid and a 4-substituted analogue in the urea process (or by a similar mixture of phthalonitriles and a metal salt). Along with mono-, tri- and a small amount of

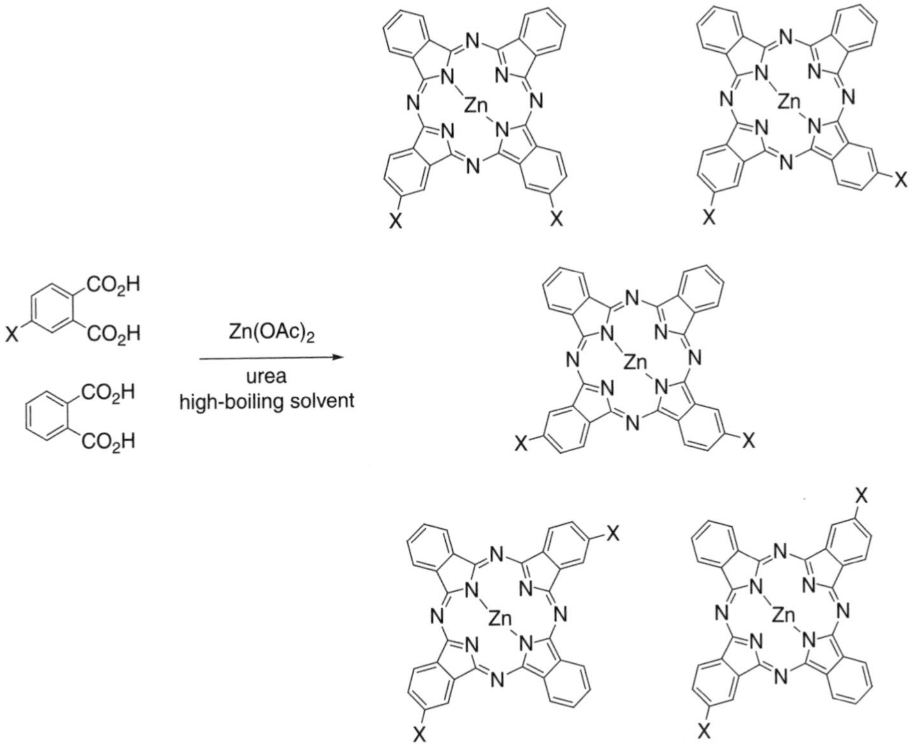

Figure 7.6 Disubstituted zinc phthalocyanines from mixed phthalic acid starting materials.

tetrasubstituted isomers, there are five possible disubstituted structures (Figure 7.6). Plainly the purification of substituted phthalocyanine isomers is thus not a straightforward matter, and exhaustive chromatography is normally required (Zimcik *et al.* 2006).

A simple method circumventing this problem is to have no substituents – with the resulting solubility problems associated with phthalocyanine pigments – or to have all α/β ring positions substituted. 4,5-Disubstituted phthalonitriles have been synthesised in order to furnish octasubstituted products (Woehrle *et al.* 1993; Durmus and Nyokong 2007a; Akkurt and Hamuryudan 2008). Alternatively, axial (i.e. non-ring) substitution is also possible via silicon as the central atom (see below).

While side products from functional group are an expected part of organic synthesis, a less obvious purity problem was also apparent fairly quickly in early PDT research: Chloroaluminium species are unstable in aqueous media. Thus, acid treatment or aqueous work-up of chloroaluminium phthalocyanines, in order to produce aqueous-soluble derivatives, results in the loss of chlorine from the central aluminium, leaving either the corresponding hydroxyaluminium phthalocyanine or the dimers thereof (Figure 7.7). Conversely, gallium(III) phthalocyanines prepared from gallium chloride maintain the central Ga—Cl portion, thus providing examples that exhibit low levels of aggregation and thus effective photosensitisation (Durmus and Nyokong 2007b).

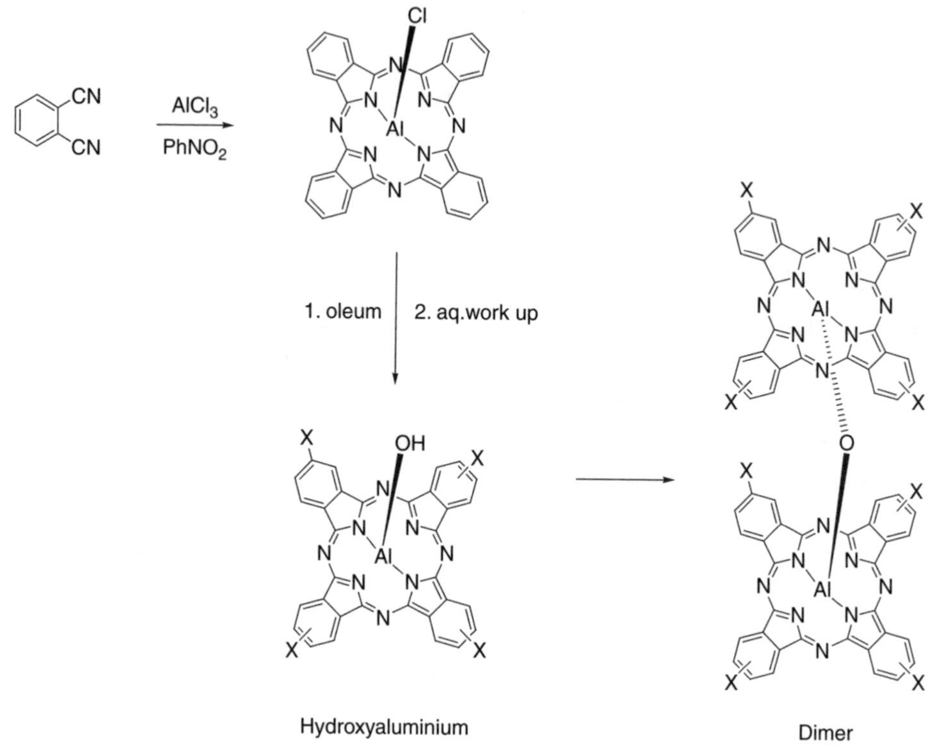

Figure 7.7 Reactions of sulphonated chloroaluminium phthalocyanine in aqueous media (X = H or SO₃H).

7.3 Photosens

Given the requirement for chemical and isomeric purity in clinically useful photosensitisers, it is perhaps surprising that a mixture of phthalocyanine isomers is in fact widely
used. This is Photosens, produced via the sulphonation of chloroaluminium phthalocyanine and purified, to a certain degree, via high-performance liquid chromatography. The
mixture is employed, for PDT purposes, only in the former Soviet Union.

Aside from Photosens, neither tetrasubstitution of the phthalocyanine nucleus nor
polyfunctionalisation has so far resulted in clinically acceptable compounds. Due to
the considerable efforts required in purification of the higher analogues, monosubstituted materials have also been investigated. The synthesis of monosubstituted
phthalocyanines originally relied on the statistical condensation of phthalic acid
derivatives – instead of reaction of the pre-formed ring system – thus requiring
chromatographic separation as a necessary purification step. However, the aqueous
solubility of monosubstituted derivatives is usually low, thus limiting their utility, and
the main use of such compounds has been for the attachment of tumour-targeting
biomolecules in research (Chen *et al.* 2006).

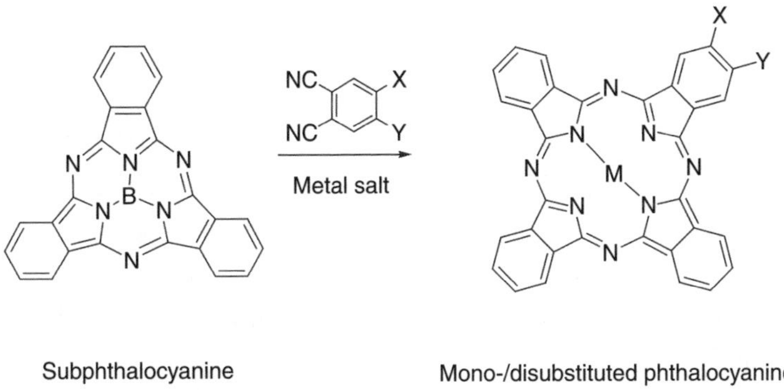

Subphthalocyanine   Mono-/disubstituted phthalocyanine

Figure 7.8 Mono-/disubstituted phthalocyanine synthesis from a sub-phthalocyanine precursor.

A major breakthrough was made via the use of subphthalocyanine precursors (Figure 7.8). Smaller analogues, based on a boron-centred trigonal-planar rather than a square-planar symmetry, can be ring-opened and another (substituted) phthalonitrile unit inserted, leading to asymmetric monosubstituted or symmetrical (same ring) disubstituted products (Matlaba and Nyokong 2002). Clearly, the use of either a 4-substituted or a 4,5-disubstituted phthalonitrile with an unsubstituted subphthalocyanine yields a product without the possibility of isomer formation.

This asymmetrical positioning of substituents on one ring has been shown to improve photodynamic activity against tumour cells *in vitro*. Thus, for example, zinc dodecafluorophthalocyanine (12 fluorines), synthesised from the fully fluorinated subphthalocyanine, exhibited increased PDT activity in culture in comparison to the wholly fluorinated analogue (16 fluorines) (Sharman and van Lier 2005).

The approach can also be used to add chromophoric asymmetry via the reaction of a 2,3-naphthalonitrile derivative with a trisulphonated subphthalocyanine, the resulting water-soluble benzophthalocyanine exhibiting a considerable bathochromic shift of *c.* 40 nm at 728 nm (Kudrevitch, Gilbert and van Lier 1996).

7.4 Naphthalocyanines

The fusion of a benzene ring on each of the pyrrole residues of the tetraazaporphyrin nucleus gives a bathochromic shift of approximately 90 nm, with small variations depending on the central metal of the resulting phthalocyanine. This is repeated by subsequent linear benzene fusion to the naphthalocyanine and anthracocyanine chromophores (Figure 7.9), the latter class having λ_{max} values in the 850 nm region (Kobayashi, Nakajima and Osa 1993).

The octabutoxynaphthalocyanine shown in Figure 7.9 has a long-wavelength absorption at 850 nm. Photoexcitation of this molecule in the cellular milieu causes cytotoxicity via a thermal, rather than a photodynamic, mechanism (Camerin *et al.* 2005). While the presence

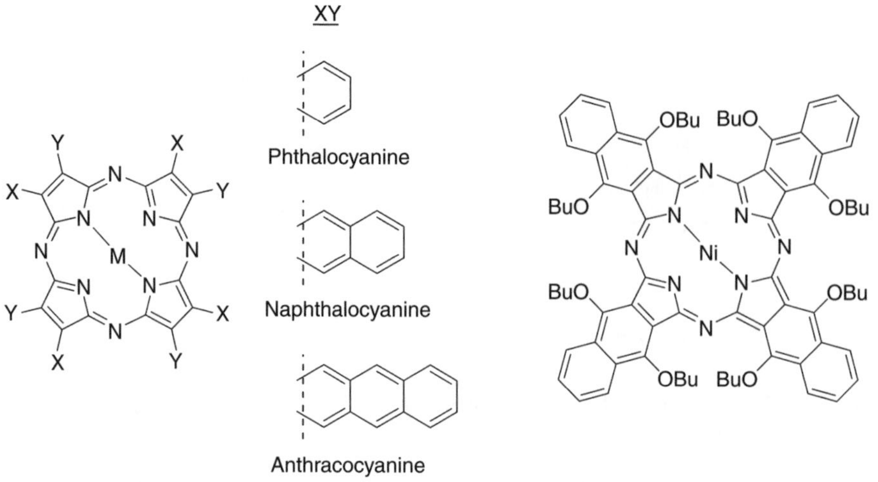

Figure 7.9 Benzene-fused tetraazaporphyrins and nickel octabutoxynaphthalocyanine.

of the central Ni(II) would, in any case, preclude efficient photosensitisation (see below), such thermal transfer in this wavelength region has also been reported for indocyanine green (see Cyanines, Chapter 8).

As it has long been established that the extension of a chromophore with a fused benzene ring generally leads to an increase in its maximum wavelength of absorption, this strategy was applied to the phthalocyanine system in order to extend photosensitisation further into the near-infrared region, allowing greater potential for therapeutic tissue penetration. A typical naphthalocyanine (Figure 7.9) has an intense, long-wavelength absorption in the region of 750–800 nm. While this satisfies the penetration depth requirement, *in vitro* testing of naphthalocyanines usually demonstrates lower singlet oxygen efficiencies in comparison to analogous phthalocyanines, for example zinc naphthalocyaninetetrasulphonic acid, $\Phi_\Delta = 0.25$, zinc phthalocyaninetetrasulphonic acid, $\Phi_\Delta = 0.68$ (Nyokong 2007). Such a decrease in photosensitising activity is in line with similar findings in other chromophoric types, such as the benzo[*a*]phenothiaziniums (Nile blue derivatives, Chapter 4).

Concerning precursors to the naphthalocyanine system, as with functionalised phthalonitriles, the preparation of substituted naphthalene analogues is not a straightforward matter, usually requiring *de novo* ring synthesis rather than ring functionalisation. Thus, the bromination of *o*-xylene derivatives leads to pro-dienes, which may then be reacted with fumaronitrile to furnish the required naphthalonitrile starting material (Figure 7.10) (Ford *et al.* 1992).

An alternative route, furnishing mono- or disubstituted 2,3-naphthalonitriles, uses furan as a starting point and its functionalisation at position 2 or both positions 2 and 5, via reaction with 1,2,4,5-tetrabromobenzene and butyllithium, reduction of the oxygen bicyclic system and subsequent replacement of the remaining *ortho*-bromines with cyanide, as shown in Figure 7.11 (Polley and Hanack 1995).

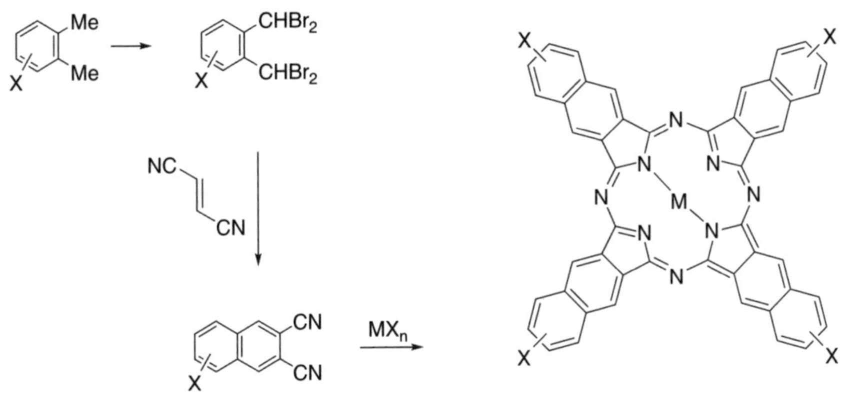

Figure 7.10 Naphthalocyanine synthesis from o-xylenes (X = alkyl, alkoxyl, halogen, nitro, etc. and may be originally in position 3 or 4 of the xylene ring).

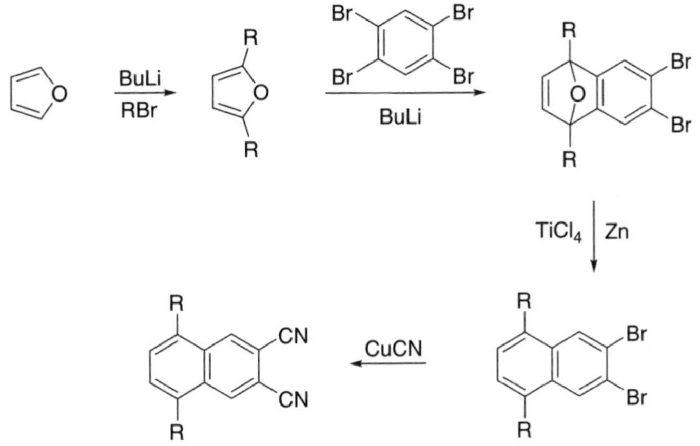

Figure 7.11 Synthesis of 2,3-naphthalonitrile derivatives from furan; R = alkyl.

Functionalisation at positions 1 and 4 of the naphthalonitrile may be afforded by using dichloronaphthoquinone as starting material (Figure 7.12). Such alkoxy-substitution affords derivatives exhibiting long-wavelength absorbance, such as the nickel octabutoxy derivative cited above as a photothermal agent (Polley and Hanack 1995).

Naphthalocyanine synthesis starting from the 1,2-dinitrile, rather than the standard 2,3-isomer, leads to a mixture of angular products, as shown in Figure 7.13. However, little has been reported concerning the utility of these analogues from the point of view of photosensitisation.

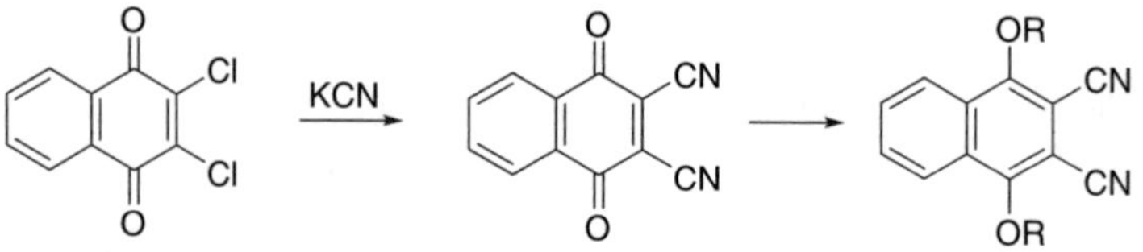

Figure 7.12 Synthesis of 2,3-naphthalonitrile derivatives from dichloronaphthoquinone.

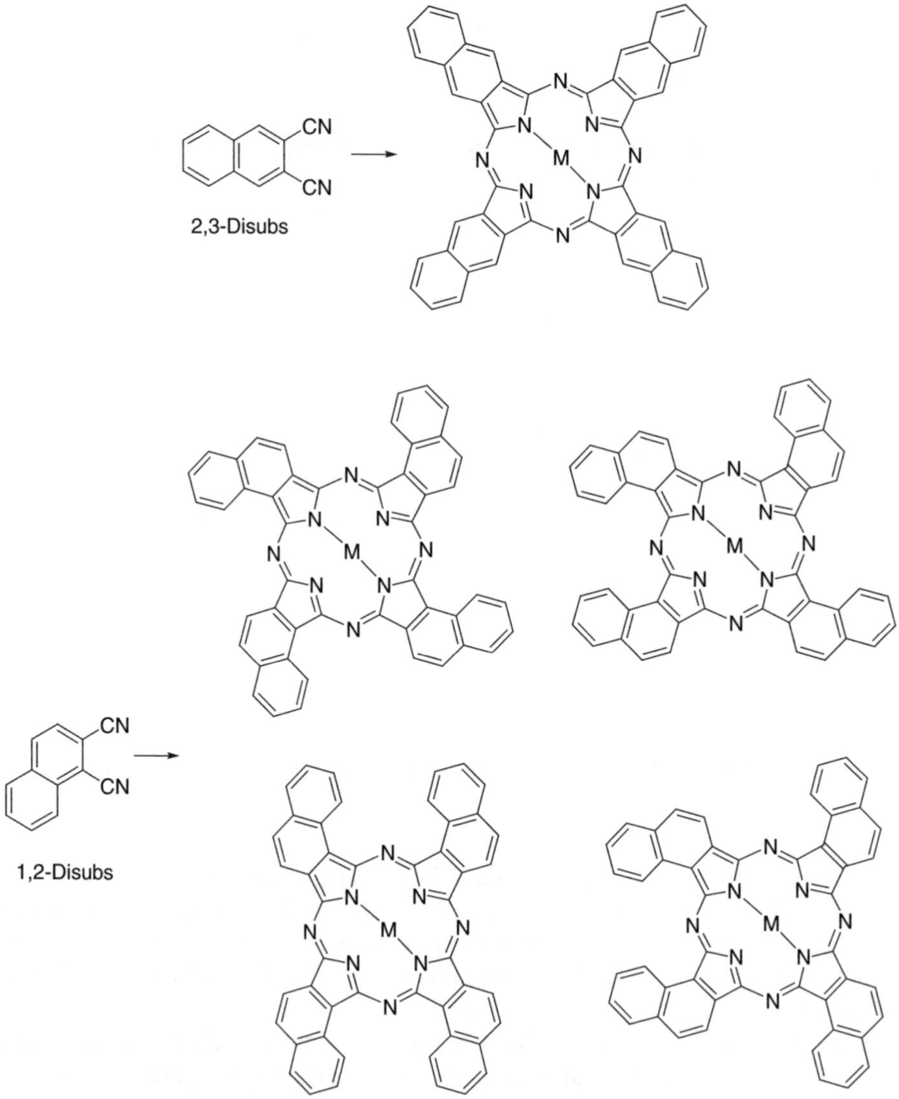

Figure 7.13 Benzo-fused phthalocyanines (naphthalocyanines).

7.5 Hetero-fused systems

The basic method of synthesis of phthalocyanine derivatives requires an aromatic 1,2-dicarboxylic acid, 1,2-dinitrile or diiminoisoindoline. This approach is widely applicable, allowing the inclusion of a range of molecular types in place of the original fused benzene, i.e. smaller/larger and/or hetero-substituted derivatives. Plainly not all of these are suitable photosensitiser candidates, but much can be learned in terms of structure function, which can be used to inform subsequent drug design.

The term 'azaphthalocyanine' covers a considerable range of different possibilities, dependent on the position and number of nitrogens substituted for aromatic carbon in the benzo-fused moieties of the phthalocyanine chromophore (Figure 7.14). In turn, this depends on the availability of suitable aromatic starting materials, and the stoichiometry of the reaction. For example, the use of 2,3-dicyanopyrazine alone furnishes an azaphthalocyanine containing eight ring nitrogens, whereas the analogous pyridine precursor would result in four. Obviously, mixed condensations of phthalic acid/phthalonitrile with pyridine/pyrazine, etc. analogues may be employed to give mono-, di- and triaza derivatives, again with similar problems regarding isomer constituency, as discussed above.

Substitution of nitrogen for $-CH=$ in the phthalocyanine system has electronic effects, since aromatic nitrogen is electron withdrawing in nature, but also provides sites for hydrophilic improvement, i.e. via quaternisation, and also potential targeting since quaternary derivatives have been shown to intercalate DNA (Dougherty 1988).

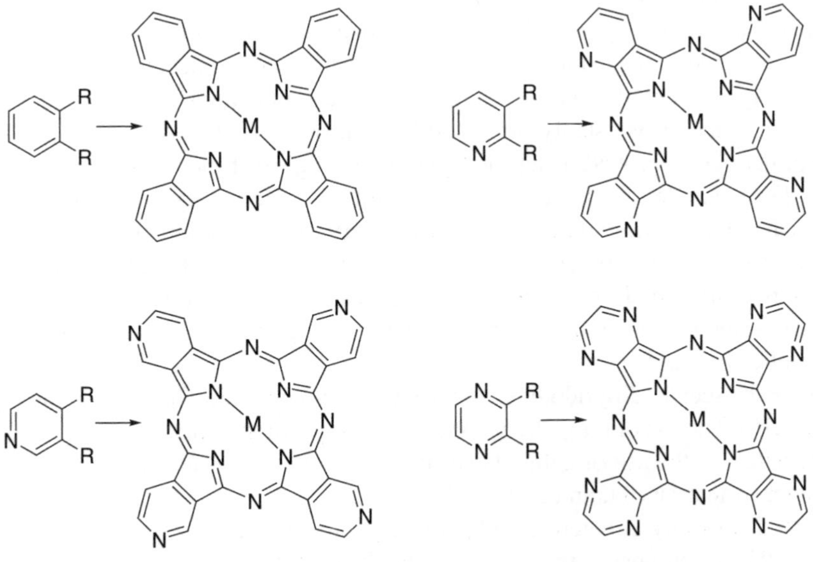

Figure 7.14 Azaphthalocyanines (R = CO_2H and CN; regular isomers only shown).

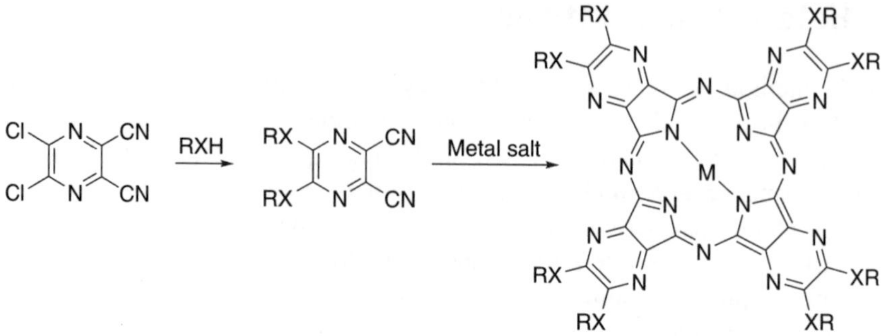

Figure 7.15 Synthesis of tetrapyrazinoporphyrazines (R = alkyl; X = O, S and N).

Tetrapyrazinoporphyrazines, i.e. derived from 2,3-dicyanopyrazines, are useful analogues of the phthalocyanine system, due to the ease of nucleophilic substitution of the chromophore, via the reactivity of the precursor 2,3-dicyanopyrazine hetero-cycles. Straightforward substitution reactions allow the synthesis of various disubstituted dicyanopyrazine derivatives, which may thus be converted into octasubstituted products (Figure 7.15). However, when compared to the phthalocyanines, singlet oxygen yields of many derivatives are low, although these may be recovered by the inclusion of sulphur-containing side chains (Kostka *et al.* 2006).

Groups containing electron-releasing sulphur directly attached to the macrocycle also exhibit improved long-wavelength absorption (*c.* 660 nm in DMF) and singlet oxygen yields (*c.* 0.6) (Musil *et al.* 2007). However, the inclusion of alkylamino groups is associated with low yields of singlet oxygen.

As noted above, the aggregation of photosensitisers is often a reason for low singlet oxygen yields and is due to the ease of stacking of the planar molecules (indeed, this disc-like planarity is essential in their use in liquid crystal applications). This tendency to stack is important, since it is usually promoted for planar molecules by aqueous media. The situation may be improved using multiple bulky ring substituents or similarly charged groups. For example, complete monomerisation has been achieved in zinc azaphthalo-cyanines having at least eight cationic charges (tertiary ammonium groups) per molecule (Zimcik *et al.* 2006). Obviously, the presence of such high charge may have conse-quences in terms of hydrophilicity, possibly pharmacology and thus activity. Obviously, suitable co-functionalisation of the molecule with lipophilic groups would produce more amphiphilic character.

This can be seen in dipyridodibenzoporphyrazines containing long-chain alkyl groups (Figure 7.16). The use of long-chain alkyl groups at positions 3 and 6 of the phthalonitrile intermediate (via the use of a thiophene sulphoxide precursor) also decreases the poten-tial isomer formation (Sakamoto *et al.* 2005).

As noted, a variety of hetero-fused phthalocyanine derivatives are available, although for PDT/PACT purposes, this tends to be limited to the azaphthalocyanine type. However, the range of derivative chromophores available also includes those having

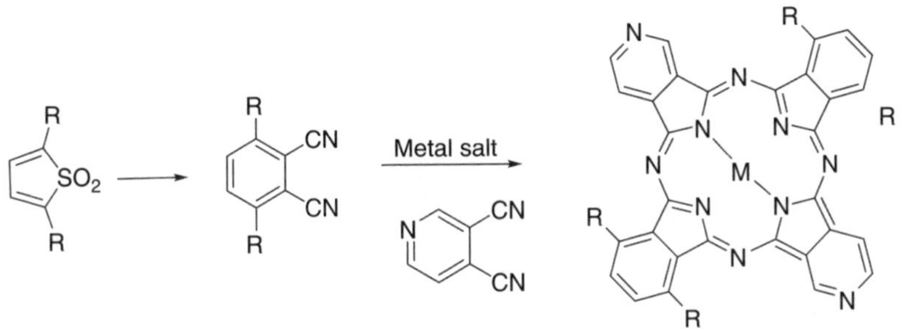

Figure 7.16 Amphiphilic azaporphyrin synthesis using long-chain alkyl substitution.

fused benzene rings replaced by five-membered hetero-aromatic systems such as imidazole, thiophene, thiazole and thiadiazole (Donzello, Ercolani and Stuzhin 2006). Such derivative chromophores generally have long-wavelength Q-band absorptions at higher energies than standard phthalocyanines, at around 600 nm.

7.6 Silicon derivatives

As already mentioned, in terms of photosensitising activity, chromophore aggregation offers significant deactivation potential for excited-state chromophores. Logically, this is less likely in structures that approximate three-dimensional rather than planar shapes; thus, the discovery of the photosensitising capabilities of silicon phthalocyanines was significant. This was particularly advantageous from the angle of photosensitiser design, due to the lack of isomer formation via axial (i.e. above and below the plane of the molecule about the central silicon) rather than peripheral functionalisation, in addition to the absence of aggregation that endows concomitant high yields of singlet oxygen (Zhu *et al.* 2006). Given the high affinity of silicon for oxygen, it is a simple proposition to replace, e.g. chlorines attached to the central silicon (i.e. from a silicon tetrachloride/phthalonitrile reacting couple) with oxygen nucleophiles. Silicon phthalocyanine functionalisation via this methodology is thus a much more straightforward proposition, solubilising groups, etc. being pre-formed and attached to the phthalocyanine via an oxygen bridge (Figure 7.17).

The premier example, Pc4, is a silicon phthalocyanine having basic dialkylaminoalkyl functionality (Figure 7.17), this endowing considerable photobiological activity in both the anticancer and the antimicrobial applications (Miller *et al.* 2007). Pc4 has been included in clinical trials for skin cancer and is an excellent candidate photoantimicrobial, particularly for blood product decontamination (see below).

The oxygen affinity alluded to above allows conjugation to sugar moieties, which – as with porphyrin photosensitisers – can lead to improved antitumour activity (Lee *et al.* 2005). Conversely, the conjugation of silicon phthalocyanine derivatives to serum lipoproteins (albumens) has been shown to increase uptake and photocytotoxicity, and

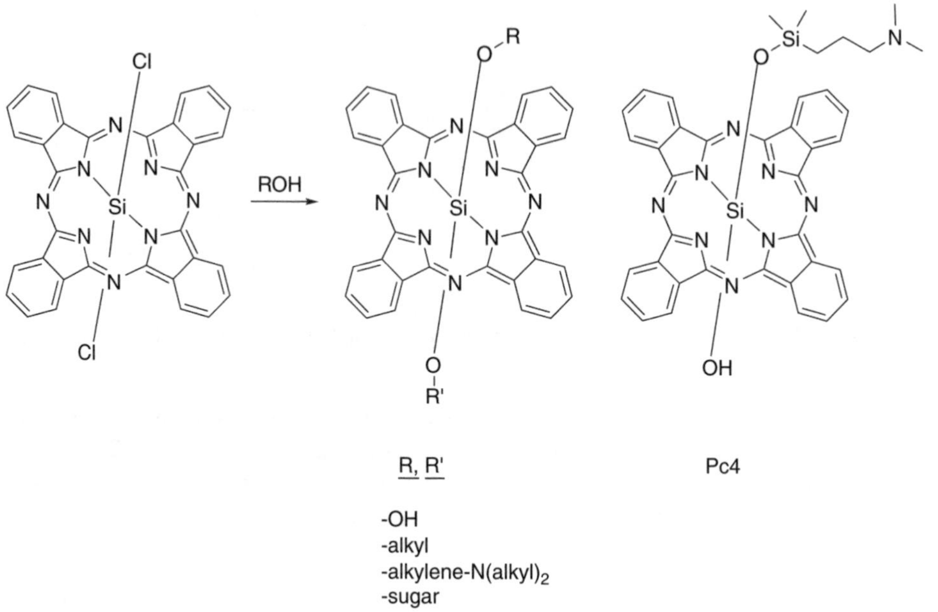

Figure 7.17 Silicon phthalocyanines.

this may be achieved, for example, in silicon phthalocyanine having axial 4-substituted phenoxy ligands (Jiang *et al.* 2006).

Considering conventionally functionalised phthalocyanines, the comparison of soluble group IV phthalocyanines (containing Si, Ge or Sn) indicated that the most effective material, each being a sulphonated mixture, was that based on germanium phthalocyanine (Kloek and Beijersbergen van Henegouwen 1996). Plainly the ring-functionalised materials have different properties (and purities) in comparison to axially substituted silicon phthalocyanines.

7.7 Photoantimicrobial activity

In terms of their photoantimicrobial application, there is a considerable literature associated with phthalocyanines. This was mainly due to the considerable investigations undertaken by Ben Hur and co-workers in the 1990s into the use of silicon phthalocyanines in blood product decontamination protocols (Zhao *et al.* 1997; Zmudzka *et al.* 1997). However, much work has also been reported subsequently by Jori and co-workers concerning peripherally rather than axially functionalised derivatives, especially those of a cationic nature (Segalla *et al.* 2002; Maisch *et al.* 2004).

Given the major requirement for blood decontamination protocols, namely that both enveloped and non-enveloped viruses should be eradicated, there is a significant problem in covering both cellular and non-cellular fractions from the point of view of collateral damage (see Chapter 11). For a range of derivatives, enveloped viruses such as the human immunodeficiency virus (HIV) and herpes simplex virus (HSV) are generally susceptible to photoinactivation, whereas non-enveloped viruses are not, indicating that the viral envelope is a target for phthalocyanine photosensitisation (Ben Hur *et al.* 1996). The use of phthalocyanines against free and cell-bound HIV has been reported and, in common with the development of these compounds in PDT, aluminium and silicon phthalocyanines offer considerable promise, both groups of compounds exhibiting light absorption in the near-infrared and efficient sensitisation of singlet oxygen. As noted above, in terms of chemical make-up, the silicon phthalocyanines used as photodynamic agents are functionalised axially through the silicon atom, rather than conventionally in the periphery of the aromatic system, and several highly active compounds have resulted from this approach. Thus, for example, silicon phthalocyanine bearing a cationic dialkylaminoalkylsilyloxy residue on the central silicon (Figure 7.17) was not only active against cell-free HIV but also against the actively replicating virus and latently infected red blood cells (Margolis-Nunno *et al.* 1996). Such an activity profile against viruses is obviously highly desirable, although there remains the problem of red blood cell damage, which requires the addition of an antioxidant. Although little work has been reported concerning the antimicrobial effects of naphthalocyanines, some antiviral examples exist, such as the sulphonated aluminium derivatives (Smetana *et al.* 1994).

In terms of structure–activity relationships for the phthalocyanines, there is no general correlation between antiviral potency and the central atom of the phthalocyanine. Conversely, the degree of phthalocyanine sulphonation and butylation has been reported to affect both the antiviral activity and the extent of haemolysis. This is logical, given the effect on the hydrophilic/lipophilic balance engendered by functionalisation of the phthalocyanine ring periphery and also supports the viral envelope mode of action alluded to above.

Unlike the successful methylene blue-plasma treatment, the photoantimicrobial phthalocyanines have not yet gained clinical approval. However, results from the photoantiviral testing of phthalocyanines have been so impressive that several derivatives have been tested as anti-HIV agents *per se*. Thus, sulphonated copper, nickel and vanadyl derivatives, as well as the metal-free analogue have all been found to be inherently virucidal, suggesting their use, for example, in blocking the sexual transmission of HIV (Vzorov *et al.* 2003).

Although widely publicised, viral contamination is not, of course, the sole factor in disease transmission through donated blood. It has been shown that blood-borne pathogens involved in tropical diseases may be photoinactivated using phthalocyanines. For example, the cationic silicon phthalocyanine mentioned above (Figure 7.17) has been shown to inactivate the malarial and trypanosomal parasites *Plasmodium falciparum* and *Trypanosoma cruzi*, respectively. Such activity is important in blood product decontamination protocols in many non-temperate parts of the world.

In line with other activities concerning blood products, the photobactericidal testing of phthalocyanines has also been carried out, *in vitro*. However, several groups have synthesised derivatives aimed more at conventional bacterial and fungal infection

(i.e. intended for use in hospital infection control). Thus, activity has been demonstrated for phthalocyanines against multidrug-resistant organisms (Wilson and Pratten
1995) and bacteria in biofilms (Wilson, Burns and Pratten 1996), while photobactericidal materials have been produced from phthalocyanines incorporated into polymer
films (Bonnett *et al.* 1993). Testing of anionic, cationic and neutral zinc phthalocyanines (Figure 7.18) against both Gram-positive and Gram-negative bacteria again
confirmed that only the positively charged phthalocyanine (a pyridinium salt) was
active (Minnock *et al.* 1996). This is a similar effect to that observed with cationic
meso-porphyrins (Chapter 6).

Similarly to other classes of photosensitiser, the acceptance of the photosensitiser
charge/Gram-class activity paradigm has led to a considerable increase in the synthesis of
(broad-spectrum) cationic derivatives. In the majority of cases, such derivatives featured
quaternary nitrogens in either pyridine or aminoalkyl moieties, often employing nucleophilic nitro-group replacement in phthalonitrile derivatives to provide suitable starting
materials (Figure 7.18).

Phthalocyanine use in both PDT and PACT is due to the structural similarity of the
chromophore with porphyrins, alongside the improved photoproperties associated with
the extended aromatic system, particularly from the point of view of increasing wavelength. That there are as yet no examples in clinical use outside the former Soviet Union
may be due to the problem of isomer formation and purification. However, functionalisation via the axial positions, as shown for silicon phthalocyanines, offers considerable
biological activity without the isomer problem and represents a logical pattern for future
development.

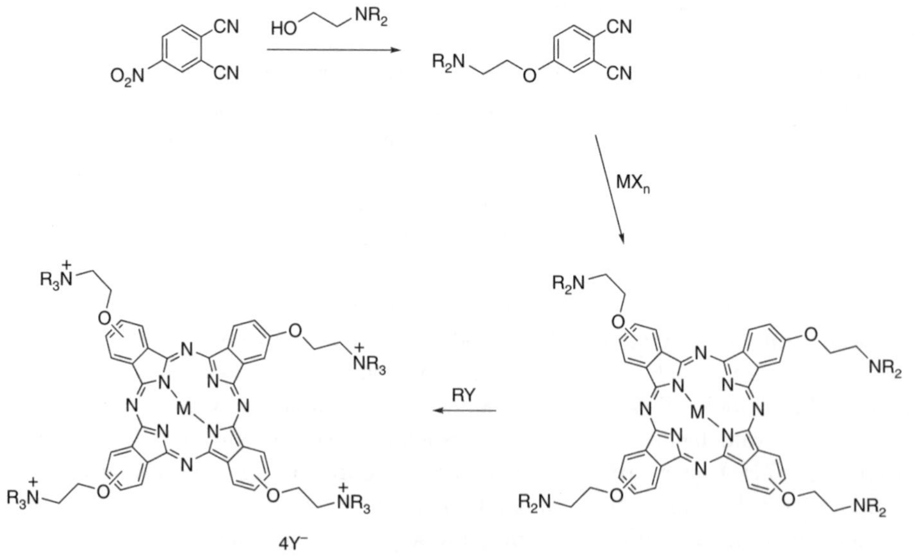

Figure 7.18 Cationic photoantimicrobial phthalocyanines (e.g. M = Zn; R = Me; Y = I).

References

Akkurt B, Hamuryudan E. (2008) Enhancement of solubility via esterification: synthesis and characterization of octakis (ester)-substituted phthalocyanines. *Dyes and Pigments* **79**: 153–158.

Ben Hur E, Moor ACE, Margolis-Nunno H, *et al*. (1996) The photodecontamination of blood components: mechanisms and use of photosensitization in transfusion medicine. *Transfusion Medicine Reviews* **10**: 15–22.

Bilton JA, Linstead RP, Wright JM. (1937) Phthalocyanines. X. Experiments in the pyrrole, isoxazole, pyridazine, furan and triazole series. *Journal of the Chemical Society* 922–929.

Bonnett R, Buckley DG, Burrow T, Galia ABB, Saville B, Songca SP. (1993) Photobactericidal materials based on porphyrins and phthalocyanines. *Journal of Materials Chemistry* **3**: 323–324.

Byrne GT, Linstead RP, Lowe AR. (1934) Phthalocyanines. II. The preparation of phthalocyanine and some metallic derivatives from *o*-cyanobenzamide and phthalimide. *Journal of the Chemical Society* 1017–1022.

Camerin M, Rello S, Villanueva A, *et al*. (2005) Photothermal sensitization as a novel therapeutic approach for tumours: studies at the cellular and animal level. *European Journal of Cancer* **41**: 1203–1212.

Chan VS, Marshall JF, Svensen R, Phillips D, Hart IR. (1987) Photosensitising activity of phthalocyanine dyes screened against tissue culture cells. *Photochemistry and Photobiology* **45**: 757–761.

Chen J, Chen N, Huang J, Wang J, Huang M. (2006) Derivatizable phthalocyanine with single carboxylate group: synthesis and purification. *Inorganic Chemistry Communications* **9**: 313–315.

Cook AH, Linstead RP. (1937) Phthalocyanines. XI. The preparation of octaphenylporphyrazines from diphenylmaleinitrile. *Journal of the Chemical Society* 929–933.

Darwent JR, Douglas P, Harriman A, Porter G, Richoux MC. (1982) Metal phthalocyanines and porphyrins as photosensitizers for reduction of water to hydrogen. *Coordination Chemistry Reviews* **44**: 83–126.

Donzello MP, Ercolani C, Stuzhin PA. (2006) Novel families of phthalocyanine-like macrocycles – porphyrazines with annulated strongly electron-withdrawing 1,2,5-thia/selenodiazole rings. *Coordination Chemistry Reviews* **250**: 1530–1561.

Dougherty G. (1988) Intercalation of tetracationic metalloporphyrins and related compounds into DNA. *Journal of Inorganic Biochemistry* **34**: 95–103.

Durmus M, Nyokong T. (2007a) Synthesis, photophysical and photochemical properties of tetra- and octa-substituted gallium and indium phthalocyanines. *Polyhedron* **26**: 3323–3335.

Durmus M, Nyokong T. (2007b) The synthesis, fluorescence behaviour and singlet oxygen studies of new water-soluble cationic gallium(III) phthalocyanines. *Inorganic Chemistry Communications* **10**: 332–338.

Ford WE, Rodgers MAJ, Schechtman LA, Sounik JR, Rihter BD, Kenney ME. (1992) Synthesis and photochemical properties of aluminum, gallium, silicon and tin naphthalocyanines. *Inorganic Chemistry* **31**: 3371–3377.

Griffiths J, Schofield J, Wainwright M, Brown SB. (1997) Some observations on the synthesis of polysubstituted zinc phthalocyanine sensitisers for PDT. *Dyes and Pigments* **33**: 65–78.

Jiang XJ, Huang JD, Zhu YJ, Tang FX, Ng DKP, Sun JC. (2006) Preparation and in vitro photodynamic activities of novel axially substituted silicon(IV) phthalocyanines and their bovine serum albumin conjugates. *Bioorganic & Medicinal Chemistry Letters* **16**: 2450–2453.

Kloek J, Beijersbergen van Henegouwen GMJ. (1996) Prodrugs of 5-aminolevulinic acid for photodynamic therapy. *Photochemistry and Photobiology* **64**: 994–1000.

Kobayashi N, Nakajima S, Osa T. (1993) Spectroscopic comparison of tetra-tert-butylated tetraazaporphyrin, phthalocyanine, naphthalocyanine and anthracocyanine cobalt complexes. *Inorganic Chimica Acta* **210**: 131–133.

Kostka M, Zimcik P, Miletin M, Klemera P, Kopecky K, Musil Z. (2006) Comparison of aggregation properties and photodynamic activity of phthalocyanines and azaphthalocyanines. *Journal of Photochemistry and Photobiology A: Chemistry* **178**: 16–25.

Kudrevitch SV, Gilbert S, van Lier JE. (1996) Synthesis of trisulfonated phthalocyanines and their derivatives using boron(III) subphthalocyanines as intermediates. *Journal of Organic Chemistry* **61**: 5706–5707.

Lee PPS, Lo PC, Chan EYM, Fong WP, Ko WH, Ng DKP. (2005) Synthesis and in vitro photodynamic activity of novel galactose-containing phthalocyanines. *Tetrahedron Letters* **46**: 1551–1554.

Linstead RP. (1934) Phthalocyanines. I. A new type of synthetic coloring matters. *Journal of the Chemical Society* 1016–1017.

Linstead RP, Lowe AR. (1934) Phthalocyanines. III. Preliminary experiments on the preparation of phthalocyanines from phthalonitrile. *Journal of the Chemical Society* 1022–1027.

Linstead RP, Noble EG, Wright JM. (1937) Phthalocyanines. IX. Derivatives of thiophene, thionaphthene, pyridine and pyrazine, and a note on the nomenclature. *Journal of the Chemical Society* 911–921.

Maisch T, Szeimies RM, Jori G, Abels C. (2004) Antibacterial photodynamic therapy in dermatology. *Photochemical and Photobiological Sciences* **3**: 907–919.

Margolis-Nunno H, Ben Hur E, Gottlieb P, Robinson R, Oetjen J, Horowitz B. (1996) Inactivation by phthalocyanine photosensitization of multiple forms of human immunodeficiency virus in red cell concentrates. *Transfusion* **36**: 743–750.

Matlaba P, Nyokong T. (2002) Synthesis, electrochemical and photochemical properties of unsymmetrically substituted zinc phthalocyanine complexes. *Polyhedron* **21**: 2463–2472.

Miller JD, Baron ED, Scull H, *et al.* (2007) Photodynamic therapy with the phthalocyanine photosensitizer Pc4: the case experience with preclinical mechanistic and early clinical-translational studies. *Toxicology and Applied Pharmacology* **224**: 290–299.

Minnock A, Vernon DI, Schofield J, Griffiths J, Parish JH, Brown SB. (1996) Photoinactivation of bacteria – use of a cationic water soluble zinc phthalocyanine to photoinactivate both Gram negative and Gram positive bacteria. *Journal of Photochemistry and Photobiology B: Biology* **32**: 159–164.

Musil Z, Zimcik P, Miletin M, Kopecky K, Petrik P, Lenco J. (2007) Influence of electron-withdrawing and electron-donating substituents on photophysical properties of azaphthalocyanines. *Journal of Photochemistry and Photobiology A: Chemistry* **186**: 316–322.

Nyokong T. (2007) Effects of substituents on the photochemical and photophysical properties of main group metal phthalocyanines. *Coordination Chemistry Reviews* **251**: 1707–1722.

Polley R, Hanack M. (1995) Synthesis of alkyl- and alkyloxy-substituted 2,3-naphthalocyanines. *Journal of Organic Chemistry* **60**: 8278–8282.

Sakamoto K, Kato T, Ohno-Okumura E, Watanabe M, Cook MJ. (2005) Synthesis of novel cationic amphiphilic phthalocyanine derivatives for next generation photosensitizer using photodynamic therapy of cancer. *Dyes and Pigments* **64**: 63–71.

Segalla A, Borsarelli CD, Braslavsky SE, *et al.* (2002) Photophysical, photochemical and antibacterial photosensitizing properties of a novel octacationic Zn(II)-phthalocyanine. *Photochemical and Photobiological Sciences* **1**: 641–648.

Sharman WM, van Lier JE. (2005) Synthesis and photodynamic activity of novel asymmetrically substituted fluorinated phthalocyanines. *Bioconjugate Chemistry* **16**: 1166–1175.

Smetana Z, Mendelson E, Manor J, *et al.* (1994) Photodynamic inactivation of herpes viruses with phthalocyanine derivatives. *Journal of Photochemistry and Photobiology B: Biology* **22**: 37–43.

Wilson M, Burns T, Pratten J. (1996) Killing of *Streptococcus sanguis* in biofilms using a light-activated antimicrobial agent. *Journal of Antimicrobial Chemotherapy* **37**: 377–381.

Wilson M, Pratten J. (1995) Lethal photosensitization of *Staphylococcus aureus* in vitro: effect of growth phase, serum and pre-irradiation time. *Lasers in Surgery and Medicine* **16**: 272–276.

Vzorov AN, Marzilli LG, Compans RW, Dixon DW. (2003) Prevention of HIV-1 infection by phthalocyanines. *Antiviral Research* **59**: 99–109.

Woehrle D, Eskes M, Shigehara K, Yamada A. (1993) A simple synthesis of 4,5-disubstituted 1,2-dicyanobenzenes and 2,3,9,10,16,17,23,24-octasubstituted phthalocyanines. *Synthesis* 194–196.

Zhao XJ, Lustigman S, Kenney ME, Ben-Hur E. (1997) Structure-activity and mechanism studies on silicon phthalocyanines with *Plasmodium falciparum* in the dark and under red light. *Photochemistry and Photobiology* **66**: 282–287.

Zhu YJ, Huang JD, Jiang XJ, Sun JC. (2006) Novel silicon phthalocyanines axially modified by morpholine: synthesis, complexation with serum protein and in vitro photodynamic activity. *Inorganic Chemistry Communications* **9**: 473–477.

Zimcik P, Miletin M, Musil Z, Kopecky K, Kubza L, Brault D. (2006) Cationic azaphthalocyanines bearing aliphatic tertiary amino substituents – synthesis, singlet oxygen production and spectroscopic studies. *Journal of Photochemistry and Photobiology A: Chemistry* **183**: 59–69.

Zmudzka BZ, Strickland AG, Beer JZ, Ben-Hur E. (1997) Photosensitized decontamination of blood with the silicon phthalocyanine Pc4: no activation of the human immunodeficiency promoter. *Photochemistry and Photobiology* **65**: 461–464.

8

Cyanines

The basis of cyanine-type construction is the presence of several, linked methine ($-CH=$) units. These allow the transmission of charge via delocalisation from one end to the other, and the greater the delocalisation, the longer the maximum wavelength of absorption of the resulting chromophore. More specifically, the hydrocarbon bridging unit has an odd number of methine units. Thus, proper cyanine structures have one unit, carbocyanines have three, dicarbocyanines five, etc. (Figure 8.1).

The structure of cyanine dyes is such that analogue formation is relatively straightforward, e.g. via variation in heteroatom, *N*-alkylation, length of polymethine chain, etc., and has led to a large number of compounds being synthesised. Indeed, given the huge number of cyanine dyes of various types produced during the last 90 years or so, it is surprising that the cyanine class is probably the least explored in terms of photosensitising drug discovery and development. As in other classes, there was a significant amount of clinically oriented work based on what might be termed 'dye therapy' in the first half of the last century, notably by Carl Browning and his co-workers. Various cyanine and related styryl candidates were examined as conventional antibacterial and antitrypanosomal therapeutics (Browning *et al.* 1924; Wainwright and Kristiansen 2003), although this is seldom mentioned in relation to more modern endeavours. In common with other

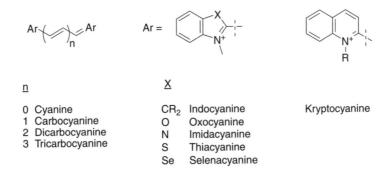

n	X	
0 Cyanine	CR$_2$ Indocyanine	Kryptocyanine
1 Carbocyanine	O Oxocyanine	
2 Dicarbocyanine	N Imidacyanine	
3 Tricarbocyanine	S Thiacyanine	
	Se Selenacyanine	

Figure 8.1 Common structural nomenclature for 'symmetrical' cyanine compounds.

Photosensitisers in Biomedicine   Mark Wainwright
© 2009 John Wiley & Sons, Ltd

'old' dye types, it would be sensible to take account of such work in building a rational basis for drug design. The very limited number of cyanine compounds used in photo-dynamic applications is normally exemplified in three examples: merocyanine 540 (MC540), kryptocyanine (N,N'-bis(2-ethyl-1,3-dioxolane) kryptocyanine, EDKC) and indocyanine green (see below).

It is well established that there are many structural types within this class that are unsuitable for the production of reactive oxygen species – this is reflected not only in the numerous examples of photographic sensitisers but also in the developing area of fluorescent probes – but there remains considerable potential outside such applications, even were this limited to analogue preparation based on the active photosensitisers mentioned above. In addition, there is increasing evidence in the literature of cyanine synthesis based on established chromophores, in order to extend the active spectrum of the parent (e.g. acridines, below).

As has been mentioned, visible light absorption by organic molecules relies on the possession of a conjugated π-system, the wavelength of light absorbed depending on charge distribution within the system. The cyanine dye class constitutes an excellent example of this, particularly in the extension of a system to provide longer-wavelength absorption.

Generally, the λ_{max} for a given cyanine dye is increased by such lengthening of the polymethine chain. This may, however, lead to problems with photoisomerisation, a major deactivation route for cyanine dyes – indeed deactivation processes for the excited cyanine include isomerisation and singlet oxygen production (Delaey et al. 2000). Increasing the polymethine chain length often yields compounds that are less chemically stable and causes decreased aqueous solubility (due to increased hydrocarbon content). A careful balance of these factors is thus required when designing likely candidates for testing as photosensitisers in biological systems. However, poor singlet oxygen production by cyanines in vitro is often much increased in cellular challenges due to external structural rigidification, e.g. via chromophore binding to biomolecules.

The photoisomerisation (cis–trans) of several oxa-, thia- and selena-cyanine analogues (Figure 8.2) has been postulated to proceed via a triplet-state mechanism, since the rate of isomerisation increases with increase in atomic mass (Cooper and Rome 1974). Obviously, the large increases in singlet oxygen yield seen in other systems due to the heavy atom effect are thus compromised in this case.

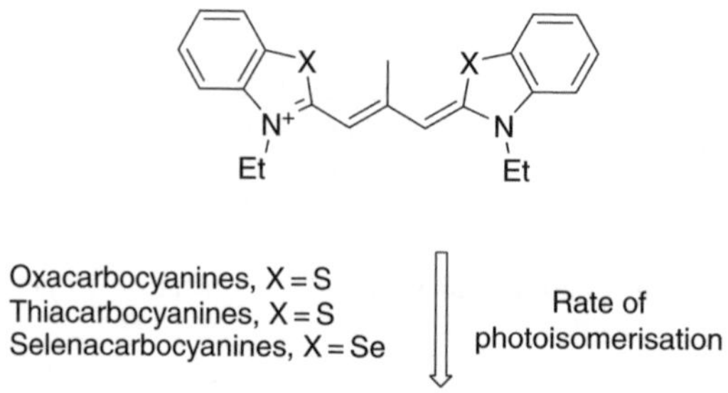

Oxacarbocyanines, X = S
Thiacarbocyanines, X = S
Selenacarbocyanines, X = Se

Rate of
photoisomerisation

Figure 8.2 Group VI cyanine derivatives.

On consideration of the basic structures involved in styryl and cyanine dyes, the main difference lies in the number of methine units present between the aromatic/heteroaromatic residues. In styryl compounds this is even, and in cyanines, odd. Both types obey the rule of increasing λ_{max} by extending the π-system of the parent chromophores. In both cases, there is a 'push–pull' scenario, with one aromatic region usually (the merocyanines are often an exception) carrying a positive charge due to a quaternary nitrogen being linked via delocalisation to an electron-rich centre, electrons flowing between the two (Figure 8.3).

For example, while typical acridine absorption is seen in the blue region (350–450 nm), this may be extended into the near infrared by cyanine-type synthesis, e.g. 9-styrylacridines (Figure 8.4). The straightforward condensation of 9-methylacridine with N,N-dialkylaminobenzaldehyde derivatives produces acidochromic dyes. These, in addition to absorbing light at or around 700 nm, can exist in neutral or protonated (N-10) forms at approximately physiological pH, since they have pK_a values of around 5. Such compounds may be useful in antitumour work due to the pH drop in tumours. This would

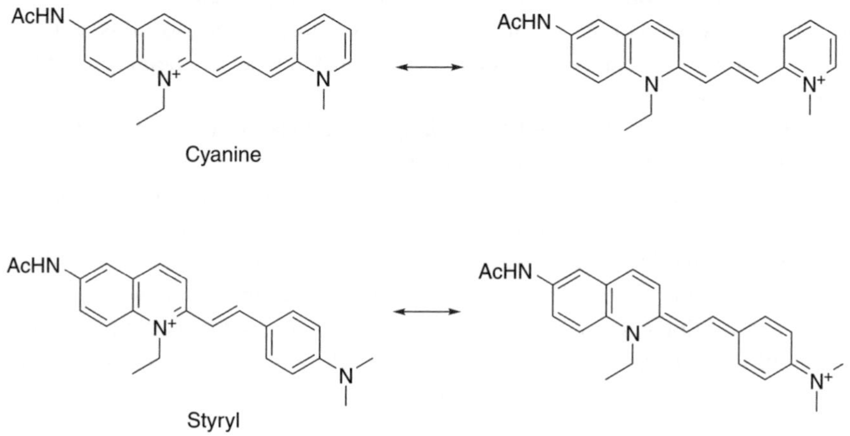

Cyanine

Styryl

Figure 8.3 Structural comparison of cyanine and styryl compounds.

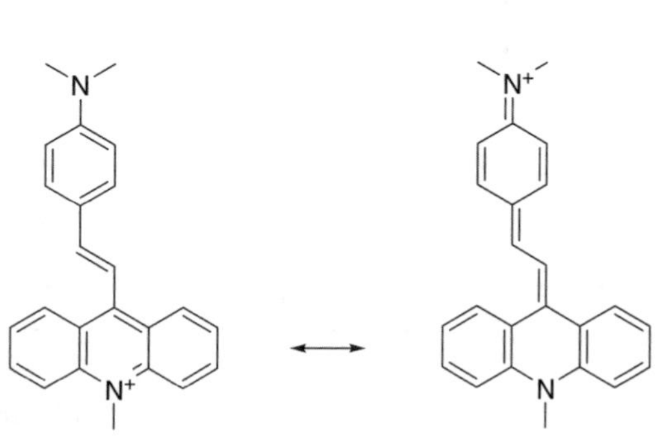

Figure 8.4 Styrylacridine compound showing extensive delocalisation.

give a higher degree of protonation inside the tumour, thus adding to the selectivity of the system, since the protonated dye is the photoactive species and has an absorption that is bathochromically shifted by around 200 nm. Extrapolating to the idea of tumour treatment *in vivo*, this would mean that peritumoural tissues would contain mainly neutral dye, which would of course remain unaffected by near-infrared light. However, a potential drawback to such compounds is that their associated singlet oxygen yields are apparently low ($\leq 5\%$ of the photosensitising activity of methylene blue *in vitro*) (Lindauer, Czerney and Grummt 1994). Their behaviour in cells has not been reported, but similar singlet oxygen results in thiacyanines, for example, have not precluded significant photocytotoxicity (see below).

8.1   Synthesis

Cyanines normally employ heteroaromatic starting materials with active carbon atoms, for example a quaternised pyridine derivative with a methyl group at position 2, 4 or 6. The electron-withdrawing effect of the quaternary nitrogen allows loss of a methyl proton, effectively providing a carbanion. Nucleophilic attack of this anion at an aldehyde or ketone with subsequent dehydration provides an elementary cyanine-type molecule (Figure 8.5).

Similarly, this may be extended to nucleophilic attack by the activated methylene species as a nucleophile, as shown in Figure 8.6.

This scenario is applicable across a wide range of active methylene heteroaromatics, although the 'acceptor' molecule is often an orthoester rather than an aldehyde or ketone. This is shown for 2-methylpyridine and triethyl orthoformate in Figure 8.7, the orthoformyl carbon constituting the 'odd' carbon of the trimethine chain in the product.

Plainly charge delocalisation requires a difference in electron density in the moieties at the termini of the polymethine bridge in order for 'flow' to occur. This is present in the examples of Figure 8.8. Of the two examples exhibiting delocalisation, the latter is a

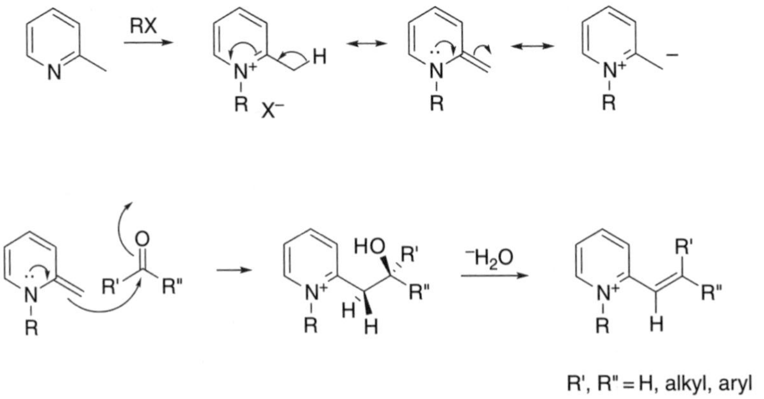

R', R" = H, alkyl, aryl

Figure 8.5   Elementary cyanine/styryl formation.

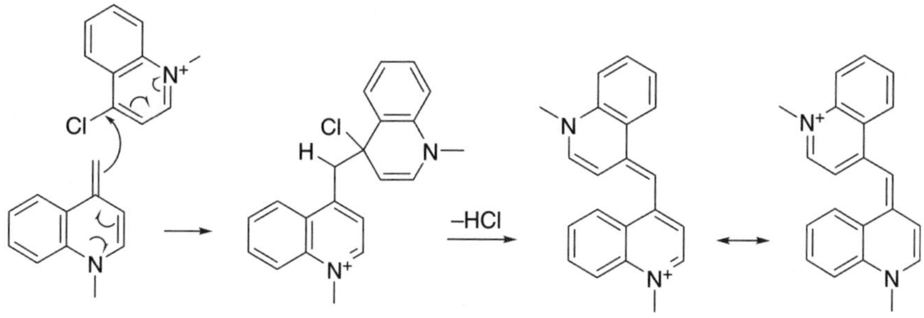

Figure 8.6 Cyanine formation from 4-quinoline units.

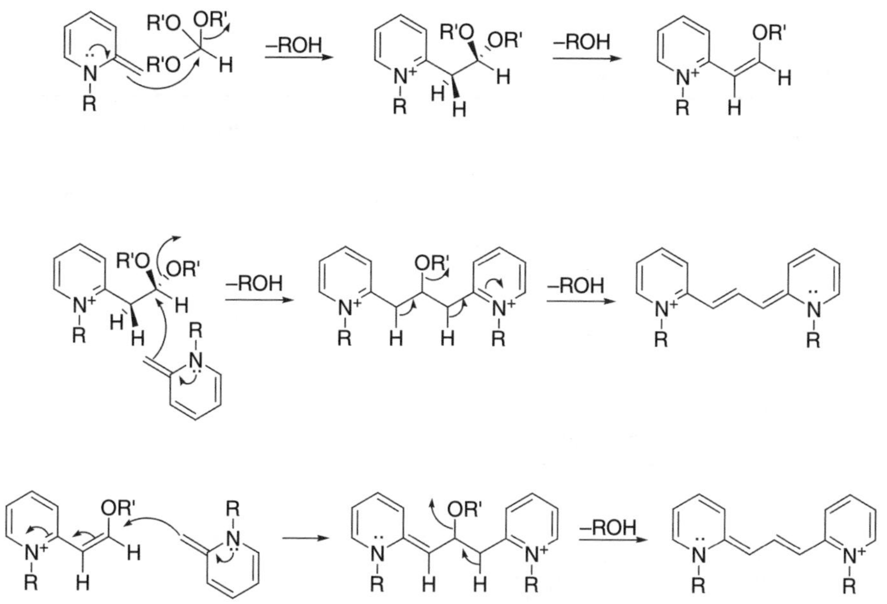

Figure 8.7 Mechanism of cyanine formation using triethyl orthoformate.

cyanine, the former is a styryl compound. However, this class is included here due to its close chemical similarity with cyanines.

As noted above, many different cyanines and styryls have been synthesised and tested for various purposes. Structure–function relationships are relatively straightforward and rational design is therefore possible. Some simple examples of carbocyanines are shown in Table 8.1.

As can be seen from Table 8.1, given a constant length of polymethine chain, the λ_{max} value may be varied by attaching different nuclei, usually heteroaromatic. For the examples shown, singlet oxygen yields *in vitro* were very low. However, in cell culture, both the quinoline and benzothiazole derivatives were phototoxic (Kassab 2001). It

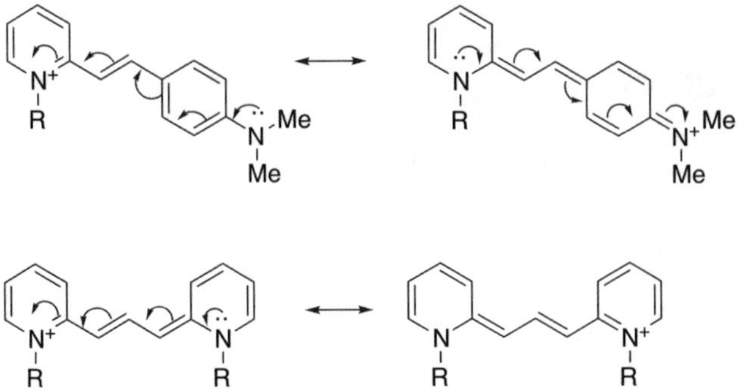

Figure 8.8 Delocalisation in cyanine/styryl structures.

Table 8.1 Simple heterocyclic carbocyanines

Heterocycle	Cyanine	λ_{max}(nm,EtOH)	Φ_Δ
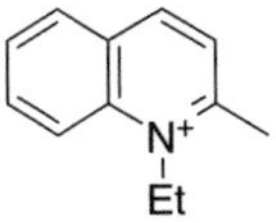	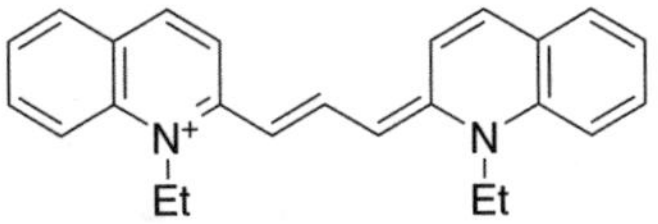	733	0.04
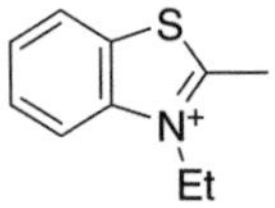	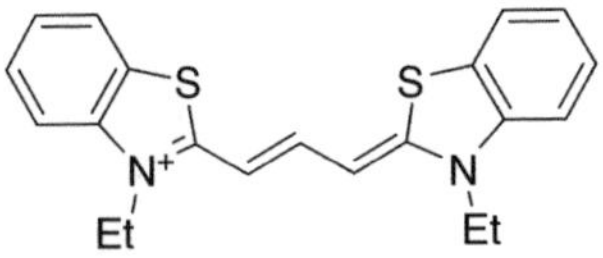	560	0.03
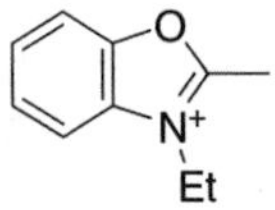	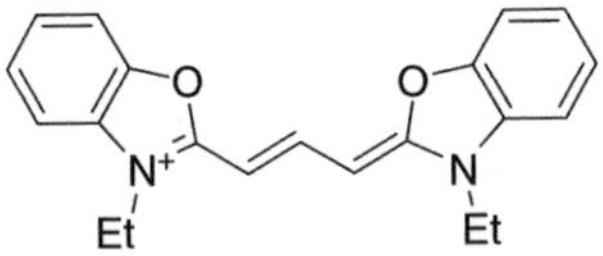	483	0.02

Source: Kassab (2001).

should be noted that, under the same singlet oxygen test conditions, indocyanine green exhibited an even poorer result.

Interestingly, while standard styryl derivatives also give typically low results, significant yields of singlet oxygen are associated with simple nitrostyryl compounds such as those shown in Figure 8.9 (Görner 1999).

There is a long history of the application of cyanine dyes in the area of colour photography, the dyes acting as colour sensitisers. Photographic sensitisation involves electron transfer from the excited dye to silver(I) ions in the photographic film, i.e. similar to the Type I photosensitisation route, causing reduction (i.e. $Ag^+ + e^- \rightarrow Ag^0$).

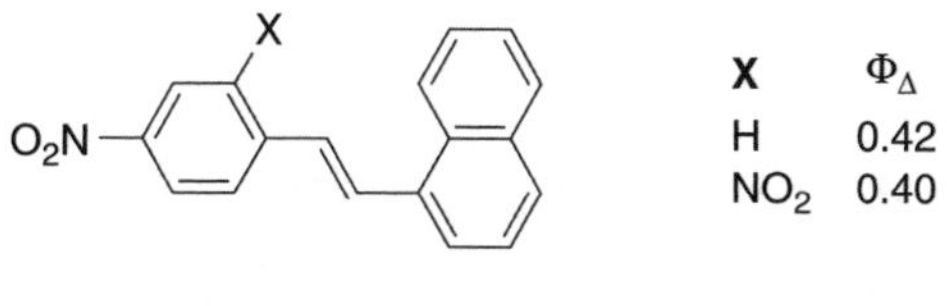

Figure 8.9 Nitrostyryl photosensitisers.

X	Φ_Δ
H	0.42
NO_2	0.40

The cyanines offer a wide range of visible absorption, including the near-infrared region. Given the huge variety of heteroaromatic precursors having acidic carbon groups, this range of absorption is unsurprising. However, as with other areas of biomedical photosensitiser research, full series of compounds have not yet been screened in order to provide optimal candidates, although, as mentioned, their use as a photographic sensitisers infers, at least, potential Type I activity.

One large survey of the suitability of cyanine dyes as anticancer photosensitisers was carried out by Delaey and co-workers (2000). Within the study, several promising candidates were observed, but, again, singlet oxygen yields and cellular photosensitisation results did not agree, supporting the suggestion that candidates should always be tested in a cellular model (Table 8.2). Interestingly, the eight kryptocyanine derivatives examined showed no photoactivity in the HeLa cell model employed in the study.

In terms of the potential application of useful activity, several rhodacyanines (e.g. Figure 8.10) have been shown to be selective for and to exert a cytotoxic effect

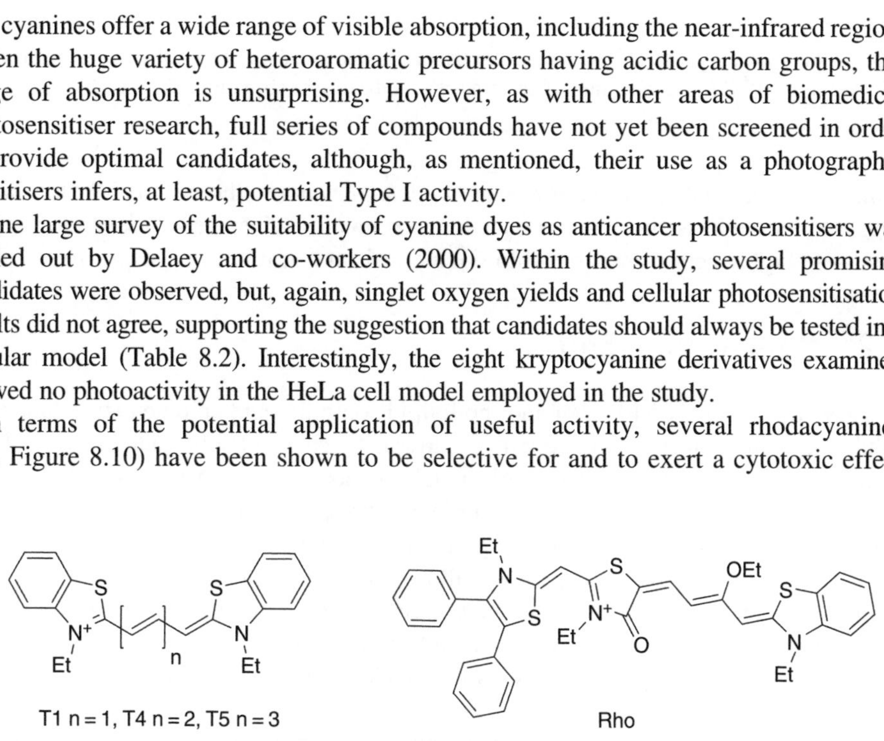

T1 n = 1, T4 n = 2, T5 n = 3

Rho

Table 8.2 Sample of cyanine candidates used by Delaey *et al.*

Cyanine	λ_{max} (nm)	$^1O_2$ yield	Toxicity (dark)	(IC$_{50}$, μM) Light	Dark:light toxicity at IC$_{50}$
T1	560	0.005	9.1	0.41	22
T4	655	0.033	3.5	0.64	5.5
T5	764	0.013	6.5	4.5	1.4
Rho	655	0.005	4.2	0.39	11
New indocyanine green	824	0.077	3.1	2.1	1.5
Hypericin	598	0.73	>10	0.83	>12
Photofrin	630	0.89	>25	5.82	>4.3

Cytotoxicity measured against HeLa cells.

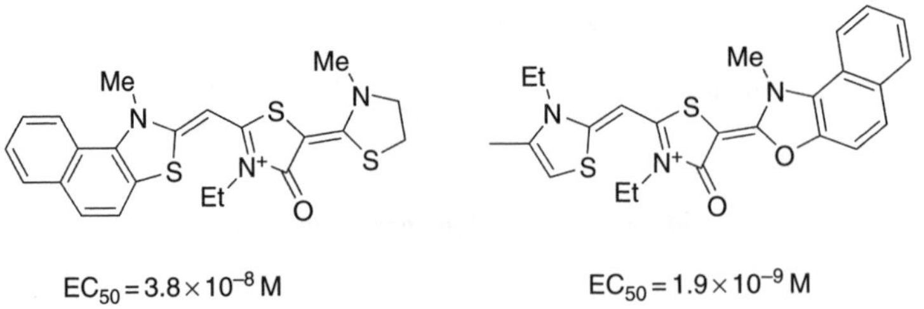

$$EC_{50} = 3.8 \times 10^{-8} \, M \qquad\qquad EC_{50} = 1.9 \times 10^{-9} \, M$$

Figure 8.10 Rhodacyanine derivative active *versus P. falciparum*, EC_{50} = concentration toxic to 50% of cells (=1.5 $\times$ 10^{-6} M for chloroquine). Compare with rhodacyanine structure in Table 8.2.

against the malarial parasite, *Plasmodium falciparum* (Pudhom *et al.* 2006), which suggests a possible use in blood product decontamination (Chapter 11). Several examples here were highly active against a chloroquine-resistant strain of *P. falciparum*.

8.2   Merocyanine 540

Merocyanine 540 (MC540) combines both lipid solubility and anionic nature (due to its sulphonic acid group). Such properties allow it to pack into the outer leaflet of cell membranes, where there are low levels of anionic head groups (Mateašik, Šikurová and Chorvát 2002). The affinity of MC540 for leukaemic cells relative to hematopoietic stem cells is, as with other stains/diseases, a perfectly logical basis for its use in diagnosis. The associated photosensitising capability allows its use in leukaemic cell destruction (Sieber, Spivak and Sutcliffe 1984), this having been extended to demonstrations in glioblastoma, neuroblastoma, melanoma and colorectal carcinoma cell lines (Yow *et al.* 2000). MC540 thus supplies an excellent demonstration of the 'stain and kill' paradigm mentioned at various points in this book.

There are many merocyanine structures possible (and available), and it is possible to tailor the structure in order to emphasise photosensitising or fluorescence behaviour, depending on product requirements.

MC540 may, in fact, be used in two different ways against target tumour cells. The conventional mode of action is that of membrane photo-oxidation, the lack of intracellular targeting reflecting the anionic character of the photosensitiser, although additional lysosomal localisation has been reported (Chen *et al.* 2000). Conversely, the idea of pre-activation of MC540 uses photoirradiation of the photosensitiser prior to its delivery to the tumour, etc., the reaction of the dye with singlet oxygen yielding stable photoproducts such as merodantoin and merocil (Figure 8.11). The target site for these photoproducts is topoisomerase II, an enzyme involved in DNA replication, and thus inhibition of macromolecular synthesis is achieved via conventional chemotherapeutic routes. One advantage of this technique lies in the lack of toxicity of the photoproducts to healthy cells, allowing attack on, for example, metastatic disease since no further light delivery is required (Sharma, Arnold and Gulliya 1995).

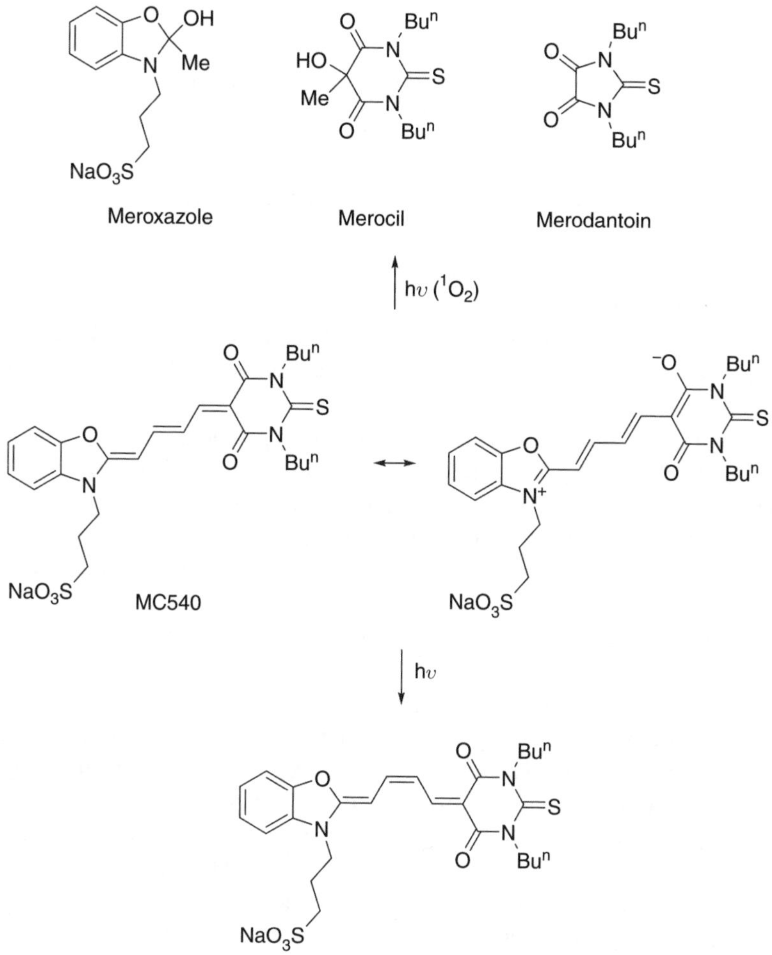

Figure 8.11 Delocalised forms of merocyanine 540, including deactivation and singlet oxygen-mediated breakdown products.

MC540 may be considered to be amphipathic. In combination with its overall negative charge, this underlies the targeting of the photosensitiser at cell membranes. Consequently, it has been found that MC540 preferentially photosensitises enveloped viruses and virus-infected cells, which led to its evaluation in preclinical models as a blood-sterilising agent (O'Brien *et al.* 1992). Plainly such an approach is logical in infection control, given the photodamage caused to viral surface antigens.

The structure of MC540 allows wide scope for the design of analogous series. In the current context, this may be necessary due to its low singlet oxygen yield (0.002). As with other photosensitiser types, use can be made of the heavy atom effect in increasing photosensitising effects, for example by replacing either the benzoxazole ring oxygen or

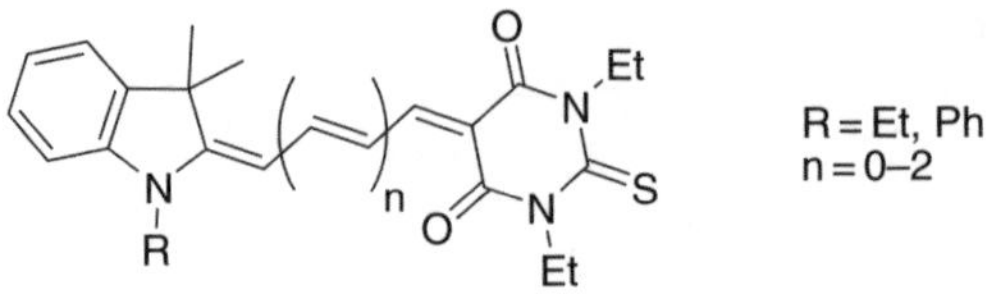

Figure 8.12   Merocyanine 540 antecedents.

the thiobarbiturate sulphur in the MC540 structure with selenium. Photoisomerisation of the central double bond of the polymethine chain, a major deactivation pathway for MC540, is also absent in the seleno-analogue (Redmond *et al.* 1994). Both of these factors obviously contribute to the increased singlet oxygen production *in vitro* (compare to thia-/selena-cyanines above).

Conversely, the presence of oxygen atoms in these positions yields a compound that undergoes relatively rapid photoisomerisation, resulting in much lowered triplet/singlet oxygen yields (Benniston, Harriman and McAvoy 1998). Obviously, methine chain rigidification inhibits such photoisomerisation (Benniston, Harriman and McAvoy 1997).

Perhaps the main drawback to the clinical use of merocyanines is their inactivation by plasma and serum components (Anderson *et al.* 1996), although this effect may be inhibited by the replacement of the ring oxygen, again with sulphur or selenium (Redmond *et al.* 1994).

Currently, the main use of MC540 lies, as noted above, in the purging of leukaemic cells from fractions required for autologous bone marrow transplants (Günther, Searle and Sieber 1992; Chen *et al.* 2000). However, as mentioned previously, there are many possible derivatives available from previous work, which might have improved properties relative to MC540. Indeed there are indole analogues in the original merocyanine publication from the 1950s (Figure 8.12), which might be seen as hybrids of MC540 and indocyanine green (Brooker, Keyes and Sprague, 1951a,b). Although photosensitising efficiencies were not investigated at the time, sufficient data are available to show that red and near-infrared absorbers were synthesised in the study.

8.3   N,N′-Bis(2-ethyl-1,3-dioxolane)kryptocyanine

Many cyanines have been derived from quinoline. *N,N′*-bis(2-ethyl-l,3-dioxolane) kryptocyanine (EDKC, a kryptocyanine; Figure 8.13) was in the vanguard of the synthetic non-porphyrin photosensitisers evaluated in the mid–late 1980s, and, as such, the understanding of its mode of action is now extensive. The importance of the 4-quinolyl unit (as opposed to 2-quinolyl derivatives such as pinacyanol) lies in the much-longer-absorption wavelengths achievable with the former (Figure 8.13).

The cationic nature of EDKC indicated its localisation in the mitochondria of malignant cells, thus providing a platform for the development of improved photosensitisers (Oseroff *et al.* 1986). The chalcogenapyrylium series (see Chapter 5) were discovered as active agents for malignant disease as a result of this work, and the interference with

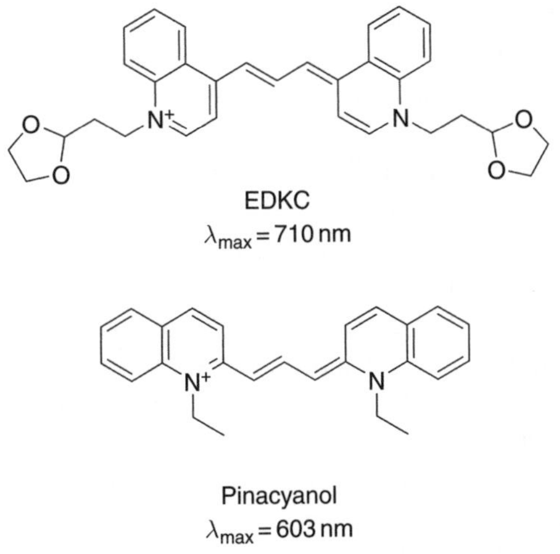

EDKC
$\lambda_{max} = 710\,nm$

Pinacyanol
$\lambda_{max} = 603\,nm$

Figure 8.13 *N,N'-bis(2-ethyl-l,3-dioxolane)kryptocyanine, EDKC, and pinacyanol.*

mitochondrial function (leading to dark toxicity) has led to further, non- photodynamic therapy (PDT), applications of cyanines (e.g. rhodacyanines; Kawakami *et al.* 1998). However, EDKC cannot be considered to be a conventional photosensitiser as it does not produce singlet oxygen, either *in vitro* or in cellular systems – this is perhaps not surprising in hindsight, considering its structure (Figure 8.13), as there is no heavy atom involvement to aid excited-state stabilisation. Indeed, unlike MC540, EDKC exerts its effects under both aerobic and anaerobic conditions and the only efficient photoprocess appears to be rapid internal conversion from the excited state to the ground state via polymethine chain isomerisation. This causes heat transfer to the surrounding cells and may thus lead to disruption. Similarly, structural alteration due to photoisomerisation might lead to considerable local membrane alterations and concomitant leakage of ions through the membrane. Additionally, electron transfer processes in the mitochondria may be disturbed by EDKC (Ara *et al.* 1987; Valdes-Aguilera, Ara and Kochevar 1988). As already noted, the large study carried out by Delaey and co-workers found no Type I or Type II activity from kryptocyanine derivatives (Delaey *et al.* 2000).

8.4 Indocyanine green

Due to its extended π-system, indocyanine green is a long-wavelength absorbing cyanine (tricarbocyanine; Figure 8.1). It is also essentially non-toxic in humans, and this has led to its intraoperative use in vital staining of malignancy (Tagaya *et al.* 2008) and blood flow tracing (Unno *et al.* 2008). More recently, this use has been extended into photodynamic areas, for example in the angiography of subjects treated for choroidal neovascularisation in age-related macular degeneration (Gomi *et al.* 2008).

As a cyanine derivative, it was originally expected that indocyanine green might act as a photosensitiser, and indeed phototoxic effects have been reported. However, it appears likely that these effects are photothermal, rather than the traditional Type I/Type II photosensitisation routes. The long-wavelength band, centred on the λ_{max} of 780 nm, allows excitation by 808 nm diode laser, and the photoexcited cyanine releases the excitational energy to its surroundings, causing thermal damage to cells (Chen *et al.* 1995). As with normal PDT, the effect is selective due to the required combination of chromophore localisation and light direction. The technique can, of course, be employed in grosser fashion to ablate tissue (photothermal ablation) (Diven, Pohl and Motamedi 1996) and has also led to the use of indocyanine green in 'tissue soldering', as an alternative to suturing (Kirsch *et al.* 1995), the heat produced on illumination increasing the rate of reaction of other 'solder' components, such as fibrin, albumin, etc.

Derivatised indocyanine green has been synthesised on a similar basis, employing the succinimide moiety (Figure 8.14) and others as reactive groups for amino group targeting (Ito *et al.* 1998; Tadatsu *et al.* 2003).

New indocyanine green has a measured singlet oxygen quantum yield *in vitro* of 0.077 (Delaey *et al.* 2000). This is less than a fifth of the accepted figure for the standard photosensitiser methylene blue (0.44; see Chapter 4). However, the photosensitising properties of indocyanine green *in vivo* are noteworthy and provide an excellent illustration of the point made earlier in the book, *viz.* that too much reliance should not be placed on singlet oxygen yield as a performance indicator in drug design and development (Chapter 3).

Comparison of the structure–property relationships of indocyanine green and new indocyanine green (Figure 8.15) also underlines the difference made by rigidification of the polymethine moiety. Indocyanine green has three *trans*-alkene groups in the straightforward chain. The isomerisation (*trans* → *cis*) of such groups is a major deactivation route from the excited state in cyanines, resulting in the low singlet oxygen yields normally associated with these compounds. The presence of the central chlorocyclohexenyl unit in the new indocyanine green molecule extends the wavelength range of indocyanine green itself (805 nm) by around 15 nm, while the resulting molecule has only three olefinic groups in the polymethine chain, as opposed to four in the original molecule.

Some interesting analogues of indocyanine green/new indocyanine green have been synthesised (Figure 8.16), both having indole, rather than benzindole nuclei, and with

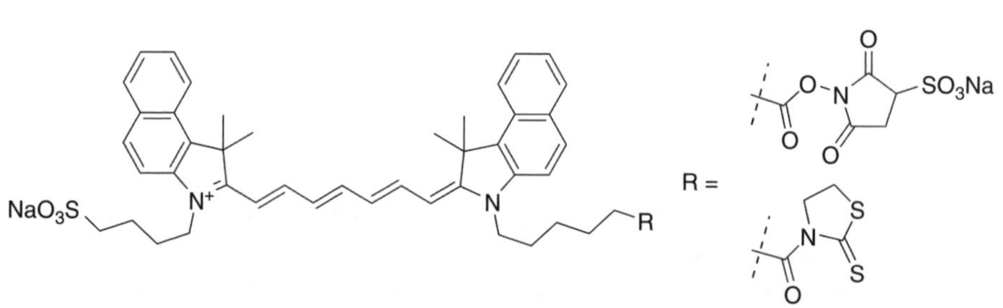

Figure 8.14 Biomolecule-reactive indocyanine green analogues.

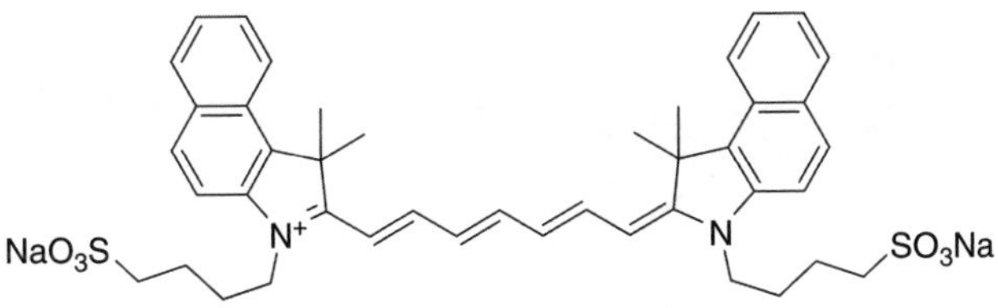

Indocyanine green

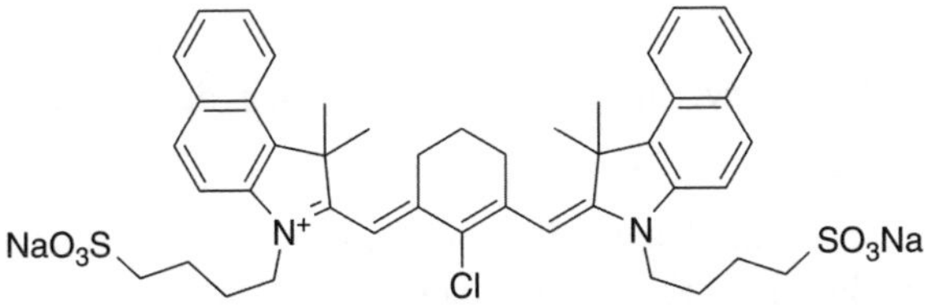

New indocyanine green

Figure 8.15 Indocyanine green and new indocyanine green.

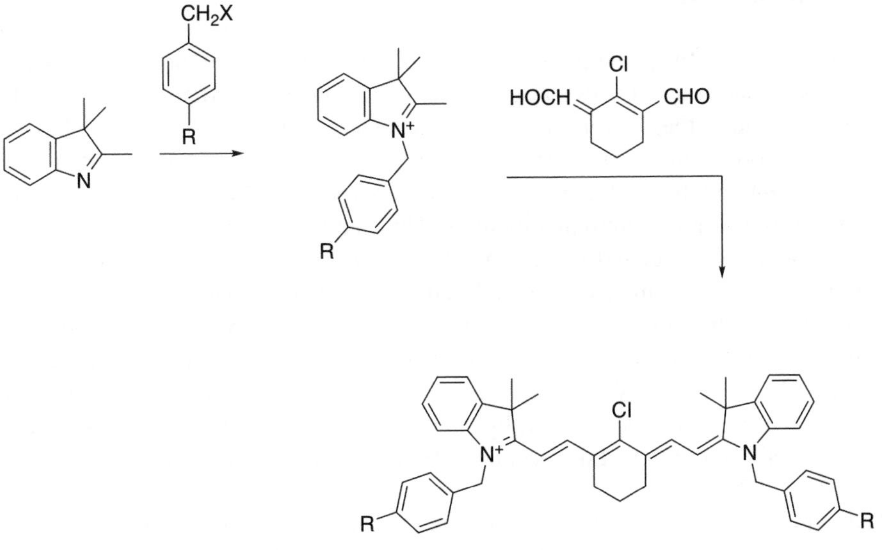

R = H, Me, CO₂H, F, NO₂

Figure 8.16 Structural variation in new indocyanine green derivatives.

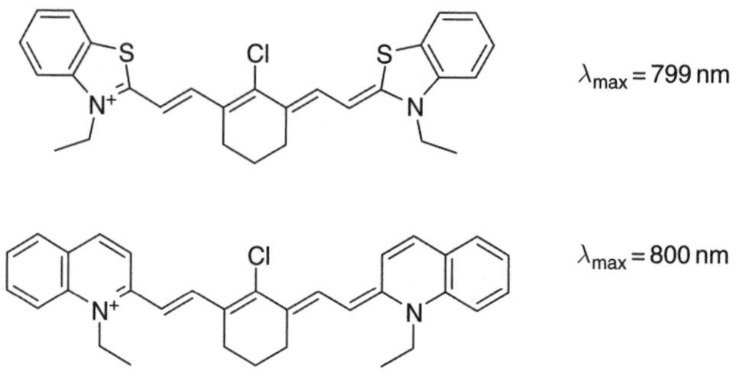

$\lambda_{max} = 799$ nm

$\lambda_{max} = 800$ nm

Figure 8.17   Novel, long-wavelength absorbing tricarbocyanines.

variously functionalised benzyl pendants attached to the ring nitrogens (Chen *et al.* 2006). The resulting bridged tricarbocyanines still absorb in the 780 nm region and demonstrate the potential of the molecular skeleton in drug discovery.

Extension of this idea to include alternative heterocyclic nuclei has produced cyanines with increased λ_{max} values (Ramos *et al.* 2002). Such research also increases the potential for new photothermal agents (Figure 8.17).

8.5   Structural improvement

On consideration of the general cyanine structure, i.e. heterocycle–polymethine–hetero-cycle, it is apparent that in several cases the development of effective *in vitro* photo-sensitisers requires the alteration of both the heterocycle and the polymethine unit. Typically this takes the form of rigidification of the chain (e.g. via the inclusion of a cyclic moiety) and the halogenation of the heterocycle. Plainly such requirements impinge on the design of optimal structures. This is discussed below.

Carbocyanines and related compounds have also been proposed as PDT agents. In common with other examples discussed here, their levels of photoactivity suffer from photoisomerisation. This problem has been approached in two ways: the introduction of substituents into the *meso* (i.e. central) position of the polymethine chain or the use of long-chain *N*-alkylation. While the former approach rigidifies the polymethine chain directly, it has been demonstrated in liposomes that the long-chain *N*-alkyl substituents cause physical anchoring and thus rigidification of the molecule within the lipid bilayer (Krieg *et al.* 1994).

The synthesis of more stable cyanines is possible using a ring system, effectively containing a pent-2-enedial moiety. This can be bridged as part of, for example, a cyclohexenyl molecule. Further stability is provided by the inclusion of a halogen (typically chlorine) in the central portion of the bridge (X, Figure 8.18). Such bridged systems are available as water-soluble compounds either by the inclusion of hydrophilic groups in the ring nitrogen side chain, as in new indocyanine green, or as a part of the annelated aromatic ring (R[1]; Figure 8.18).

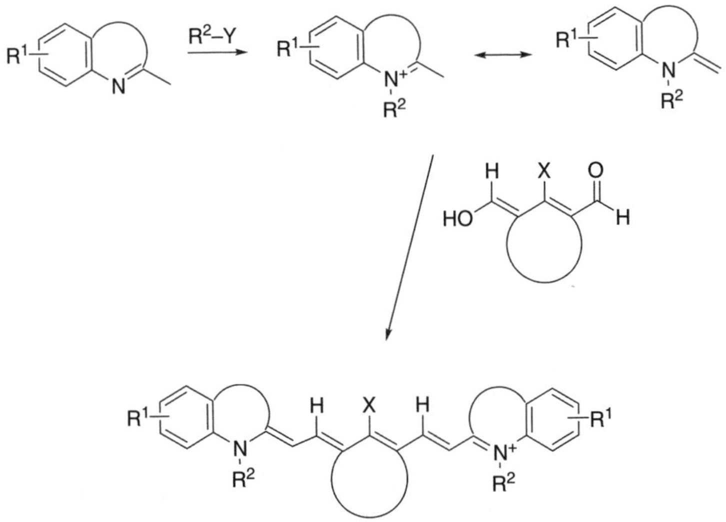

Figure 8.18 Synthesis of stabilized cyanines via incorporation of central ring system (X = halogen; R^1 = functional group; R^2 = alkyl or substituted alkyl).

8.6 Squaric and croconic acid derivatives

Squarylium and croconium dyes are 1,3-disubstituted compounds (based on the central residue) resulting from the condensation of one equivalent of squaric or croconic acid, respectively, with two molar equivalents of electron-donating aromatic or heterocyclic methylene bases (Figure 8.19).

Effectively, this endows greater rigidity to the polymethine chain, since the central double bond is incorporated in a ring system, as mentioned in the previous section. The resulting dyes are typically stable and exhibit intense absorption in the visible and near-infrared regions. Examples of squarylium (or squaraine) structures produced using benzothiazole-type heterocycles have been investigated (Santos *et al.* 2004), although in terms of use in photosensitisation, recourse must be made to the heavy atom effect in order to achieve reasonable singlet oxygen yields. Thus, the replacement of sulphur in the thiazole moiety with selenium reportedly results in up to a 10-fold increase in $^1O_2$ yield (Table 8.3).

It is noticeable from the table that the singlet oxygen yields are highest with longer alkyl pendants on the azole moieties. These inhibit photoisomerisation (Santos *et al.* 2004), although this behaviour was not recorded in the analogous squaric acid parents shown in Table 8.4 (Santos *et al.* 2003).

The influence of the central squarylium or croconium unit is, as mentioned above, to inhibit the non-radiative decay of the singlet excited state due to the *trans–cis* photoisomerisation typical of the cyanines, with a consequently increased

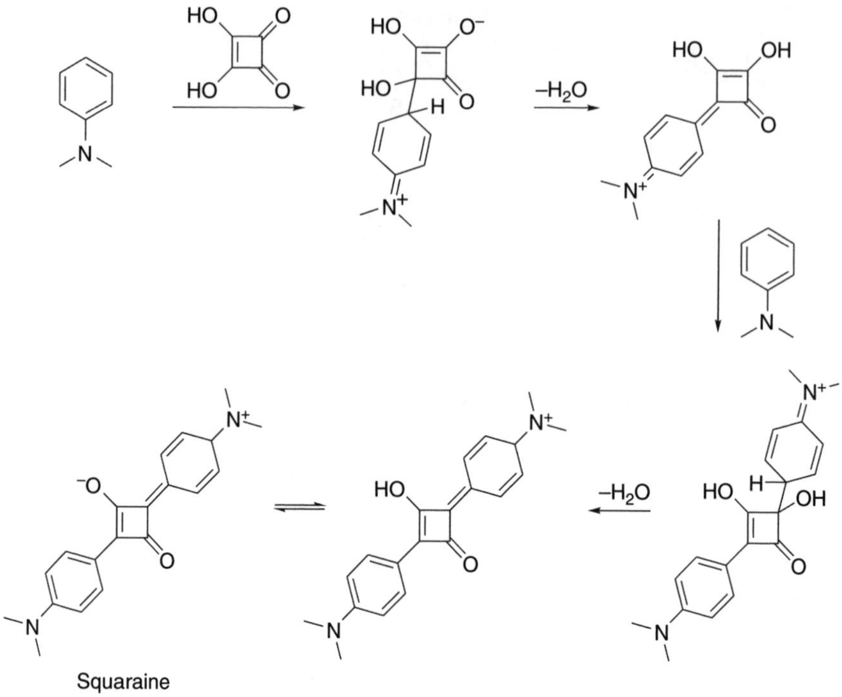

Figure 8.19 Synthesis of simple squaraine compounds.

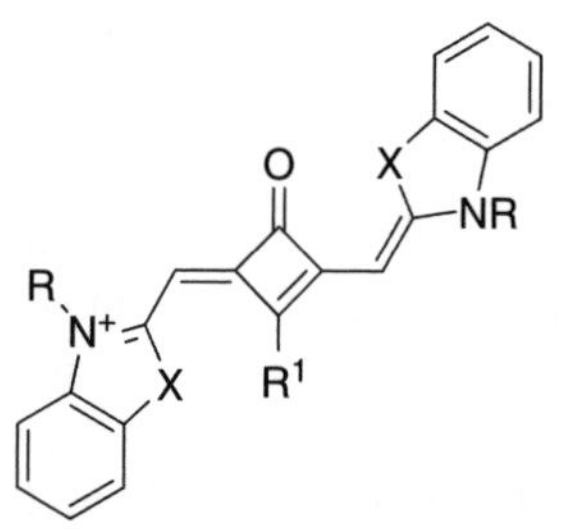

Table 8.3 Aminosquarylium derivatives based on benzothiazole and benzoselenazole

R	R^1	X	λ_{max} (nm)	Φ_Δ
Et	NHMe	S	659	0.04
Et	NHMe	Se	674	0.45
n-Hexyl	NHMe	S	659	0.14
n-Hexyl	NHMe	Se	677	0.68
Et	NEt$_2$	S	668	0
Et	NEt$_2$	Se	686	0.26

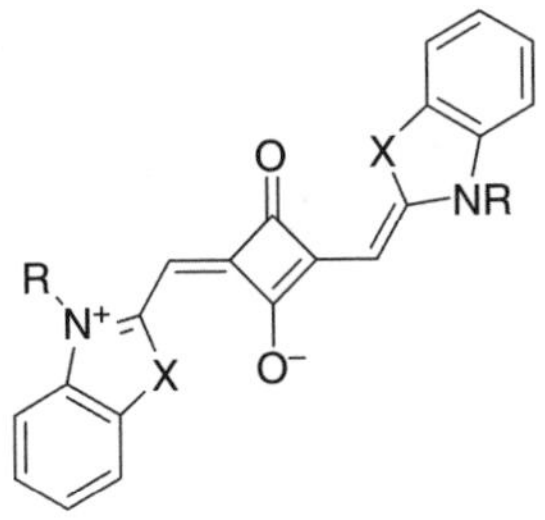

Table 8.4 Squaraine derivatives based on benzothiazole and benzoselenazole

R	X	λ_{max} (nm)	Φ_Δ
Et	S	650	0.26
Et	Se	665	0.31
n-Hexyl	S	650	0.26
n-Hexyl	Se	668	0.31

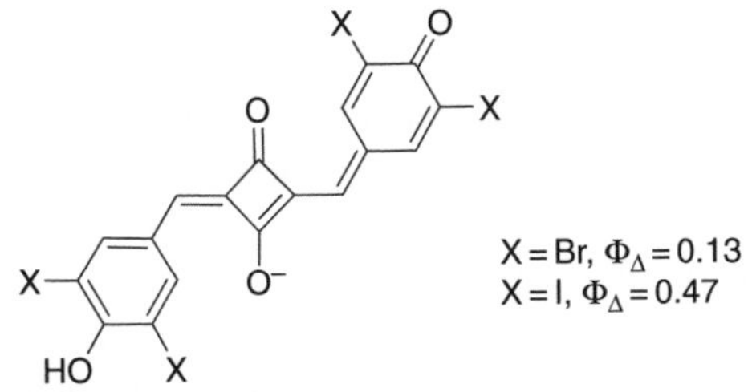

X = Br, $\Phi_\Delta = 0.13$
X = I, $\Phi_\Delta = 0.47$

Figure 8.20 Phenolic squaraines capable of singlet oxygen production.

efficiency of intersystem crossing. Other heavy atom effects have been reported for simpler squaraine cyanines having pendant halogenated phenols in place of hetero-aromatics. Again, only brominated or iodinated derivatives (Figure 8.20) appear to have sufficient effect on relaxation to allow singlet oxygen production (Ramaiah *et al.* 1997).

Derivatives of the six-membered analogue, rhodizonic acid, are fewer, but, given the same aryl/heteroaryl substitution pattern, absorb at longer wavelengths (*c.* 800 nm) than the squarylium or croconium analogues (Figure 8.21).

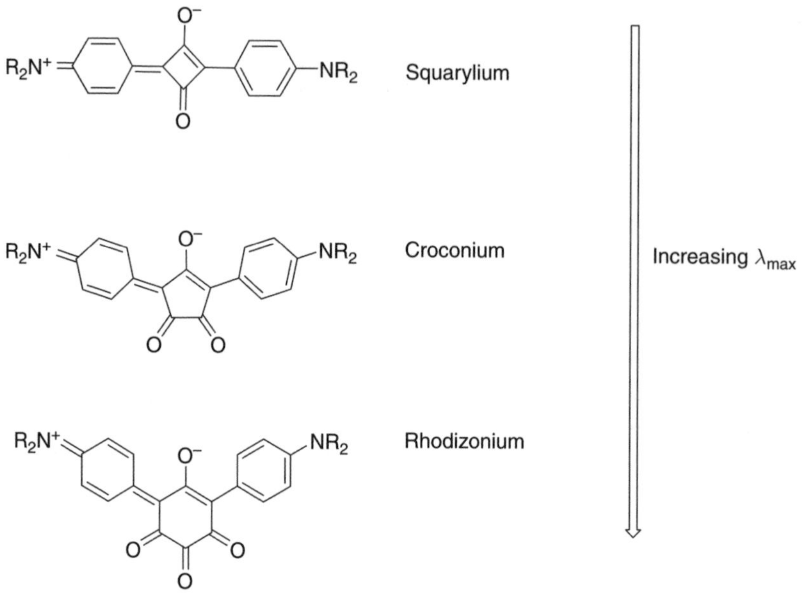

Figure 8.21 Structural comparison of squarylium-type bridged compounds.

8.7 Functional cyanines

As can be seen throughout the current chapter, the structure of cyanines and the related styryl compounds may be produced as positive, negative or neutral species, depending on the synthetic route employed. In addition, the delocalised charge may be altered in the molecule as a whole by the addition of ionic side chains. For example, the original merocyanines were electronically neutral (although zwitterion formation is possible via resonance), but this situation is altered by the inclusion of a pendant alkylenesulphonic acid, for example on the benzoxazole nitrogen in MC540 (Figure 8.11). Theoretically, the use of a quaternary ammonium moiety in this site would alter the biological activity profile, e.g. extending the photobactericidal spectrum to include Gram-negative strains. MC540 is itself usually only photobactericidal towards Gram-positive organisms (Dunne and Slater 1998). However, the photoactivity is apparent against species such as *Staphylococcus aureus* both in planktonic and in biofilm stages (Lin, Chen and Huang 2004).

Compared to MC540, the different charge on the carbocyanines obviously has ramifications regarding sites of action. Thus, whereas MC540 acts at the plasma membrane, cationic cyanines may be internalised, typically in the mitochondria. However, as more examples are synthesised, new targets are being identified. For example, the 3,3′-dihexyloxacarbocyanine derivative causes specific photodamage to microtubules (Lee, Wu and Chen 1995).

As with other classes of photosensitiser, functionalisation of cyanines to include biomolecules has been carried out in order to improve specificity, although this has usually been aimed at improving the performance of fluorescent probes (Chipon *et al.* 2006). Similar chemistry is applicable, of course, to photosensitiser optimisation.

Similarly to the phthalocyanines of the previous chapter, cyanine derivatives offer considerable advantages over porphyrins from the point of view of improved photoproperties, particularly where significantly longer-absorption wavelengths are required. In common with other dye classes, there are large numbers of previously synthesised derivatives to act as structure–function templates for improved photosensitisers. In addition, compounds absorbing in the 800 nm region offer photothermal capabilities.

References

Anderson GS, Gunther WHH, Searle R, Bilitz JM, Krieg M, Sieber F. (1996) Inactivation of photosensitizing merocyanine dyes by plasma, serum and serum components. *Photochemistry and Photobiology* **64**: 683–687.

Ara G, Aprille JR, Malis CD, *et al.* (1987) Mechanisms of mitochondrial photosensitization by the cationic dye, *N,N'*-bis (2-ethyl-1,3-dioxylene) kryptocyanine (EDKC): preferential inactivation of the electron transport chain. *Cancer Research* **47**: 6580–6585.

Benniston AC, Harriman A, McAvoy C. (1997) Photoisomerization of sterically hindered merocyanine dyes. *Journal of the Chemical Society, Faraday Transactions* **93**: 3653–3662.

Benniston AC, Harriman A, McAvoy C. (1998) Effect of resonance polarity on the rate of isomerization of merocyanine dyes. *Journal of the Chemical Society, Faraday Transactions* **94**: 519–525.

Brooker LGS, Keyes GH, Sprague RH, *et al.* (1951a) Studies in the cyanine dye series. XI. The merocyanines. *Journal of the American Chemical Society* **73**: 5326–5332.

Brooker LGS, Keyes GH, Sprague RH, *et al.* (1951b) Color and constitution. X. Absorption of the merocyanines. *Journal of the American Chemical Society* **73**: 5332–5350.

Browning CH, Cohen JB, Ellingworth S, Gulbransen R. (1924) The antiseptic action of the apocyanine, carbocyanine and isocyanine series. *Proceedings of the Royal Society: B* **96**: 317–333.

Chen JY, Cheung NH, Fung MC, Wen JM, Leung WN, Mak NK. (2000) Subcellular localization of merocyanine 540 (MC540) and induction of apoptosis in murine myeloid leukemia cells. *Photochemistry and Photobiology* **72**: 114–120.

Chen WR, Adams RL, Bartels KE, Nordquist RE. (1995) Chromophore-enhanced *in vivo* tumor cell destruction using an 808[hs]nm diode laser. *Cancer Letters* **94**: 125–131

Chen X, Peng X, Cui A, Wang B, Wang L, Zhang R. (2006) Photostabilies of novel heptamethine 3*H*-indolenine cyanine dyes with different *N*-substituents. *Journal of Photochemistry and Photobiology A: Chemistry* **181**: 79–85.

Chipon B, Clavé G, Bouteiller C, Massonneau M, Renard PY, Romieu R. (2006) Synthesis and post-synthetic derivatization of a cyanine-based amino acid. Application to the preparation of a novel water-soluble NIR dye. *Tetrahedron Letters* **47**: 8279–8284.

Cooper W, Rome KA. (1974) The external heavy atom effect on the photoisomerization of cyanine dyes. *Journal of Physical Chemistry* **78**: 16–21.

Delaey E, vanLaar F, DeVos D, Kamuhabwa A, Jacobs P, de Witte P. (2000) A comparative study of the photosensitizing characteristics of some cyanine dyes. *Journal of Photochemistry and Photobiology, B: Biology* **55**: 27–36.

Diven DG, Pohl J, Motamedi M. (1996) Dye-enhanced diode laser photothermal ablation of skin. *Journal of the American Academy of Dermatology* **35**: 211–215.

Dunne WM, Slater WA. (1998) Antimicrobial activity of merocyanine 540: a photosensitizing dye. *Diagnostic Microbiology and Infectious Disease* **32**: 101–105.

Gomi F, Ohji M, Sayanagi K, *et al.* (2008) One-year outcomes of photodynamic therapy in age-related macular degeneration and polypoidal choroidal vasculopathy in Japanese patients. *Ophthalmology* **115**: 141–146.

Görner H. (1999) Photoprocesses in 4-nitro- and 2,4-dinitro-substituted trans-1-styrylnaphthalene, trans-9-styrylanthracene and related systems. *Journal of Photochemistry and Photobiology A: Chemistry* **126**: 15–21.

Günther WHH, Searle R, Sieber F. (1992) Structure-activity relationships in the antiviral and antileukemic photoproperties of merocyanine dyes. *Seminars in Hematology*. **29**: 88–94.

Ito S, Muguruma N, Hayashi S, *et al.* (1998) Development of agents for reinforcement of fluorescence on near-infrared ray excitation for immunohistological staining. *Bioorganic & Medicinal Chemistry* **6**: 613–618.

Kassab K. (2001) Photophysical and photosensitizing properties of selected cyanines. *Journal of Photochemistry and Photobiology B: Biology* **68**: 15–22.

Kawakami M, Koya K, Kai TU, *et al.* (1998) Structure-activity of novel rhodacyanine dyes as antitumour agents. *Journal of Medicinal Chemistry* **41**: 71–75.

Kirsch AJ, Miller MI, Hensle TW, *et al.* (1995) Laser tissue soldering in urinary tract reconstruction: first human experience. *Urology* **46**: 261–266.

Krieg M, Bilitz JM, Srichai MB, Redmond RW. (1994) Effects of structural modifications on the photosensitizing properties of dialkylcarbocyanine dyes in homogeneous and heterogeneous solutions. *Biochemica et Biophysica Acta* **1199**: 149–156.

Lee C, Wu SS, Chen LB. (1995) Photosensitization by 3,3'-dihexyloxacarbocyanine iodide: specific disruption of microtubules and inactivation of organelle motility. *Cancer Research* **55**: 2063–2069.

Lin HY, Chen CT, Huang CT. (2004) Use of merocyanine 540 for photodynamic inactivation of *Staphylococcus aureus* planktonic and biofilm cells. *Applied and Environmental Microbiology* **70**: 6453–6458.

Lindauer H, Czerney P, Grummt UW. (1994) 9-(4-Dialkylaminostyryl)acridines – a new class of acidochromic dyes. *Journal fuer Praktische Chemie/Chemiker-Zeitung* **336**: 521–524.

Mateašik A, Šikurová L, Chorvát D. (2002) Interaction of merocyanine 540 with charged membranes. *Bioelectrochemistry* **55**: 173–175.

O'Brien JM, Gaffney DK, Wang TP, Sieber F. (1992) Merocyanine 540-sensitized photoinactivation of enveloped viruses in blood products: site and mechanism of phototoxicity. *Blood* **80**: 277–285.

Oseroff AR, Ohuoha D, Ara G, McAuliffe D, Foley J, Cincotta L. (1986) Intramitochondrial dyes allow selective in vitro photolysis of carcinoma cells. *Proceedings of the National Academy of Sciences of the USA* **83**: 9729–9733.

Pudhom K, Kasai K, Terauchi H, *et al.* (2006) Synthesis of three classes of rhodacyanine dyes and evaluation of their in vitro and in vivo antimalarial activity. *Bioorganic & Medicinal Chemistry* **14**: 8550–8563.

Ramaiah D, Joy A, Chandrasekhar N, Eldho NV, Das S, George MV. (1997) Halogenated squaraine dyes as potential photochemotherapeutic agents. Synthesis and study of photophysical properties and quantum efficiencies of singlet oxygen generation. *Photochemistry and Photobiology* **65**: 783–790.

Ramos SS, Santos PF, Reis LV, Almeida P. (2002) Some new symmetric rigidified triheterocyclic heptamethinecyanine dyes absorbing in the near infrared. *Dyes and Pigments* **53**: 143–152.

Redmond RW, Srichai MB, Bilitz JM, Schlomer DD, Krieg M. (1994) Merocyanine dyes: effect of structural modifications on photophysical properties and biological activity. *Photochemistry and Photobiology* **60**: 348–355.

Santos PF, Reis LV, Almeida P, Oliveira AS, Vieira Ferreira LF. (2003) Singlet oxygen generation ability of squarylium cyanine dyes. *Journal of Photochemistry and Photobiology A: Chemistry* **160**: 159–161.

Santos PF, Reis LV, Almeida P, Serrano JP, Oliveira AS, Vieira Ferreira LF. (2004) Efficiency of singlet oxygen generation of aminosquarylium cyanines. *Journal of Photochemistry and Photobiology A: Chemistry* **163**: 267–269.

Sharma R, Arnold L, Gulliya KS. (1995) Correlation between DNA topoisomerase II activity and cytotoxicity in MC540 and merodantoin sensitive and resistant human breast cancer cells. *Anticancer Research* **15**: 295–303.

Sieber F, Spivak JL, Sutcliffe AM. (1984) Selective killing of leukemic cells by merocyanine 540-mediated photosensitization. *Proceedings of the National Academy of Sciences of the United States of America* **81**: 7584–7587.

Tadatsu T, Ito S, Muguruma N, *et al.* (2003) A new infrared fluorescent-labeling agent and labeled antibody for diagnosing microcancers. *Bioorganic & Medicinal Chemistry* **11**: 3289–3294.

Tagaya N, Yamazaki R, Nakagawa A, *et al.* (2008) Intraoperative identification of sentinel lymph nodes by near-infrared fluorescence imaging in patients with breast cancer. *The American Journal of Surgery* **195**: 850–853.

Unno N, Suzuki M, Yamamoto N, *et al.* (2008) Indocyanine green fluorescence angiography for intraoperative assessment of blood flow: a feasibility study. *European Journal of Vascular and Endovascular Surgery* **35**: 205–207.

Valdes-Aguilera O, Ara G, Kochevar IE. (1988) Phototoxicity mechanism of a kryptocyanine dye in human red cell membranes and isolated murine mitochondria. *Cancer Research* **48**: 6794–6798.

Wainwright M, Kristiansen JE. (2003) Quinoline and cyanine dyes – putative anti-MRSA drugs? *International Journal of Antibiotic Agents* **22**: 479–486.

Yow CMN, Mak NK, Szeto S, *et al.* (2000) Photocytotoxic and DNA damaging effect of Temoporfin (mTHPC) and merocyanine 540 (MC540) on nasopharyngeal carcinoma cell. *Toxicology Letters* **115**: 53–61.

9

Natural product photosensitisers

As mentioned in the main introduction to this book, light is essential in nature and many organisms (including *Homo sapiens*) use light energy to survive, e.g. via photosynthesis. Light-harvesting chemicals (e.g. in photosystems I and II) require chromophores that are adapted to absorb specific wavelengths, usually coincident with natural sunlight, which thus allows the electronic transitions involved in energy absorption. While it is well established that porphyrins are the major light-absorbing pigments in photosynthesis, there are many other uses of chromophores in nature, usually in pigmentation or colouration phenomena, that have been evolved to the organisms' advantage, for example as insect or bird attractants in plants.

The use of naturally occurring compounds, or their derivatives, as drugs appears to be more acceptable to both the pharmaceutical industry and to their end-users, the rationale being that, in many cases, a substance of natural origin is probably less harmful to the human subject than one that has been produced synthetically. Logically, the natural products tested are under investigation for a similar purpose to that exhibited in the wild, for example the antibiotic activity of fungal metabolites from soil.

Outside chlorophylls and haematoporphyrin, the use of natural chromophores in photosensitisation is less well appreciated. Obviously, as with synthetic photosensitisers, light absorption leads to electronic transitions, which result in chemical reaction or further energy transfer to the immediate environment, with the production of reactive oxygen species that may result in cytotoxicity. In nature, such processes occur in both photochemical attack and defence, for example in the parasitisation of plant leaves by the mould family *Cercospora* and in insecticidal secretions by the plant bloodroot (*Sanguinaria canadensis*), respectively.

The chemicals or chromophoric types responsible for such natural photosensitisation are by now well established among the photodynamic community and have been investigated as putative photosensitisers for various applications, both in the natural state and, increasingly, as semi-synthetic or purely synthetic preparations, in a similar process to that developed from the naturally occurring porphyrins. There are thus growing collections of photosensitisers based on, for example, the perylenequinonoid, anthraquinone, isoquinoline and furocoumarin chromophores.

Photosensitisers in Biomedicine   Mark Wainwright
© 2009 John Wiley & Sons, Ltd

Natural product photosensitisers have been investigated in both the anticancer and the antimicrobial areas. The proposed use of riboflavin as a photodisinfecting agent for blood products is a suitable case in point, since this photosensitiser is effective against bacteria and viruses yet has no significant toxicity in the human host, as it is an essential vitamin (B_2) (Ruane *et al.* 2004; Reddy *et al.* 2008).

Given the variety of photoactive molecules available from natural sources, it is simplest to discuss them by chemical class rather than by potential clinical use.

9.1 Condensed aromatic pigments: perylene- and phenanthroperylenequinones

Lead compounds in this area are the condensed aromatic structures associated with the hypericin and hypocrellin families, both of which exhibit photosensitising properties in nature (Diwu and Lown 1994). Hypericism is a condition of skin photosensitisation exhibited by ruminants after feeding on plants containing the pigment hypericin, the pigment being photoexcited in sub-dermal capillaries. Hypocrellins are structurally related photosensitisers (they have fewer fused benzene rings; Figure 9.1) secreted by moulds such as *Hypocrellae bambusae*, in order to break down plant cell walls prior to digestion of nutrients from the cell interior.

Both hypericins and hypocrellins have been shown to be pharmacologically active in terms of malignant disease, lipoprotein binding and membrane activity being due to the combination of high, positive log$P$ values coupled with formal negative molecular charge. Both classes were thus originally investigated for anticancer photodynamic therapy (PDT) properties, but antiviral activity (in some cases without light) led to examination in terms of antimicrobial potential, the main area of interest for such photosensitisers in this case being the photodecontamination of blood products.

Clearly, both the hypericin and hypocrellin chromophores have considerable quinonoid character, and this gives rise to significant redox behaviour among the various derivatives. This access to facile electron transfer also promotes Type I photosensitisation, although most derivatives appear to exhibit both Types I and II behaviour *in vitro* (Diwu and Lown 1993).

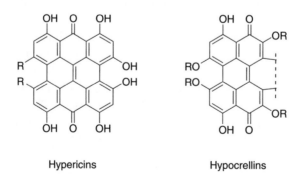

Hypericins   Hypocrellins

Figure 9.1 Hypericin and hypocrellin chromophores (R is usually Me).

9.2   Hypericin-type photosensitisers

There are few natural examples of the hypericin (phenanthroperylene) chromophore, the major one being hypericin itself (Figure 9.2), found in the leaves, stems and petals of St John's wort (*Hypericum perforatum*). Stentorin (from the ciliate *Stentor coeruleus*) is reportedly more photoactive than hypericin in some cell culture tests and also offers a longer λ_{max} (619 nm *versus* 590 nm) (Dai *et al.* 1995; Miyake, Harumoto and Iio 2001). Another natural analogue which is also photoactive, is blepharismin, a red pigment isolated from the protozoan *Blepharisma japonicum*. This differs from hypericin and stentorin in having an apical phenyl residue (Spitzner *et al.* 1998), i.e. located above the plane of the molecule (Figure 9.2).

Blepharismin exhibits effective antibacterial activity against Gram-positive bacteria including MRSA (Pant *et al.* 1997).

As with the porphyrins, there have been no naturally occurring cationic derivatives of hypericin discovered to date, derivatives usually occurring in either the neutral or the anionic form, since the only ionisable groups present are phenolic (i.e. ionised as phenoxide). The absence of any positive charge, despite the planar or pseudoplanar

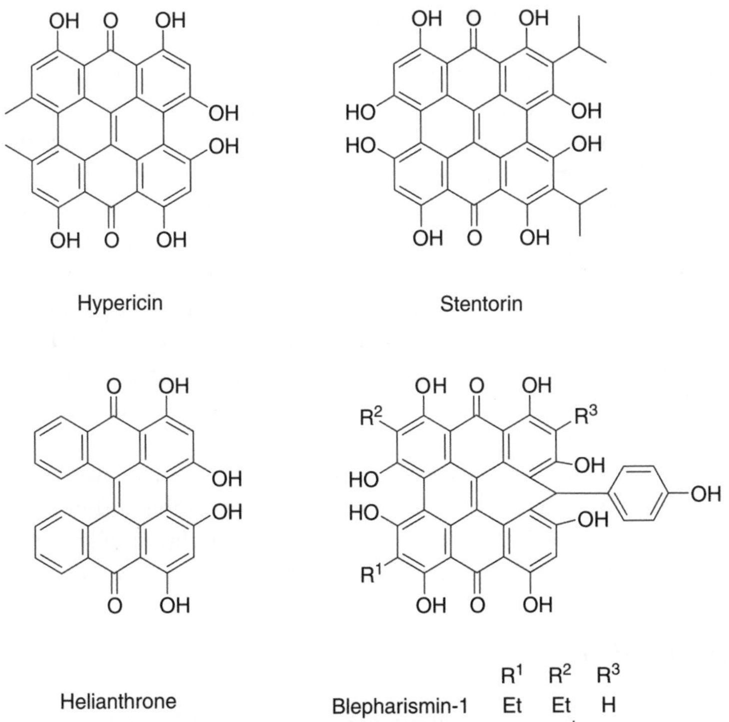

Hypericin

Stentorin

Helianthrone

	R¹	R²	R³
Blepharismin-1	Et	Et	H
Blepharismin-2	Et	Prⁱ	H
Blepharismin-5	Prⁱ	Prⁱ	Me

Figure 9.2   Hypericin derivatives.

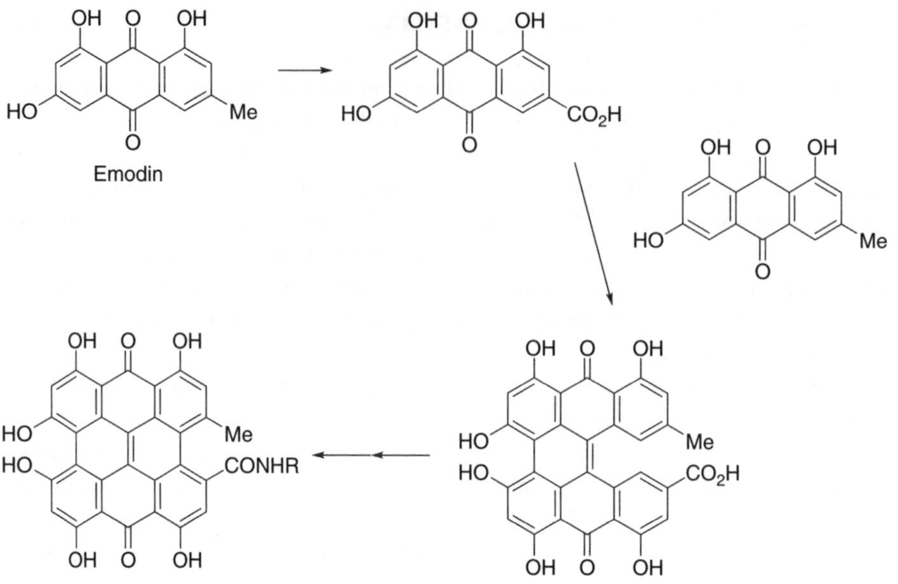

Figure 9.3 Synthesis of novel hypericin derivatives from emodin.

molecular nature, here means that conventional antinucleic acid activity (i.e. via an intercalative mechanism) is unlikely. However, several examples have been reported to be highly active antivirals, for example in being effective in the photodisinfection of media containing HIV-1 (Hudson, Harris and Towers 1993). Sulphonation of either hypericin or helianthrone chromophores achieves greater aqueous solubilisation but also provides compounds with decreased singlet oxygen efficiencies (Roslanice *et al.* 2000).

There is little reported work concerning photoactive synthetic derivatives of hypericin, although these may be produced via oxidation of the emodin (anthraquinone) precursor to the corresponding acid, followed by oxidative coupling, as shown in Figure 9.3, and amide production, furnishing the new derivatives (Crnolatac *et al.* 2005). Such approaches could be usefully employed in the production of cationic derivatives for photoantimicrobial research.

The singlet oxygen yield of hypericin (in ethanol) is much lower than that of hypocrellin A (0.35 *versus* 0.88 in ethanol) due to structural differences. However, photosensitising activity can be raised – in line with other chromophoric types – by the use of halogen substitution (Guedes and Eriksson 2006; Shen, Ji and Zhang 2006).

9.3 Hypocrellins

Members of this class of photosensitiser are usually mould metabolites, produced by *Cercospora* spp., *Hypocrella bambusae*, etc. The main function of these pigments in nature appears to be in plant parasitisation, utilising the photodynamic effect to break down cell walls. As with hypericin, the antiviral and anti-protein kinase C activity of various naturally occurring hypocrellins (Figure 9.4) suggested their screening for the photodecontamination application, as well as for cancer PDT (Diwu *et al.* 1994). Also, similarly to hypericin, the lead compounds exhibit suboptimal light absorption properties, maximum wavelengths being in the 580–590 nm region, and this has led to considerable synthetic efforts to improve matters.

While the chemical reactivity of hypericin is limited, lead perylenequinone compounds such as hypocrellin B (Figure 9.4) are more labile, and the provision of new derivatives, usually via nucleophilic substitution or condensation reactions, is consequently simpler, both the methoxy groups attached to the quinonoid portions and the acetyl carbonyl being sufficiently reactive (Figure 9.5). Reaction of hypocrellin B with amines produces derivatives with improved, i.e. red-shifted, photoproperties (Lee *et al.* 2006), with similar singlet oxygen yields to the parent compound. Type I photosensitisation is also possible, this being in line with the redox behaviour of hypericin mentioned above.

The production of improved photoantimicrobials via this route should also be possible, particularly with the provision of cationic derivatives. Weakly basic derivatives have been synthesised, for example via similar reaction with 1,3-propanediamine, while the use of diaminoethane gives rise to suitable steric conditions for bridging substitution across the neighbouring carbonyl and methoxy-bearing positions (Xu *et al.* 2004). The resulting piperidino-fused derivative is significantly red-shifted and has improved singlet oxygen efficiency (0.82) compared to the parent compound (Figure 9.5).

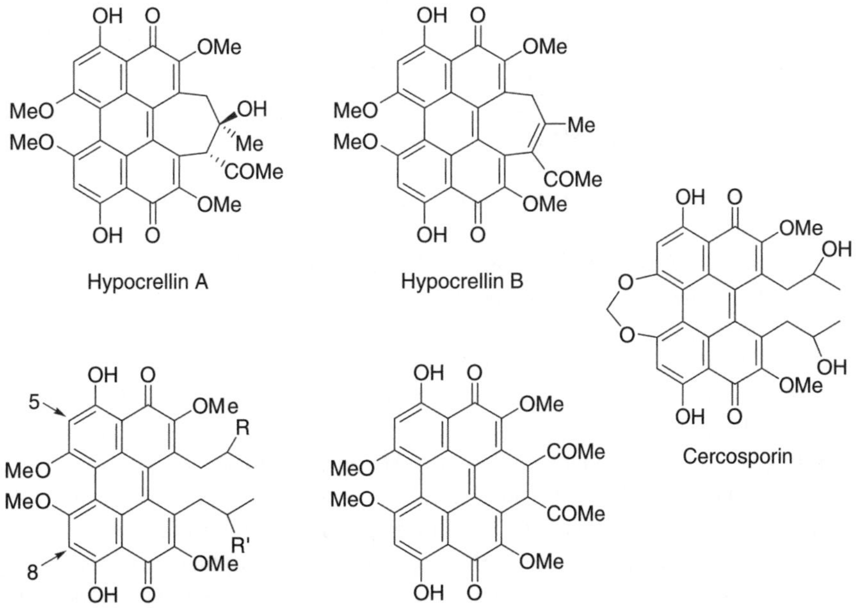

Hypocrellin A

Hypocrellin B

Cercosporin

Calphostin C, R = COPh, R' = CO$_2$(4-C$_6$H$_4$OH)
Calphostin D, R = R' = H

Elsinochrome A

Figure 9.4 Perylenequinone derivatives.

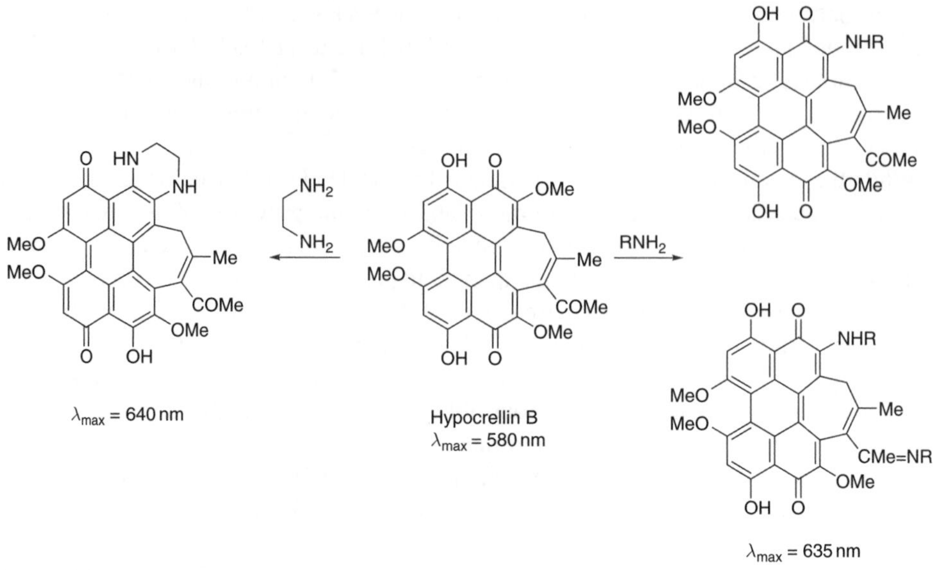

Figure 9.5 Novel amino derivatives of hypocrellin B.

As mentioned, it is well established that hypocrellins are inhibitors of protein kinase C (PKC), a key enzyme in the proliferation of tumour cells. Various semi-synthetic approaches to new hypocrellin-type photosensitisers, involving their action against PKC, have been attempted utilising as starting material the known natural product, cercosporin via conjugate addition of nucleophiles such as thiophenol at positions 5 and 8. Calphostin C (Figure 9.4) was apparently more active in PKC inhibition than any of the newer derivatives, leading to the suggestion that the increased photoactivity against PKC arises via the addition of cysteine residues in active sites of the protein (through −SH) at positions 5 and 8 of calphostin C (see Figure 9.4). Thus, when these positions are blocked, the photoactivity decreases (Diwu *et al.* 1994).

Several new derivatives of the hypocrellins have exhibited improved levels of photoactivity in cell culture. The hypocrellin structure is such that the phenolic groups (positions 4 and 9) may be replaced by nucleophiles, and alkylamino substitution has recently been shown to have dramatic effects on the phototoxic nature of the prepared derivatives. Thus, the 4,9-bis(*n*-butylamino) derivative of hypocrellin B (Figure 9.4) has a much higher ε_{max} value than the parent compound, leading to far greater phototoxicity against EMT-6 cells in culture. At the same time, the introduction of the *n*-butylamino side chains decreases the dark toxicity of the derivative relative to the parent, thus yielding a compound reportedly having a high light:dark toxicity ratio (≥ 167). This particular study demonstrated the potential of the hypocrellin skeleton for functionalisation, leading to changes in physico-chemical properties and improved characteristics for PDT (Estey *et al.* 1996).

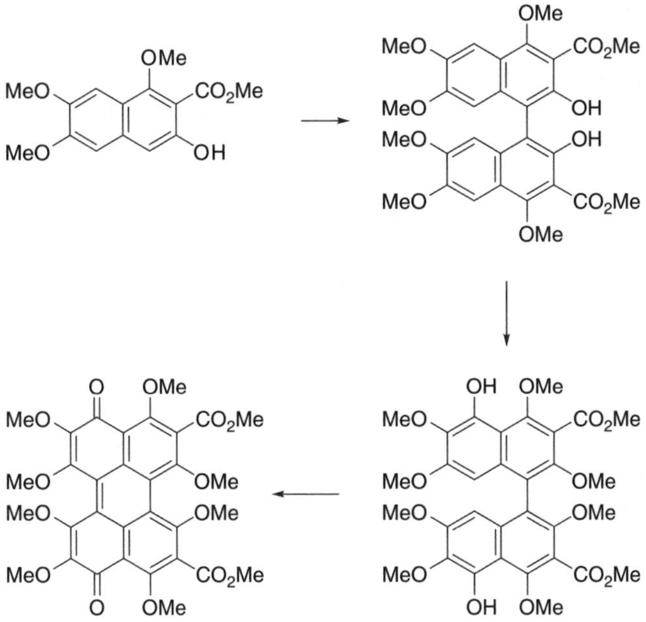

Figure 9.6 Perylenequinone synthesis from a naphthol derivative.

A simpler derivative of hypocrellin B may be produced via the simple expedient of oxidation of its 14-methyl group to the corresponding carboxylic acid. The product is more hydrophilic than the parent compound ($\log P = +0.63$ *versus* $+1.62$), but with a decreased singlet oxygen yield (0.26 *versus* 0.76).

Heavy atom effects have not been investigated significantly with the hypocrellins, not surprisingly given the high-end singlet oxygen yields *in vitro*. However, the chelation of heavy metal ions by this class has been demonstrated to increase the singlet oxygen yield of hypocrellin A (Zhou 2005). Since metal ion chelation by the adjacent carbonyl and β-hydroxyl moieties is important in pharmacology and indeed in the action of drugs such as the tetracyclines, similar chelating activity in related photosensitisers, such as the hypericins and precursor anthraquinones (below), may be a fruitful avenue of research in the future.

As shown previously with hypericin, *de novo* synthesis of perylenequinone derivatives is possible via oxidative coupling of precursors, in this case naphthols (Figure 9.6) (Mulrooney *et al.* 2003).

9.4 Anthraquinones

In line with the above, the antimicrobial activity of the hypericin progenitors, the anthra-quinones, is also significant. In addition, anthraquinones represent a ubiquitous class of compounds in nature, so the reported activities here offer considerable future potential. The antiviral capabilities of this class has been known for some time (Schinazi *et al.* 1990), and these capabilities, in combination with their photosensitising properties, support more concerted investigation of the suitability of this class as photoantimicrobials.

That photosensitising effects can be associated with anthraquinones is well known from both the dye and pharmaceutical industries. Synthetic anthraquinones produced as textile dyes may act as causative agents in skin phototoxicity, e.g. 'bikini dermatitis', the photoactive dye leaching from the textile due to the warm, moist environment. Anthraquinone drugs such as the anticancer agent doxorubicin (Figure 9.7) also exhibit phototoxicity *in vitro*, although to a lesser degree than mainstream PDT agents. However, whereas blue anthraquinone textile dyes normally absorb in the 600 nm region, the λ_{max} values for anthraquinone-based drugs are usually not much greater than 500 nm, lessening their utility in conventional PDT. These factors may not be too important, given that the proposal on anthracycline PDT would presumably be to employ the photoactivity as an additive therapy, rather than as a single means of tumour destruction (Saxton *et al.* 1995), i.e. to use an anthraquinone having a favourable combination of tumour selectivity and reasonable light-absorption properties. In addition, although drugs such as doxorubicin show a reasonable selectivity for tumours, they also exert adverse effects

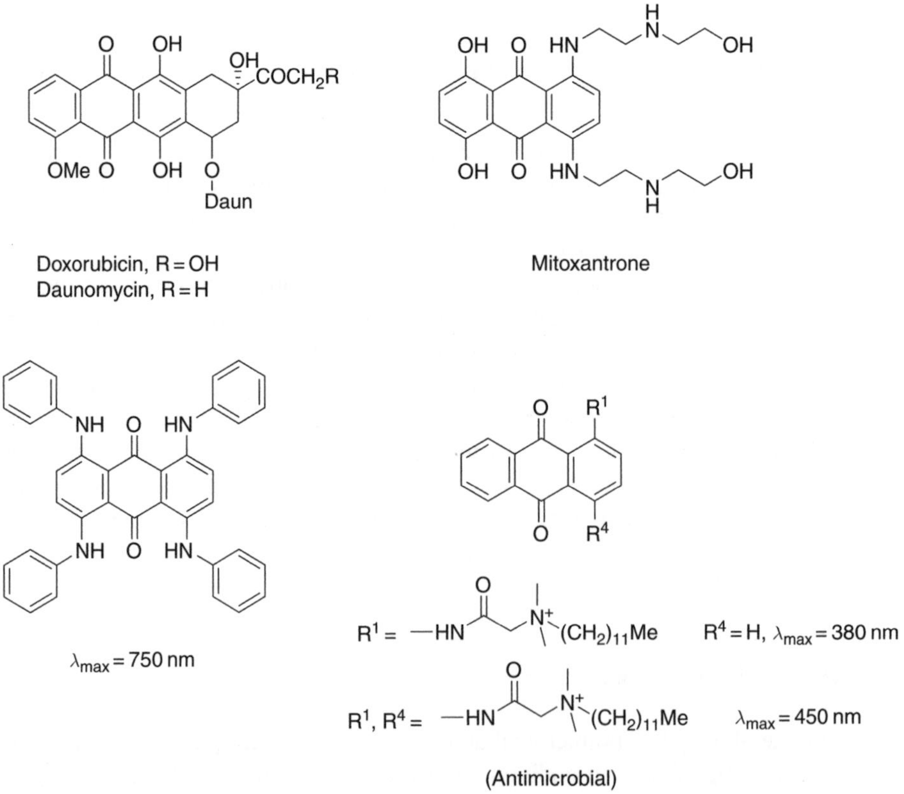

Doxorubicin, R = OH
Daunomycin, R = H

Mitoxantrone

$\lambda_{max} = 750$ nm

$R^1 = $ —HN...N^+(CH$_2$)$_{11}$Me $R^4 = H$, $\lambda_{max} = 380$ nm

$R^1, R^4 = $ —HN...N^+(CH$_2$)$_{11}$Me $\lambda_{max} = 450$ nm

(Antimicrobial)

Figure 9.7   Anthraquinone anticancer drugs and dye derivatives (Daun = daunosamine sugar moiety).

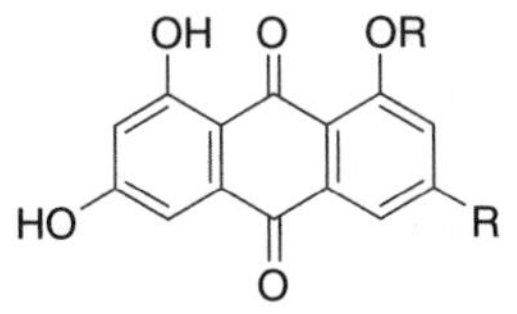

Emodin, R = Me
Aloe emodin R = CH₂OH

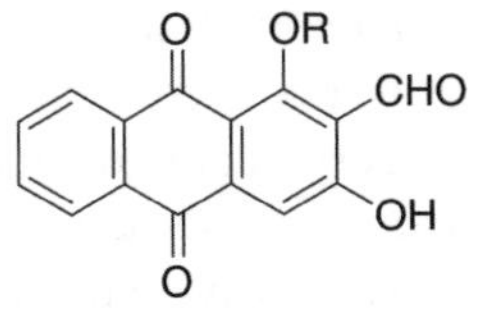

Damnacanthal, R = Me
Nordamnacanthal R = H

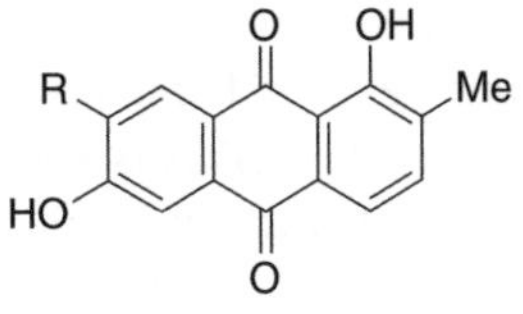

Heterophylline, R = Me
Soranjidiol, R = H

Figure 9.8 Natural product anthraquinone photosensitisers.

on healthy cells, in particular cardiotoxicity. The utilisation of the extra photodynamic activity associated with such drugs may make it possible to produce the same levels of overall antitumour activity at lower drug doses, with a concomitant decrease in side effects. Long clinical use also means that the pharmacology of the anthracycline drugs is very well understood.

Extrapolation of this reasoning suggests that it should be possible to synthesise blue anthraquinone dye derivatives, which are suitable for PDT and photodynamic antimicrobial chemotherapy *per se*. Long-wavelength absorbers (>750 nm; Figure 9.7) have been reported (Chao 1992), as have dye derivatives that exhibit antimicrobial action (Ma, Sun and Sun 2003).

Among proper natural products, good examples of this potential are also given by emodin and its hydroxymethyl analogue, aloe emodin (Figure 9.8). Both of these anthraquinones are reportedly photoactive against Gram-positive bacteria, including MRSA (Hatano *et al.* 2005), although not against Gram-negative bacteria. Interestingly, closely related anthraquinones, such as damnacanthal (Figure 9.8), have been reported to possess significant antifungal properties, e.g. against *Candida* and *Trichophyton* spp. (Singh *et al.* 2006). Both damnacanthal and nordamnacanthal, and the related soranjidiol and heterophylline exhibit significant levels of singlet oxygen production on illumination at 337 nm (Comini *et al.* 2007).

Naturally occurring anthraquinones such as damnacanthal and nordamnacanthal are associated with significant yields of singlet oxygen – up to 70% efficiency – depending on conditions, again with the ability to produce radical species via the Type I route (Nunez Montoya *et al.* 2005).

9.5 Psoralens (furocoumarins)

While photodynamic therapy and photoantimicrobial disinfection are gradually becoming accepted in the clinic, photoactive drugs have actually been employed for millennia. Psoralen phototherapy (or, more precisely, photo*chemotherapy*) is a well-established treatment for the skin disorder psoriasis, employing long-wavelength ultraviolet (UV) light (thus PUVA (psoralen-UV-A)), and also more recently in the treatment of malignant

diseases such as lymphoma. In the latter treatment, the blood fraction containing malignant cells and the psoralen drug is removed from the body and irradiated, before
replacement. This is known as extracorporeal photochemotherapy or photopheresis.

Psoralen and angelicin derivatives – linear and angular furocoumarins, respectively
(Figure 9.9) – have been known since the early civilisations of the East and Middle East in
the treatment of skin disorders, employing the extracts of common plants such as
bergamot in conjunction with natural sunlight. More recently, the structure–activity
relationships and sites of action of the key constituents of such extracts have been
elucidated and synthetic analogues prepared. The field of psoralen photomedicine, and
in particular photopheresis, has become incredibly active in recent years, with a
bewildering array of newly synthesised compounds as well as the *in vitro* testing of
natural congeners and structural isomers of the furanocoumarin unit. A representative
sample of the more promising compounds appears in Figure 9.10.

Due to their size and shape, psoralens are effective nucleic acid intercalators.
Linear furocoumarins (e.g. 8-methoxypsoralen [8-MOP] and 4,5′,8-trimethylpsoralen;

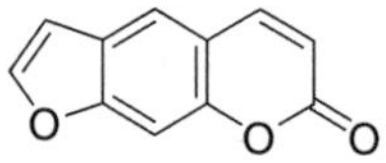

Psoralen – linear fusion 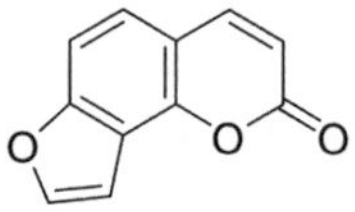   Angelicin – angular fusion

Figure 9.9 Psoralens and angelicin chromophores.

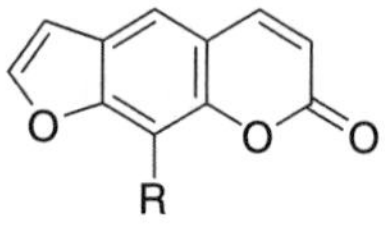

R = H, Psoralen
R = OMe, 8-methoxypsoralen(8-MOP)

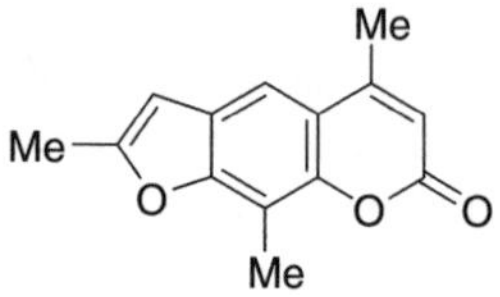

4, 5', 8-Trimethylpsoralen (TMP)

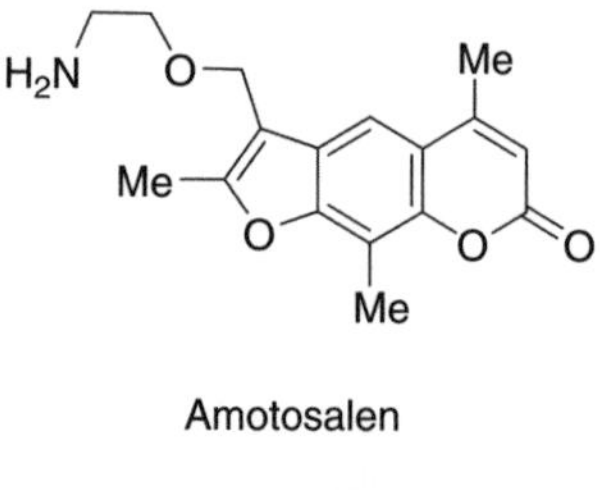

Amotosalen

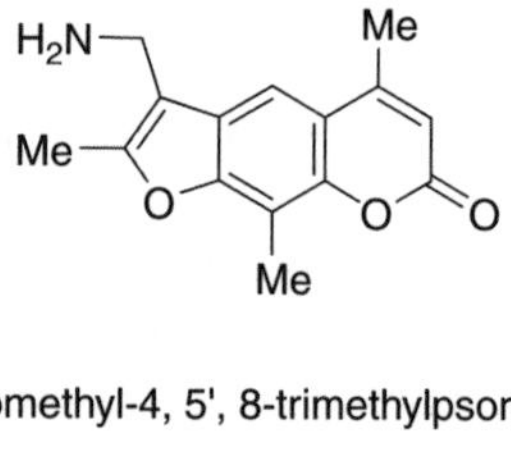

4'-Aminomethyl-4, 5', 8-trimethylpsoralen (AMT)

Figure 9.10 Clinically relevant psoralen derivatives.

Figure 9.10) absorb long-wavelength UV light, having wavebands in the region of 300–400 nm, which are used for practical illumination (absorption of UV light by nucleic acids occurs at considerably lower wavelengths). Intercalated psoralens may thus be photoexcited *in situ*, causing the promotion of [2 + 2] cycloaddition reactions with olefinic moieties in nucleotide bases (e.g. cytosine; Figure 9.11). The formation of mono-adducts from the furan side of the molecule result exclusively from the singlet excited state, and pyrone adducts mainly from the triplet (Cimino *et al.* 1985). Cycloaddition may occur at both the furan and the pyrone moieties of the intercalated psoralen, and this produces the internal cross-linking of the DNA helix.

The accepted mode of action of furocoumarins is their photocycloaddition to pyrimidine bases in target cell DNA, i.e. via a photochemical method. However, it is also established that the furocoumarins have associated singlet oxygen production capabilities. While the yields of singlet oxygen tend to be somewhat lower than standard photosensitisers such as rose Bengal (Φ_Δ c.0.8), they may nevertheless contribute a photodynamic component to the activity of psoralens or angelicins, the process being activated at the same wavelengths as the photochemical process. The range of singlet oxygen quantum yields for furocoumarins has been reported to be 0.02–0.4 (Jones, Young and Truscott 1993).

The structure of the psoralen molecule may be altered in order to increase the selectivity of photochemical reaction, rather than giving the mixture of mono- and bis-adduct

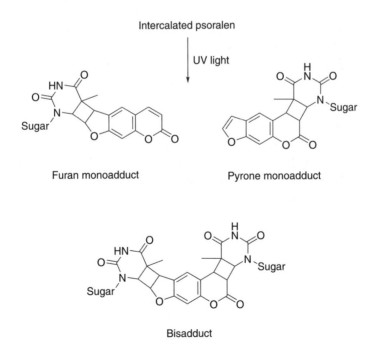

Figure 9.11 Cycloaddition reaction products of the psoralen nucleus with cytosine in DNA (sugar = deoxyribose).

formation outlined above. In addition, the idea of a cross-linking agent may be less than attractive to many clinicians, particularly if the disorder under treatment is non-malignant. Psoralen functionalisation in the pyrone (six-membered oxygen) ring can yield compounds that are unable to form cycloadducts via this ring due to steric factors. However, as it is well established that methylation in the furanocoumarin nucleus furnishes compounds, e.g. 4,5′,8-trimethylpsoralen (Figure 9.10) or 4,6,4′-trimethylangelicin (Figure 9.12), which cause DNA cross-linking (Wamer *et al.* 1995), the steric factor is plainly important in its inhibition. The 3-ethoxycarbonyl analogue has been reported to be sufficiently hindered to give rise only to mono-adducts with DNA.

As mentioned above, the major clinical use for psoralens remains in the treatment of psoriasis. The extracorporeal approach, photopheresis, is obviously more related to PDT, being based on the selectivity of psoralen derivatives for malignant cells, such as the lymphocytes implicated in cutaneous T-cell lymphoma. It should also be noted that there is also a strong link here to the photodecontamination of blood fractions/products covered below and in Chapter 11. In photopheresis, oral administration of the psoralen drug leads to its uptake by malignant T cells in the bloodstream, i.e. there is a selectivity requirement. Sequential removal of 0.5 l aliquots of blood, component separation and illumination of the white cell fraction with the relevant wavelength of UV light lead to DNA damage via photoadduct formation as shown above (Figurer 9.11) and thus to direct cytotoxicity. Protein damage at the cell membrane leads to cell death or causes sufficient changes in cellular morphology such that the reintroduction of the white cell fraction into the bloodstream may cause an auto-vaccination effect, i.e. the malignant cells are not recognised by the body's immune system and are thus destroyed (Schmitt, Chimenti and Gasparro 1995). Psoralen–protein interaction has been much overlooked throughout the development of useful clinical products, yet the widely reported mechanism of psoralen–DNA adduct formation cannot explain the immunotherapeutic basis of the treatment.

In addition to the photochemical reaction of furocoumarins, they can also cause the hydroxylation of guanosine in nucleic acids, a mechanism often associated with the intermediacy of singlet oxygen. Although it is suggested that psoralen photodamage to the cell membrane may be due to psoralen photoadducts with biomolecules other than nucleic acids, this sort of damage can certainly be envisaged as occurring via a photo-dynamic rather than a photochemotherapeutic route, since the production of singlet

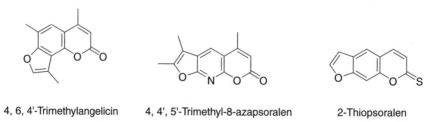

4, 6, 4'-Trimethylangelicin   4, 4', 5'-Trimethyl-8-azapsoralen   2-Thiopsoralen

Figure 9.12   Improved angelicin and psoralen derivatives.

oxygen by psoralen derivatives in solution is now well established. If this is indeed the case, the ongoing efforts in new drug design and synthesis in this area, and particularly those involving increasing the Φ_Δ values, are well justified.

When a drug is established for the treatment of a particular disorder, analogues are synthesised as a matter of course in order to improve on the activity of the lead compound. There are two main reasons for the development of analogues of psoralen: increasing the selectivity for target cells or increasing the photoactivity at the target site, for example via the heavy atom effect. Thus, many bioisosteres and structural isomers of the furanocoumarin nucleus have been isolated or synthesised *de novo* (Figure 9.12). The replacement of the furan oxygen with sulphur or selenium gives rise to compounds having much improved photoactivity. In addition, the 8-azapsoralens, i.e. analogues arising from replacement of carbon-8 with nitrogen, exhibit lower incidences of DNA cross-link formation. The activity of 4,4',5'-trimethyl-8-azapsoralen in terms of its inhibition of DNA synthesis in Ehrlich cells has been reported to be six times that of 8-MOP (Figure 9.10) and it was efficacious in the clearance of psoriasis in clinical trials. Conversely, 8-MOP showed slightly higher activity than the azapsoralen (Guiotto *et al.* 1995).

Earlier work concerning the use of the naturally occurring 8-MOP (Figure 9.10) against tumour cell DNA led to the investigation of 8-MOP and some of its congeners, targeted at the development of photoantivirals, especially those used for the control of HIV infection arising from blood donation. Improvements were reported for both trimethyl derivatives of psoralen (4,5',8-trimethylpsoralen, TMP; Figure 9.10) and angelicin (4,6,4'-trimethylangelicin, TMA; Figure 9.12) against HIV-1, in comparison to the activity of 8-MOP (Miolo *et al.* 1994).

Thus, the application of psoralen derivatives as photoantivirals in blood product decontamination began with 4,5',8-trimethylpsoralen as a lead compound. However, newer compounds such as the aminomethyl derivative of trimethylpsoralen (AMT; Figure 9.10) offered improved aqueous solubility and nucleic acid binding. AMT was also an excellent lead candidate for plasma and platelet fractions, exhibiting high activity against a range of viruses (Corash 2000). The improved activity of AMT in comparison to trimethylpsoralen is thought to be due to the cationic nature of the former. Amotosalen hydrochloride is currently marketed as part of a long-wavelength UV-activated system for pathogen reduction in blood plasma, the psoralen photosensitiser being active against a broad range of microbial species – viruses, bacteria and protozoa (Solheim and Seghatchian 2006). Although Amotosalen has broad-spectrum photobactericidal activity, it has lower than desirable activity against the important Gram-negative pathogen, *Klebsiella pneumoniae*. However, elsewhere, the range of activity also extends to protozoan species such as *Trypanosoma cruzi*, *Plasmodium falciparum* and *Leishmania mexicana* (Pelletier, Transue and Snyder 2006).

Although the shape of angelicin may not be as efficient in terms of DNA binding as its linear psoralen congener, it has been employed as a lead structure for synthetic furocoumarin development. The associated antifungal activities of angelicin derivatives have been reported against *Candida albicans*, *Cryptococcus neoformans*, *Saccharomyces cerevisiae* and *Aspergillus niger* (Sardari *et al.* 1999). The corresponding levels of

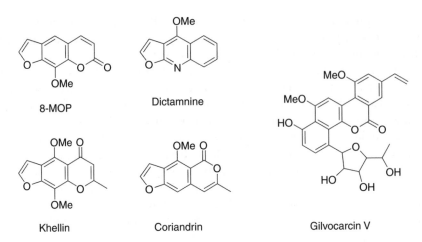

Figure 9.13 Photosensitiser types related to psoralen.

toxicity in cultured human cell lines are sufficiently low to suggest the use of the new derivatives in antifungal therapy.

It should be noted that related systems such as the psoralen isomers the furoisocoumarins, furochromones and benzannelated gilvocarcins, e.g. coriandrin, khellin and gilvocarcin V, respectively (Figure 9.13), also contain photoactive members, again generally targeting cellular DNA (Towers, Page and Hudson 1997). Rather unusually, gilvocarcin V is reportedly active against both enveloped and non-enveloped viruses (Lytle, Wagner and Prodouz 1993). The quinoline derivative, dictamnine (Figure 9.13), is also highly effective against viruses, the quinoline portion of the molecule acting as a support, allowing photoaddition of the furan moiety (Hudson 1989). Another tricyclic family, the β-carbolines, are photoactive and, again, effective against viruses, but the mechanism here is not thought to be photocycloaddition.

9.6 Isoquinoline alkaloid photosensitisers

Among natural product photosensitisers there can be few better examples than the fused isoquinoline derivative sanguinarine. Isolated from the root of *Sanguinaria canadiensis* (Bloodroot), its major function in nature is in the plant's chemical defence against insects. Sanguinarine is exuded by the leaves and sticks to the insect cuticle, where the action of daylight causes hole burning. This is not sophisticated but is an excellent demonstration of the photodynamic effect. The inherent (dark) antimicrobial effects of sanguinarine have led to its use as a constituent in toothpastes and mouthwashes (Grenby 1995).

The principal photosensitising compounds among the fused isoquinolines are thus the well-established sanguinarine, and also berberine. Both of these have a considerable number of naturally occurring congeners (Figure 9.14). As described above, the photodynamic activity of such compounds is important in their use in the natural environment, usually in defence, Reported singlet oxygen quantum efficiencies are 0.16 and 0.25 for sanguinarine and berberine, respectively (Arnason *et al.* 1992).

The planar cationic nature of these molecules and their congeners suggests nucleic acid activity via an intercalative mechanism, and this is supported by the literature (Schmeller, Latz-Brüning and Wink 1997). Indeed, the DNA binding by the sanguinarine analogue nitidine (Figure 9.14) has led to its use in anticancer research (Cushman, Mohan and Smith 1984). Such activity is perhaps not surprising, given the chromophoric similarity, particularly of the sanguinarine family of compounds, with the standard DNA probe ethidium bromide (Figure 9.14, and see Chapter 4). Many analogous structures are now established, with sanguinarine and berberine congeners often occurring together (Liang *et al.* 2006).

The intercalative mode of action associated with the sanguinarine/berberine-type photosensitisers has been established using the ethidium bromide displacement assay for a range of derivatives. However, different intercalative binding affinities are apparent, and the sequence selectivity varies depending on the derivative. For example,

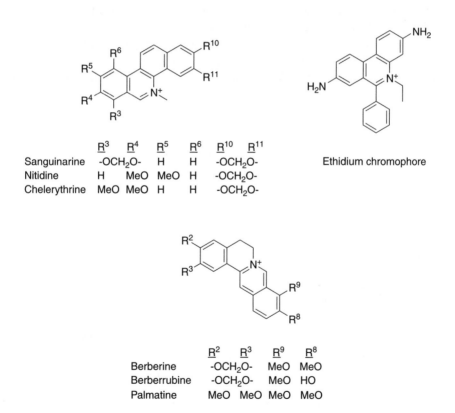

	R^3	R^4	R^5	R^6	R^{10}	R^{11}
Sanguinarine	-OCH₂O-		H	H	-OCH₂O-	
Nitidine	H	MeO	MeO	H	-OCH₂O-	
Chelerythrine	MeO	MeO	H	H	-OCH₂O-	

Ethidium chromophore

	R^2	R^3	R^9	R^8
Berberine	-OCH₂O-		MeO	MeO
Berberrubine	-OCH₂O-		MeO	HO
Palmatine	MeO	MeO	MeO	MeO

Figure 9.14 Sanguinarine and berberine derivatives and the ethidium chromophore.

sanguinarine and nitidine have been found to bind preferentially to regions of DNA containing alternating guanine–cytosine base pairs, whereas, conversely, chelerythrine (Figure 9.14) shows greater selectivity for regions having contiguous GC base pairs (Bai *et al.* 2006). Clearly, the production of free radicals or singlet oxygen in such close proximity to DNA would lead to rapid cell death, as with other photoantimicrobials.

While the activity and photoactivity of natural products such as sanguinarine against 'standard' bacteria has long been established, this has been to cover a range of clinically important pathogens, including *Mycobacterium* spp., *Candida* spp. and *Plasmodia* spp. (Iwasa *et al.* 1998; Giuliana *et al.*, 1999; Newton *et al.* 2002).

9.7　Riboflavin

Riboflavin, vitamin B2, is an essential nutrient in humans, thus the expected low toxicity. It is also a photosensitiser, Types I and II photosensitisation being possible on illumination with blue light (λ_{max} = 475 nm). Like many other natural photoactive systems, it also photodegrades. However, it is interesting that each of the breakdown products, lumiflavin and lumichrome, is also a photosensitiser, absorbing visible light in the region 390–480 nm (Hardwick *et al.* 2004). Since all three species (Figure 9.15) interact with DNA, this makes riboflavin highly attractive, particularly in the photodisinfection of blood and blood products (Dardare and Platz 2002), especially from the viewpoint of low toxicity, i.e. making treated products safe for use in human recipients (Seghatchian and de Sousa 2006).

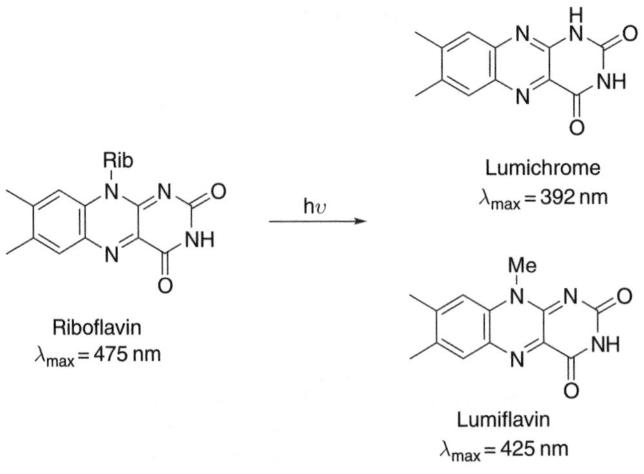

Figure 9.15　Riboflavin and its photoproducts, lumiflavin and lumichrome.

The reported activity of riboflavin as a photoantimicrobial is impressive, covering a range of Gram-positive and Gram-negative bacteria and viruses important in blood products, such as HIV-1 and West Nile virus (Goodrich *et al.*, 2006).

9.8   Terthiophenes

The thiophene-containing terthiophenes are common metabolites in *Asteraceae* spp., presumably acting as phototoxins against insects in a similar way to sanguinarine, as discussed previously. Metabolites normally consist of the linkage of two or three thiophene units in linear fashion, with conjugation and (often) a terminal acetylene group. Such molecules, normally excited at wavelengths around 350 nm, are known to produce significant quantities of singlet oxygen on illumination and have been screened both as photoantiviral and as photobactericidal agents (Ciofalo, Petruso and Schillaci 1996). The photodynamic activity of α-terthienyl (Figure 9.16) has been modelled using its photolytic effect on liposomes containing glucose (McRae, Yamamoto and Towers, 1985).

Synthetic efforts in this area have produced water-soluble amino derivatives (e.g. bisaminomethylterthiophene; Figure 9.16), which are also useful photosensitisers (Saito *et al.* 2001). The antimembrane properties of such photosensitisers makes them ideal candidates against enveloped viruses.

Given that the preceding sections do not represent an exhaustive survey, there are plainly a considerable number of natural products that have the potential both to act as photosensitisers and to provide starting materials for the semi-synthesis of further, improved series. The number of naturally occurring photosensitisers will continue to increase due to the continuing analysis of our natural resources in the search for useful materials. However, the period of time between discovery and clinical reality is long indeed for new chemical entities produced as part of an organised and funded drug discovery programme, and this delay will necessarily be much longer for a newly discovered natural product. For example, the perylene-quinones are inherently efficient anticancer agents and antivirals due to their interactions with PKC, and yet there are still no clinically useful PDT agents arising from the considerable body of research carried out on these compounds. This underlines the Cinderella status of photodynamic therapy, which remains a minority, though growing, clinical activity.

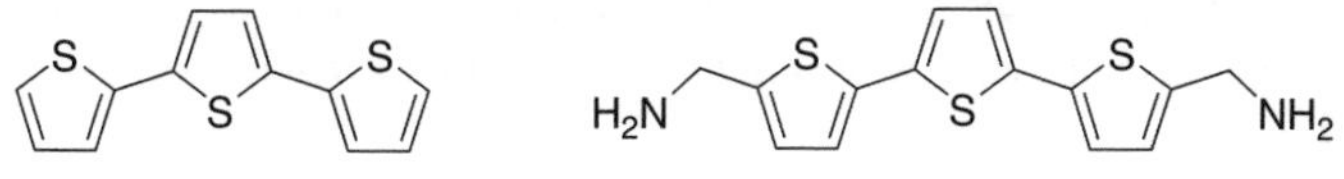

α-Terthienyl   BAT

Figure 9.16   Terthiophenes.

References

Arnason JT, Guerin B, Kraml MM, Mehta B, Redmond RW, Scaiano JC. (1992) Phototoxic and photochemical properties of sanguinarine. *Photochemistry and Photobiology* **55**: 35–38.

Bai LP, Zhao ZZ, Cai Z, Jiang ZH. (2006) DNA-binding affinities and sequence selectivity of quaternary benzophenanthridine alkaloids sanguinarine, chelerythrine, and nitidine. *Bioorganic & Medicinal Chemistry* **14**: 5439–5445.

Chao YC. (1992) 1,4,5,8-Tetrakis(arylamino)anthraquinones: near infrared absorbing dyes. *Dyes and Pigments* **19**: 123–128.

Cimino GD, Gamper HB, Isaacs ST, Hearst JE. (1985) Psoralens as photoactive probes of nucleic acid structure and function: organic chemistry, photochemistry, and biochemistry. *Annual Review of Biochemistry* **54**: 1151–1193.

Ciofalo M, Petruso S, Schillaci D. (1996) Quantitative assay of photoinduced antibiotic activities of naturally-occurring 2,2′: 5′,2′-terthiophenes. *Planta Medica* **62**: 374–375.

Comini LR, Núñez Montoya SC, Sarmiento M, Cabrera JL, Argüello G. (2007) Characterizing some photophysical, photochemical and photobiological properties of photosensitizing anthraquinones. *Journal of Photochemistry and Photobiology A: Chemistry* **188**: 185–191.

Corash L. (2000) Inactivation of viruses, bacteria, protozoa and leukocytes in platelet and red cell concentrates. *Vox Sanguinis* **78**: 205–210.

Crnolatac I, Huygens A, van Aerschot A, Busson R, Rozenski J, deWitte PAM. (2005) Synthesis, *in vitro* cellular uptake and photo-induced antiproliferative effects of lipophilic hypericin acid derivatives. *Bioorganic & Medicinal Chemistry* **13**: 6347–6353.

Cushman M, Mohan P, Smith ECR. (1984) Synthesis and biological activity of structural analogues of the anticancer benzophenanthridine alkaloid nitidine chloride. *Journal of Medicinal Chemistry* **27**: 544–547.

Dai R, Yamazaki T, Yamazaki I, Song PS. (1995) Initial spectroscopic characterization of the ciliate photoreceptor stentorian. *Biochimica Biophysica Acta – Bioenergetics* **1231**: 58–68.

Dardare N, Platz MS. (2002) Binding affinities of commonly employed sanitizers of viral inactivation. *Photochemistry and Photobiology* **75**: 561–564.

Diwu Z, Lown JW. (1993) Photosensitization with anticancer agents 17. EPR studies of photodynamic action of hypericin: Formation of semiquinone radical and activated oxygen species on illumination. *Free Radical Biology and Medicine* **14**: 209–215.

Diwu Z, Lown JW. (1994) Phototherapeutic potential of alternative photosensitizers to porphyrins. *Pharmacology and Therapeutics* **63**: 1–35.

Diwu Z, Zimmermann J, Meyer T, Lown JW. (1994) Design, synthesis and . *Biochemical Pharmacology* **47**: 373–385.

Estey EP, Brown K, Diwu Z, *et al.* (1996) Hypocrellins as photosensitizers for photodynamic therapy: a screening evaluation and pharmacokinetic study. *Cancer Chemotherapy and Pharmacology* **37**: 343–50.

Giuliana G, Pizzo G, Milici ME, Giangreco R. (1999) *In vitro* activities of antimicrobial agents against *Candida* species. *Oral Surgery Oral Medicine Oral Pathology Oral Radiology and Endodontology* **87**: 44–49.

Goodrich RP, Edrich RA, Li J, Seghatchian J. (2006) The Mirasol[TM] PRT system for pathogen reduction of platelets and plasma: an overview of current status and future trends. *Transfusion and Apheresis Science* **35**: 5–17.

Grenby TH. (1995) The use of sanguinarine in mouthwashes and toothpaste compared with some other antimicrobial agents. *British Dental Journal* **178**: 254–258.

Guedes RC, Eriksson LA. (2006) Effects of halogen substitution on the photochemical properties of hypericin. *Journal of Photochemistry and Photobiology A: Chemistry* **178**: 41–49.

Guiotto A, Chilin A, Manzini P, Dall'Acqua F, Bordin F, Rodighiero P. (1995) Synthesis and antiproliferative activity of furocoumarin isosteres. *Farmaco* **50**: 479–488.

Hatano T, Kusuda M, Inada K, *et al.* (2005) Effects of tannins and related polyphenols on methicillin-resistant *Staphylococcus aureus*. *Phytochemistry* **66**: 2047–2055.

Hardwick CC, Herivel TR, Hernandez SC, Ruane PH, Goodrich RP. (2004) Separation, identification and quantification of riboflavin and its photoproducts in blood products using HLPC with fluorescence detection: a method to support pathogen-reduction technology. *Photochemistry and Photobiology* **80**: 609–615.

Hudson JB. (1989) Plant photosensitizers with antiviral properties. *Antiviral Research* **12**: 55–74.

Hudson JB, Harris L, Towers GHN. (1993) The importance of light in the anti-HIV effect of hypericin. *Antiviral Research* **20**: 173–178.

Iwasa K, Kim HS, Wataya Y, Lee DU. (1998) Antimalarial activity and structure-activity relationships of protoberberine alkaloids. *European Journal of Medicinal Chemistry* **33**: 65–69.

Jones SG, Young AR, Truscott TG. (1993) Singlet oxygen yields of furocoumarins and related molecules – the effect of excitation wavelength. *Journal of Photochemistry and Photobiology B: Biology* **21**: 223–227.

Lee HY, Zhou ZX, Chen S, Zhang MH, Shen T. (2006) New long-wavelength ethanolamino-substituted hypocrellin: photodynamic activity and toxicity to MGC803 cancer cell. *Dyes and Pigments* **68**: 1–10.

Liang M, Zhang W, Hu J, Liu R, Zhang C. (2006) Simultaneous analysis of alkaloids from *Zanthoxylum nitidum* by high performance liquid chromatography–diode array detector–electrospray tandem mass spectrometry. *Journal of Pharmaceutical and Biomedical Analysis* **42**: 178–183.

Lytle CD, Wagner SJ, Prodouz KN. (1993) Antiviral activity of gilvocarcin V plus UVA radiation. *Photochemistry and Photobiology* **58**: 818–821.

Ma M, Sun Y, Sun G. (2003) Antimicrobial cationic dyes: part 1: synthesis and characterization. *Dyes and Pigments* **58**: 27–35.

McRae DG, Yamamoto E, Towers GHN. (1985) The mode of action of polyacetylene and thiophene photosensitizers on liposome permeability to glucose. *Biochimica et Biophysica Acta* **821**: 488–496.

Miolo G, Tomanin R, De Rossi A, Dall'Acqua F, Zacchello F, Scarpa M. (1994) Antiretroviral activity of furocoumarins plus UVA light detected by a replication-defective retrovirus. *Journal of Photochemistry and Photobiology B: Biology* **26**: 241–247.

Miyake A, Harumoto T, Iio H. (2001) Defence function of pigment granules in *Stentor coeruleus*. *European Journal of Protistology* **37**: 77–88.

Mulrooney CA, Li X, DiVirgilio ES, Kozlowski MC. (2003) General approach for the synthesis of chiral perylenequinones via catalytic enantioselective oxidative biaryl coupling. *Journal of the American Chemical Society* **125**: 6856–6857.

Newton SM, Lau C, Gurcha SS, Besra GS, Wright CW. (2002) The evaluation of forty-three plant species for *in vitro* antimycobacterial activities; isolation of active constituents from *Psoralea corylifolia* and *Sanguinaria Canadensis*. *Journal of Ethnopharmacology* **79**: 57–67.

Nunez Montoya SC, Comini LR, Sarmiento M, *et al.* (2005) Natural anthraquinones probed as Type I and Type II photosensitizers: singlet oxygen and superoxide anion production. *Journal of Photochemistry and Photobiology B: Biology* **78**: 77–83.

Pant B, Kato Y, Kumagai T, Matsuoka T, Sugiyama M. (1997) Blepharismin produced by a protozoan *Blepharisma* functions as an antibiotic effective against methicillin-resistant *Staphylococcus aureus*. *FEMS Microbiology Letters* **155**: 67–71.

Pelletier JPR, Transue S, Snyder EL. (2006) Pathogen inactivation techniques. *Best Practice and Research in Clinical Haematology* **19**: 205–242.

Reddy HL, Dayan AD, Cavagnaro J, Gad S, Li J, Goodrich RP. (2008) Toxicity testing of a novel riboflavin-based technology for pathogen reduction and white blood cell inactivation. *Transfusion Medicine Reviews* **22**: 133–53.

Roslanice M, Weitman H, Freeman D, Mazur Y, Ehrenberg B. (2000) Liposome binding constants and singlet oxygen quantum yields of hypericin, tetrahydroxy helianthrone and their derivatives: studies in organic solutions and in liposomes. *Journal of Photochemistry and Photobiology B: Biology* **57**: 149–158.

Ruane PH, Edrich R, Gampp D, Keil SD, Leonard RL, Goodrich RP. (2004) Photochemical inactivation of selected viruses and bacteria in platelet concentrates using riboflavin and light. *Transfusion* **44**: 877–885.

Sardari S, Mori Y, Horita K, Micetich RG, Nishibe S, Daneshtalab M. (1999) Synthesis and antifungal activity of coumarins and angular furanocoumarins. *Bioorganic & Medicinal Chemistry* **7**: 1933–1940.

Saxton RE, Paiva MB, Lufkin RB, Castro DJ. (1995) Laser photochemotherapy: a less invasive approach for treatment of cancer. *Seminars in Surgery and Oncology* **11**: 283–289.

Schinazi RF, Chu CK, Babu JR, *et al.* (1990) Anthraquinones as a new class of antiviral agents against human immunodeficiency virus. *Antiviral Research* **13**: 265–272.

Schmeller T, Latz-Brüning B, Wink M. (1997) Biochemical activities of berberine, palmatine and sanguinarine mediating chemical defence against microorganisms and herbivores. *Phytochemistry* **44**: 257–266.

Saito TK, Takahashi M, Muguruma H, Niki E, Mabuchi K. (2001) Phototoxic process after rapid photosensitive membrane damage of 5,5'-bis(aminomethyl)-2,2': 5',2'-terthiophene dihydrochloride. *Journal of Photochemistry and Photobiology B: Biology* **61**: 114–121.

Schmitt IM, Chimenti S, Gasparro FP. (1995) Psoralen-protein photochemistry – a forgotten field. *Journal of Photochemistry and Photobiology B: Biology* **27**: 101–107.

Seghatchian J, de Sousa G. (2006) Pathogen-reduction systems for blood components: The current position and future trends. *Transfusion and Apheresis Science* **35**: 189–196.

Shen L, Ji HF, Zhang HY. (2006) Anion of hypericin is crucial to understanding the photosensitive features of the pigment. *Bioorganic & Medicinal Chemistry Letters* **16**: 1414–1417.

Singh DN, Verma N, Raghuwanshi S, Shukla PK, Kulshreshtha DK. (2006) Antifungal anthraquinones from *Saprosma fragrans. Bioorganic & Medicinal Chemistry Letters* **16**: 4512–4514.

Solheim BG, Seghatchian J. (2006) Update on pathogen reduction technology for therapeutic plasma: an overview. *Transfusion and Apheresis Science* **35**: 83–90.

Spitzner D, Höfle G, Klein I, Pohlan S, Ammermann D, Jaenicke L. (1998) On the structure of oxyblepharismin and its formation from blepharismin. *Tetrahedron Letters* **39**: 4003–4006.

Towers GHN, Page JE, Hudson JB. (1997) Light-mediated biological activities of natural products from plants and fungi. *Current Organic Chemistry* **1**: 395–414.

Wamer WG, Timmer WC, Wei RR, Miller SA, Kornhauser A. (1995) Furocoumarin-photosensitized hydroxylation of guanosine in RNA and DNA. *Photochemistry and Photobiology* **61**: 336–340.

Xu S, Chen S, Zhang M, Shen T. (2004) Hypocrellin derivative with improvements of red absorption and active oxygen species generation. *Bioorganic & Medicinal Chemistry Letters* **14**: 1499–1501.

Zhou J, Liu J, Wang X, Zhang B. (2005) Effect of chelation to lanthanum ions on the photodynamic properties of hypocrellin A. *Journal of Physical Chemistry B* **109**: 19529–19535.

PART III

Applications

10

Photodynamic therapy in oncology

As mentioned at various stages in this book, there is a long history associated with the development of clinical anticancer photodynamic therapy. That the present level of sophistication, with several photosensitisers approved for use in patients, draws mainly on clinical findings associated with porphyrin phenomena or artefacts is supported by the fact that the clinically approved drugs, porfimer sodium and temoporfin, are porphyrin derivatives, whereas 5-aminolaevulinic acid (ALA) and methyl aminolaevulinate are porphyrin precursors that are metabolised to protoporphyrin IX *in situ*. However, given the low number of approved agents and their similarity, photodynamic therapy (PDT) has an enormous amount to offer in oncology, with a broad range of suitable presentations, but more particularly in the area of patient compliance and acceptability, due to the low levels of toxicity and post-treatment scarring.

Conventional treatment for malignant disease is based on surgery, radiotherapy and chemotherapy. While these approaches have developed into fine arts during the past 50 years, there remains an underlying problem common to each: selectivity for malignant among healthy cells. When a surgeon excises a tumour, the normal practice is to remove a healthy margin also. Hopefully, this will ensure that the malignancy will not be able to re-grow. However, the removal of the associated healthy tissue often results in scarring or other irregularity at the tumour site. Although this approach is not strictly selective for the tumour, the removal of a margin is essential and justifiable in the circumstances.

Radiotherapy is the use of a beam of high-energy electromagnetic radiation, which is focused on the tumour mass, ideally causing necrosis of the cancer via ionisation of biomolecules within the aberrant cells. The energy associated with the radiation is obvious, given the lead shielding of body parts not undergoing treatment. The efficacy of this approach depends on the focusing of the irradiating beam, since there is no inherent selectivity of γ- or X-rays for tumour cells. The ionisation events required for tumour damage can occur quite as easily in normal cells. Alternative or derivative therapy based on neutron capture agents (normally based on boron) allows a greater likelihood of tumour damage, but this requires a suitable therapeutic ratio between malignant and healthy cells. As with surgery, a lack of selectivity for the tumour – unsurprising given the siting of many internal tumours, e.g. within organs – leads to the necrosis of surrounding, and sometimes distant, tissue.

Photosensitisers in Biomedicine Mark Wainwright
© 2009 John Wiley & Sons, Ltd

Chemotherapy, as mentioned in a previous chapter, properly refers to the use of chemicals to improve the diseased state and can thus refer to the chemical drug treatment of any disease or impairment. However, the common modern usage of the term normally refers to the treatment of cancer.

In the early years of cancer chemotherapy, the drugs employed were little more than crude cytotoxic (cell-killing) agents, such as mechlorethamine – very similar in structure and nature to the mustard gas on which it was based. The mode of action of the drug at the target site, then as now, entailed the chemical reaction of the drug with a biomolecule such as a nucleotide base, leading to alkylation/further rearrangement and nucleic acid damage. The targeting of such crude agents was correspondingly poor, often to the detriment of the patient.

Ideally, poisoning of the tissue identified as malignant would be a matter of delivering the toxic agent directly to the required site. In reality, this is not normally possible and relies on systemic delivery, with the potential for the reactive drug to attack biomolecules in the bloodstream or in metabolising organs before it reaches the target site. This is further complicated in non-solid or disseminated cancers, such as leukaemia, although such presentations are less amenable to surgery or radiotherapy since the target sites are single cells within the bloodstream and bone marrow. While molecular drug design has improved selectivity to a certain extent, the fact remains that cancer chemotherapy is based on reactive chemicals administered systemically.

Photodynamic therapy, in its anticancer application (Figure 10.1), has much in common with both radio- and chemotherapy. Like radiotherapy, PDT requires irradiation of the tumour site but with low-energy, non-damaging red or near-infrared light. Like the neutron capture approach and like chemotherapy, selective uptake of the agent by the tumour is required, to allow a proper therapeutic ratio with the surrounding tissue. However, the photosensitisers used here are non-toxic in the absence of light.

There are, of course, disadvantages with the use of photosensitisers in anticancer therapy. Any lack of selectivity for the tumour may cause the photosensitisation of healthy tissue, either in the tumour environment or at a more distant site, such as the skin – this has been widely observed with both porphyrin drugs following systemic administration, although skin photosensitisation is not a new phenomenon. This photosensitisation of healthy tissue has been reported for many aromatic drug types, usually sufficiently lipophilic partly to compartmentalise in the skin, where they undergo subsequent excitation by sunlight. There is a considerable experience of such behaviour with the tetracyclines, sulphonamides and phenothiazines, among others. Patients undergoing PDT are normally given a light dosimeter in order to be informed as to their daily exposure level, so that skin damage may be avoided.

A second disadvantage occurs in the treatment of solid tumours and the distribution of light therein. Logically, the photosensitisation mechanism requires the trigger of light absorption, so highly selective uptake and distribution of the proposed photosensitiser in malignant tissue is of little use if the whole of the tumour mass is not illuminated. The problem of light dosimetry, again allied to the field of radiotherapy, is not a straightforward proposition. This is mainly due to the irregular structure of the tumour mass, i.e. some areas/volumes will allow the transmission of different amounts of light.

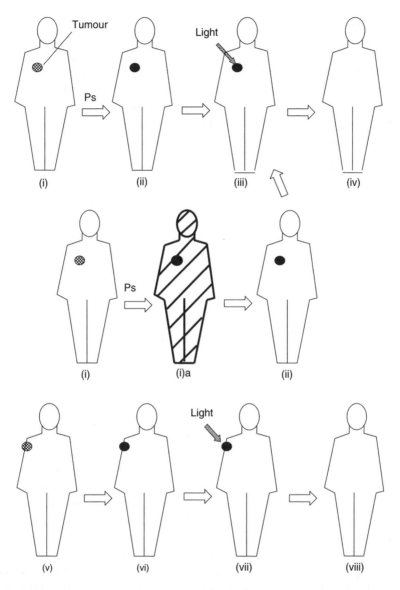

Figure 10.1 Outline approach to the photodynamic therapy of a localised tumour. Stages (i)–(iv) apply to the systemic use of a photosensitiser (Ps), which may exhibit a variable period of systemic localisation [(i)a] before concentrating in the tumour. Stages (v)–(viii) apply to local use, e.g. with ALA or one of its derivatives.

For example, the central portion of an established tumour is likely to be less vascularised, denser and thus more opaque than the outer regions. Such portions require more effort in illumination. In addition, often there is a necrotic tumour core, which has a lower associated oxygen tension, again leading to problems in photosensitiser action.

In terms of photosensitiser light absorption, whether in the anticancer or antimicrobial application, it is important to consider the endogenous light absorption associated with the proposed therapy. Normally, the main absorbers that might interfere with the light used for excitation are haem and melanin, to varying degrees, depending on the tumour type and situation. Consequently, as indicated at various stages throughout this book, photosensitisers should be designed to absorb light outside the spectra of such endogenous absorbers. Normally this means wavelengths beyond 630 nm.

Effective illumination of a tumour mass may be achieved by fibre implants, i.e. overlap of diffused light from multiple fibre optics. However, larger tumours are more likely to be difficult to treat because of the number of fibres required. This is certainly one of the reasons that PDT is still more likely to be used for small tumours.

Due to endogenous absorption and scattering, there is an approximate relationship between absorption wavelength and depth of tissue penetration. However, the penetration of red light through tissue is rarely greater than 5 mm (Alexiades-Armenakas 2006), and this is necessarily a limiting factor in tumour treatment with photosensitisers absorbing in this region.

10.1 Photosensitisers for use in photodynamic therapy

As clinical PDT is based on the initial and lasting use of haematoporphyrin derivative, it is not surprising that many new derivatives have been reported as potential improvements (Nyman and Hynninen 2004). As discussed in Chapters 6 and 7, alternative chromophores such as the chlorins, bacteriochlorins and purpurins, and also the completely synthetic phthalocyanines dominate the second- and third-generation photosensitisers for anticancer application.

Plainly there is a considerable structural similarity among such photosensitisers. Other families of compounds were also investigated from the angle of either established selectivity for tumour cells or photosensitisation, rather than any structural relationship to the porphyrins. Most lead compounds in this area were synthetic dyestuffs or biological stains, such as the azine and cyanine dyes (Chapters 4 and 8, respectively). Finally, there are other, non-porphyrin, natural products – such as the psoralens and perylenequinone derivatives – which have associated photoactivities. However, as mentioned above, the fact remains that there are currently only four drugs approved for the photodynamic treatment of cancer.

One of the main reasons for this book is to give greater coverage to the wide range of photosensitisers available as potential therapeutics. Thus, each of the classes mentioned has provided candidate compounds for screening as PDT agents. While initial choices were dictated by the availability of lead compounds, chemical synthesis has, over time, provided series of compounds to enable structure–activity relationships (SAR) to be developed, which, in turn, aid in the design of improved derivatives. While the development of a suitable drug for use in the clinic does not depend solely on chemical synthesis – drug development for PDT must be a multidisciplinary project, as with conventional drugs – the fact remains that the input of the chemist in design, synthesis and initial testing has been paramount in successful groups.

As with conventional anticancer drug development, the synthesis of analogues for PDT screening is normally based on a lead structure. However, whereas the chemical character-isation of conventional anticancers, such as *cis*-platin derivatives, usually entails structure confirmation and physicochemical profiling (log$P$, pKa, etc.), this is further complicated for photosensitisers by the inclusion of detailing Type I/Type II photosensitisation behaviour (e.g. singlet oxygen yield) and light absorption characteristics (maximum absorption wave-length λ_{max} and absorption coefficient ε_{max}). As with photoantimicrobials, there is little logic in pursuing only compounds having significant photosensitising capability in chemical or spectrophotometric tests, since it may be that such activity is only realised in cells (see discussion in Chapter 3). However, knowledge of singlet oxygen production and so on would be expected during later stages of development in order to understand mode(s) of action and cell-killing behaviour. Success in screening against a number of different tumour cell lines is normally followed by small animal testing. In addition, at this stage, highly promising candidates are tested for deleterious behaviour – toxicity and mutagenicity (Ames, Comet and micronucleus tests) in host-type cell lines – and the mammalian pharmacokinetics are established. Candidacy in a clinical trial requires a pure compound having optimal performance in this pathway, but also requires a reasonably economic background – this may eliminate photosensitisers from very low yielding synthetic routes and again provides work for the laboratory chemist in process optimisation, and scale-up of promising compounds is often contracted out to industrial concerns having Good Laboratory Practice certification. Similarly, work is usually required at each stage following analysis of the data from the various tests in terms of structure optimisation and the development, as mentioned, of SAR. It is to be hoped that the growth of clinical PDT as a treatment modality will allow sufficient funding of research groups so that computer modelling of SAR can be carried out. Some idea of the route to a clinical photosensitising drug is given in Chapter 3.

10.2   Indications for photodynamic therapy

Light delivery is the major governing factor in the use of PDT against cancer. The design and development stages outlined above should guarantee that a systemically delivered, selective photosensitiser is concentrated in malignant tissue, in the same way as a conventional chemotherapeutic agent. Rational photosensitiser design should optimise its response to long-wavelength light, but this is of little use if the tumour mass cannot be illuminated efficiently. Consequently, although there are constant developments in light delivery, deep-seated/larger tumours are usually more difficult to treat. This means that the most common types of tumour treated with PDT remain those that are skin related. This is in line with the similar treatment of hyperproliferative skin diseases such as psoriasis using psoralens and ultraviolet A (PUVA).

The normal range of indications for PDT is thus skin malignancy, except for pigmented melanoma (Braathen *et al.* 2007), oral (Meisel and Kocher 2005), head and neck (Li *et al.* 2006), gastrointestinal tract (Petersen *et al.* 2006), lung (Moghissi and Dixon 2005), genitourinary tract and prostate (McFadden and Cruickshank 2005; Senior 2005; Zhu and Finlay 2006), and brain (Stylli *et al.* 2004). Application to internal tumours normally requires the use of fibre-optical light delivery, again increasing the degree of difficulty of the approach, since larger tumours require multiple fibre-optic implantation in order to guarantee illumination of the whole volume of the tumour. Clearly, some target areas lend themselves more than others to light delivery due to established endoscopic approaches (head and neck, lung, oesophagus and stomach). Conversely, the treatment of haematological diseases (leukaemia and lymphoma) requires removal of blood containing the aberrant cell population, followed by treatment outside the body (extracorporeal photochemotherapy). The strong case for the use of PDT in salvage therapy should also be emphasised.

Discussion of the development of PDT for various indications follows.

10.3   Skin

The accessibility of the skin makes associated malignancy the simplest target for local and/ or topical cancer treatment. This accessibility also has a downside in the worldwide increase in skin cancer in modern times, normally caused by over-exposure to high-energy ultraviolet light, although the human papillomavirus and chemical exposure may some- times be implicated. Clearly, the best way to combat the increase in skin cancer occurrence is to avoid the predisposing factors, but there is little sign of this recommendation being followed, particularly with respect to unprotected exposure to ultraviolet light.

Generally, skin cancers can be divided into melanoma and non-melanoma types. The latter group comprises basal cell and squamous cell carcinoma (BCC/SCC), named for the respective skin strata and actinic keratoses (AK), which are often the precursors of the SCC type. In terms of severity, melanoma is by far the most dangerous form, mainly due to the propensity to formation and diffusion of microsatellites (metastases). Correspondingly, the clinical approach to non-melanoma skin cancer may reasonably include PDT. For melanoma this is unlikely, the preferred route being surgical excision, although PDT may have a part to play subsequently (see below).

BCC, SCC and AK would not normally be considered as, necessarily, a life-threatening disease. However, disfigurement is an unfortunately common consequence of conventional treatment, and this is a criterion on which the photodynamic approach scores very posi- tively. The surgical removal of skin tumours from the face and neck with resulting scarring is obviously less acceptable to the patient than if the tumour had been removed from a part of the body normally covered in everyday life, e.g. torso.

Initially, the photodynamic approach to non-melanoma skin cancers consisted of systemic administration of an haematoporphyrin derivative (HpD) preparation followed, after a suitable interval, by superficial illumination. With HpD, however, general skin photosensitivity was a constant side effect, leading to poor patient response. The advent of topical aminolaevulinic therapy (see Chapter 6) altered the situation radically, making the treatment of these particular skin cancers much more convenient, on an outpatient basis. Local application allows considerably improved control over skin sensitisation, since the peritumoural area can be efficiently masked, even with large presentations.

Limitations to the application of ALA-PDT to non-melanoma skin cancers relate mainly to the lesion thickness and degree of invasiveness. In turn, this is governed by the transmission of light in tissue. For shallow lesions, blue light may be employed, utilising the maximal absorption of PPIX at 409 nm, although, overall, most use appears to be made of the λ_{max} of 635 nm and red light. However, even here, it would be unlikely that topical ALA-PDT could be employed in the treatment of invasive cancer.

Thus, for non-melanoma skin cancer, there are applications for the PDT approach. However, for melanoma this is not the case, mainly due to the characteristics of this tumour type.

There are several major differences between melanomas and the remaining skin cancers, most of which are due to the presence of melanin, at various levels, in the former. The function of melanin pigment in the skin is mainly photoprotective, from the aspect of both light absorption and antioxidant activity. However, although melanin absorbs light up to and beyond 600 nm, the absorption in the red region is weak. Therefore, it is possible to photosensitise melanoma cell destruction with long-wavelength absorbing candidates, and this has been shown for both phthalocyanine and phenothiazinium derivatives (Rice, Wainwright and Phoenix 2000; Kolarova *et al.* 2007). Indeed, in the latter there is greater rationale for use due to the strong melanin binding of the phenothiazine and phenothiazinium chromophores.

Given the urgent treatment requirement of malignant melanoma, it is unlikely that the first choice option of clinical excision will change in the near future. However, the topical photodynamic treatment of the tumour bed (Figure 10.2), in order to ensure that no melanoma cells remain, post surgery, may have some merit, particularly with a melanin-selective candidate. The tracking of disseminated

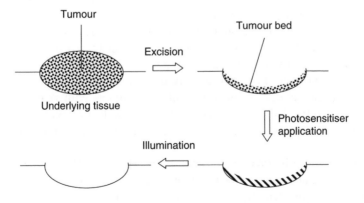

Figure 10.2 Tumour excision/treatment of tumour bed.

melanoma metastases using radio-iodinated methylene blue has previously been demonstrated in mice (Link 1999).

Another pigmented skin tumour is the violaceous Kaposi's sarcoma (KS), which was often the first sign of HIV infection early in the AIDS pandemic. Several photosensitisers proved efficacious in the treatment of KS, including tin etiopurpurin (SnET2, Chapter 6). However, since the tumour is caused by human herpesvirus 8 in the immunocompromised host, the incidence of KS has decreased considerably with the advent of highly active antiretroviral therapy (HAART). It should be realised that HAART is by no means available universally, thus there may be some requirement for the photodynamic treatment of this particular type of tumour, if only in palliation.

10.4   Head and neck

As with skin cancer, tumours of the head and neck generally offer straightforward access, either by superficial or by interstitial illumination, and are thus obvious candidates for the photodynamic approach. Similarly, the advent of ALA treatment protocols has made advances on previous systemically administered porphyrin photosensitisers possible, and, again, the excellent cosmesis offered by modern PDT is preferable in many cases to radiotherapy or surgery, e.g. in dealing with tumours of the oral cavity or the face.

In terms of the scope of treatable disease, this includes (skin) tumours of the face, oral cavity (tongue, jaw), nasopharynx, larynx and neck, although it is important to remember the propensity for metastatic disease from several of these as primary tumours, such as those of the tongue and the nasopharynx. Illumination, as mentioned, is relatively straightforward because of the proximity of the target. Haematoporphyrins were widely used in head and neck cancer at the modern outset of clinical PDT, but although they exhibited useful selectivity for tumours a short time after i.v. administration, generalised skin photosensitisation was a ubiquitous side effect. However, the excellent performance of Photofrin in cancer of the larynx should be mentioned, giving as it does high levels of tumour control and voice preservation.

The related monoaspartyl derivative of chlorin e_6, talaporfin (see Chapter 6), has also reached clinical trial for head and neck cancer. The increased hydrophilicity of this molecule aids in its relatively short-term skin photosensitivity (Yoshida *et al.* 2008).

As mentioned, ALA offers a considerably improved scenario and may be administered orally or intravenously, or may be applied locally as with skin tumour therapy. ALA has also been trialled in oral leukoplakia, a white lesion, typically of the inside of the cheek, which shows a predisposition for progression to malignancy (Kübler *et al.* 1998).

Foscan (*meso*-tetra(3-hydroxyphenyl)chlorin, *m*THPC; see Chapter 6), has been promoted with considerable energy for the treatment of head and neck cancer, and many multi-centre trials have been carried out. However, it appears that Foscan does not offer significant advantages over other clinically approved photosensitisers and has not been licensed in the USA for head and neck cancer. Foscan appears to suffer from a lack of selectivity, many occurrences of systemic photosensitisation being reported as well as tumour recurrences (Allison *et al.* 2005).

10.5 Gastrointestinal tract

The application of PDT to tumours of the stomach and colon has been investigated by various groups, the accessibility of tumour sites being aided via gastroscopy or colonoscopy. However, the major activity within this field must be in the treatment of Barrett's oesophagus (dysplasia of the tissue at the junction of the oesophagus and stomach).

Initial uses of PDT in this area were aimed at the palliation of oesophageal obstruction, i.e. in the treatment of oesophageal cancer, in order to improve swallowing capability and thus nutrition and quality of life. This was particularly aimed at the elderly patient, where the degree of extension of life via palliative care may represent a significant benefit. Porfimer sodium is again approved for use in such cases.

Traditional approaches to the treatment of malignancy here are based on surgery, radiotherapy or laser ablation, and, as in other cases, have associated complications with regards to structural tissue damage through a lack of selectivity, for example with oesophagectomy.

Barrett's oesophagus is a pre-malignant condition, mainly associated with the squamous epithelium of the lower oesophagus. As such, the depth of aberrant tissue is not great and does not truly justify a surgical approach if milder alternatives are available, particularly in view of the potential morbidity caused by surgery.

As with other areas of PDT, established vital staining methods have provided a ready-made agent. In this case, methylene blue is widely used pre-operatively to establish the presence and borders of dysplastic tissue. Subsequently, the use of ALA has been found to offer improved results.

10.6 Breast

Where PDT has been employed in breast cancer, it has usually been targeted at secondary lesions, e.g. metastatic lymph involvement, etc. For example, cutaneous metastases from the breast have been treated with PDT to the chest wall, with Photofrin as the photosensitiser (Khan, Dougherty and Mang 1993). Indeed, there have been various trials using either HpD or Photofrin, but always against recurrent or conventional refractory disease (Calzavara-Pinton *et al.* 1996). Similarly, *meso*-tetra(4-sulphophenyl) porphyrin (TPPS$_4$, Chapter 6) was employed against cutaneous breast metastases but applied locally and thus demonstrating an absence of systemic photosensitivity (Lapeš, Petera and Jirsa 1996).

It is not widely appreciated that white light imaging of the whole breast was introduced as early as the 1920s, although this technique was not particularly discriminatory as regards the type of any lesions present. The approach was advanced towards the end of the century using red and near-infrared light, although it remains of limited value for smaller lesions (Fantini and Taroni 2008). However, the demonstration of transmission of long-wavelength light through breast tissue should be viewed positively due to the potential for excitation of selective photosensitisers.

In terms of more modern photosensitisers, SnET2 has been tested clinically against breast cancer, but only against recurrent disease (Mang *et al.* 1998). Lutetium texaphyrin has also been included in clinical trails for recurrent disease (Sessler and Miller 2000).

Given the general move towards tissue sparing and breast conservation, it is likely that PDT will be allowed a larger role in breast cancer. As with the other areas of activity here, greater selectivity for tumours can only increase the degree of clinical acceptance, as will the increasing expertise in whole intact breast irradiation (Allison *et al.* 2006).

10.7 Lung

As the main cause of cancer deaths, lung cancer, logically, has been the focus of a great research effort. However, the presentations of lung cancer vary (alveolar and bronchial) and may be localised or diffuse, often being secondary metastases. Because of the essential nature of the lung in oxygen/CO_2 transport, malignancy often causes airway blockage of varying severity and consequent morbidity.

The use of PDT in lung cancer can achieve tumour necrosis and prolonged life, especially with smaller tumours. More regularly the approach is used palliatively in order to free airways, as with other morbid occlusive tumour presentations. PDT has also been used in conjunction with other conventional approaches, e.g. surgical debulking. In addition, the photodynamic approach may be more suitable than surgery due to the poor respiratory state of the patient, who may often present in old age with several smoking-related factors such as chronic obstructive pulmonary disease (COPD) – emphysema, bronchitis, etc. – and be simply unfit for surgery. However, the use of PDT in early central lung cancer remains sparse, despite the widespread and unfortunately common occurrence of the disease (Moghissi and Dixon 2008). Given the availability of bronchoscopy, both administration of the photosensitiser and subsequent illumination should be relatively straightforward in such conditions.

As in other cancers, the use of HpD and its early analogues had some degree of success, but this was limited by photosensitivity problems. Various (porphyrin-related) second-generation photosensitisers were included in clinical trials in the 1990s, such as SnET2, BPD-MA and *m*THPC (Sutedja and Postmus 1996). The intraoperative use of PDT in the related malignant mesothelioma may be another way forward, the tumour site being debulked before adjunctive treatment with Foscan, for example (Ris 2005).

10.8 Genitourinary tract and prostate

Highly hydrophilic photosensitisers may be reasonably expected to concentrate rapidly in the bladder and may also be instilled directly. Methylene blue was used in this way as an approach to the treatment of superficial bladder cancer in the early period of clinical PDT but has been superseded by ALA (Williams *et al.* 1989). When instilled into the human bladders, hypericin (see Chapter 9) also exhibits highly selective accumulation in malignant lesions and is used as a fluorescent diagnostic tool for superficial bladder cancer (Chapter 12). The phototoxicity of hypericin logically suggests its use for whole bladder wall PDT for superficial bladder cancer (Kamuhabwa *et al.* 2004). Porfimer sodium is approved for use in carcinoma of the bladder *in situ*, and the use of ALA in conjunction with BCG therapy has also been trialled (Szygula *et al.* 2004).

Cancer of the prostate often constitutes quite a problematic presentation to therapy, whether conventional or photodynamic, due to the close proximity of adjacent organs. However, since interstitial approaches exist, these may be adaptable to PDT protocols. In terms of photosensitisers used, Tookad (Chapter 6) has been investigated as a vascular-acting photosensitiser for the treatment of prostate cancer (Huang *et al.* 2007). Foscan, ALA and texaphyrin have also been included in clinical trials (Zhu and Finlay 2006).

As the conventional treatment of vulval cancer, surgery is often associated with significant morbidity, particularly due to what amounts to the patient's feelings of mutilation. The excellent post-operative cosmesis offered by PDT constitutes a positive aspect in such cases. Typically, ALA is employed in the phototreatment of both vulval and cervical cancer, as intraepithelial presentations (Booth, Poole and Moghissi 2006).

10.9 Brain

Brain tumours are perhaps the most terrifying, due to the effects of disease on this all-important organ and very poor prognosis generally. There are several types of tumour, depending on the cell progenitor – astrocytomas, gliomas/glioblastomas, etc. – or, often, the tumour is a secondary. Access to brain tumours may be very difficult, and maintenance of the border between malignant and healthy tissue is essential. Consequently, PDT has a positive role to play as an adjuvant to surgery.

HpD-mediated photodynamic therapy has been investigated as an adjuvant treatment for cerebral glioma, i.e. adjuvant PDT following surgical resection of the tumour (Stylli *et al.* 2005). It has been found that HpD uptake is higher in glioblastomas than in astrocytomas, and also higher in recurrent than in primary brain tumours (Stylli *et al.* 2004).

Foscan has also been used in the treatment of brain tumours in an intraoperative mode, photosensitiser fluorescence being used to aid guided surgical resection (see also Chapter 12). Of the small number of patients in the trial, those given intraoperative PDT showed a longer median survival time (Kostron, Fiegele and Akatuna 2006).

In most trials, there has been some incidence of post-treatment photosensitivity and also some of intracranial swelling and oedema (Stylli and Kaye 2006).

10.10   Haematological disease (leukaemia and lymphoma)

Extracorporeal photochemotherapy, or photopheresis, is based on the selectivity of psoralen derivatives for malignant cells, such as the lymphocytes implicated in cutaneous T-cell lymphoma (CTCL; see Chapter 9). Oral administration of the psoralen photosensitiser allows uptake by malignant T cells in the bloodstream. This is followed by sequential removal of 0.5 l aliquots of blood, which then undergo component separation and illumination of the white cell fraction with the relevant wavelength of ultraviolet light, resulting in DNA damage via photoadduct formation (Chapter 6) and thus to direct cytotoxicity. Protein damage at the cell membrane may cause cell death or lead to sufficient changes in cellular morphology such that when the blood fraction is reintroduced into the bloodstream an autovaccination effect results, i.e. the malignant cells are not recognised by the body's immune system and are thus destroyed (Schmitt, Chimenti and Gasparro 1995). PUVA therapy and ALA-PDT are also approved for CTCL (Knobler 2004).

Although no direct photodynamic treatment of leukaemia exists, photosensitisers are employed against leukaemic cells in the purging of autologous bone-marrow grafts. Merocyanine 540 (Chapter 8) has been employed considerably in this respect (Sieber 1987).

10.11   Targeting/formulation

As with other therapeutic areas, while the ideal approach would be to use perfectly selective agents, those involved in photodynamic therapy have not yet discovered any magic bullets. Consequently, high selectivity and activity exhibited in cell culture and even in small animal trials normally requires formulation work in order to realise similar results in the clinic.

The delivery of the photosensitiser to the tumour site has to be considered, of course. This often depends on the presentation of the malignancy, i.e. its position in the body, external or internal, localised or diffuse (malignant cells in the blood). Creams containing a photosensitiser may be used in topical application, whereas systemic administration (oral and i.v.) would be expected for internal tumours or leukaemias. Surprisingly, direct instillation of photosensitiser into the tumour mass appears to be unsuccessful (Brown, Brown and Walker 2004). Given the noted absence hitherto of a magic bullet, despite the best efforts of medicinal chemists, use has been made of delivery vehicles such as liposomes or antibodies, and other technologies exist, such as dendrimers and nanodots, each of which are discussed below. An alternative approach involves the use of metabolic activation of a pro-drug. This is currently employed in the tumour delivery of ALA (*q.v.*) via the use of its methyl ester (with subsequent *in situ* hydrolysis), which may display improved skin penetration due to greater lipophilicity (Kloek and Beijersbergen van Henegouwen 1996).

10.12 Liposomes

Rapid membrane construction by tumour cells means that their requirement for cholesterol is raised above normal, and the required cholesterol is sequestered from the blood supply via over-expressed low-density lipoprotein (LDL) receptors. While the binding of lipophilic (hydrophobic) drugs to LDL may be used logically to deliver such drugs more selectively to tumour cells, there remains a problem with drug solubility in standard media for administration. This may be overcome via the use of liposomes containing the candidate drug (Derycke and de Witte, 2004). In addition, the liposomes, depending on their construction, may concentrate in malignant tissue (interstitium) or may be so designed as to be taken up by tumour cells via standard biochemical recognition. However, simple concentration in the peritumoural region with subsequent release of the drug still offers some degree of selectivity – and certainly a different photosensitiser delivery pattern – above standard i.v. administration. In addition, the very low aqueous solubility of some candidate photosensitisers is such that a complex mixture of solvents and surfactants is required for their administration. Furthermore, as covered here in various chapters, the solubilisation of a photosensitising chromophore by chemical functionalisation often leads to equally complex separation/purification protocols. For example, the encapsulation of zinc phthalocyanine in liposomes may be considered a much more straightforward route to efficient PDT of tumours, especially where the liposome is designed to be tumour-specific, compared to the sulphonation of zinc phthalocyanine, which is followed by an extended and complex purification, even before the i.v. administration and distribution of the pure isomer. However, conventional liposomes, e.g. of the simple phospholipid/cholesterol type, exhibit half-lives that are too short for useful clinical therapy, usually due to disintegration following interaction with serum lipoproteins (Figure 10.3).

Improved targeting is given by passive-targeted liposomes, which, as mentioned, localise in the tumour insterstitium, allowing local release of the photosensitiser. Active-targeted liposomes may be designed to bind at the tumour cell surface or be internalised via receptor-mediated endocytosis (Figure 10.3). However, to ensure sufficient circulation, passive-targeted liposomes need to be disguised, i.e. via external modification, against damaging interaction with lipoproteins. Glucuronidation and PEGylation (attachment of poly(ethylene glycol)) may be used to this end. The use of glucuronide modification has been shown to increase the photodynamic efficacy of liposomal benzoporphyrin derivative monoacid ring A (BPD-MA) in a sarcoma-bearing mouse model, compared to BPD-MA in conventional liposomes (Oku et al. 1997). It should be mentioned that PEGylation is increasingly employed in photosensitiser delivery, from the angle of both disguising the photosensitiser from the body's defences (as with liposomes) and imparting greater aqueous solubility.

Tumour selectivity may be increased by the use of long-circulating liposomes (section 10.12) also having targeting moieties attached to the exterior, normally via a linker molecule. Thus, active-targeted liposomes may have, for example, folate or tumour-specific antibodies or lipoproteins attached.

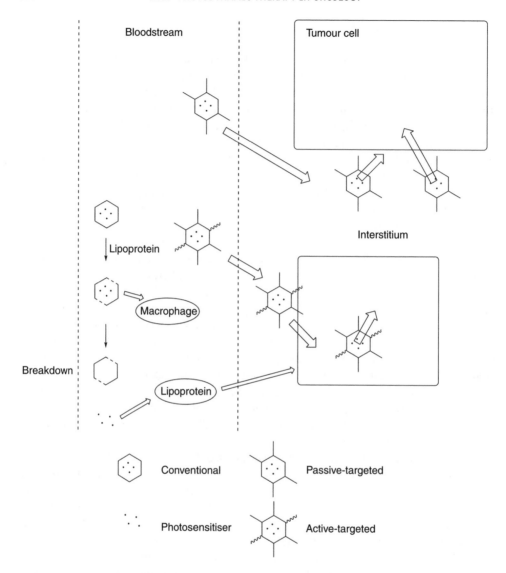

Figure 10.3   Liposomal delivery of photosensitisers.

10.13   Biomolecular conjugation

This approach to increasing photosensitiser delivery relies on the presence of certain surface-bound receptors on the tumour. Thus, a range of porphyrin and phthalocyanine photosensitiser molecules have been coupled with albumens and LDLs, normally via standard amide (peptide) coupling reactions (Sharman, van Lier and Allen 2004).

Similarly, small molecules with a cellular receptor that is over-expressed by tumour cells, may also be used to target photosensitisers. For example, an

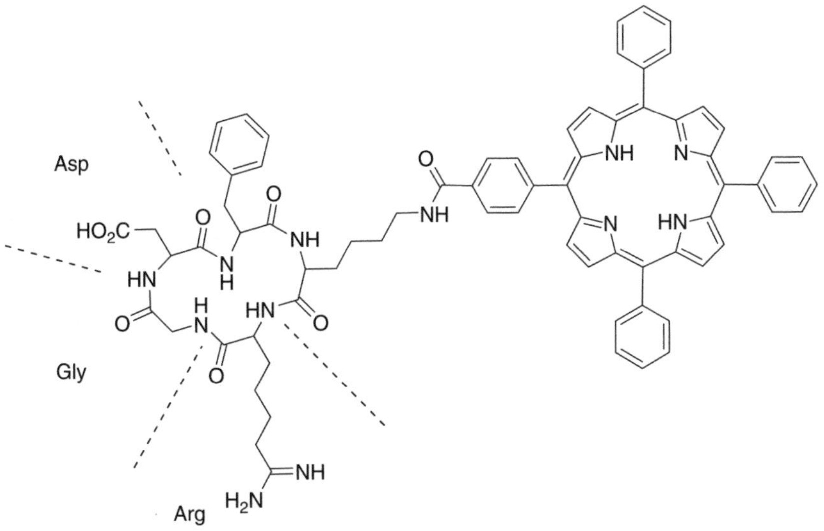

Figure 10.4 Peptide-targeted photosensitiser containing the RGD fragment.

oestradiol–*meso*-tetraphenylporphyrin conjugate, linked via an amide bond has been demonstrated to displace oestradiol, and to be taken up selectively by breast cancer cells *in vitro* (Swamy *et al*. 2002). Other porphyrin– and phthalocyanine–small biomolecule conjugates are covered in Chapters 6 and 7, respectively.

Understanding of the structure–function relationships of antibodies has lead to the use of active moieties from these complex molecules in bioconjugation, rather than the whole biomolecule. This should have the effect of allowing the photosensitiser freedom to function, rather than being swamped by the much larger peptide or protein. This approach has been applied to the RGD peptide fragment, which is involved in cell adhesion, a range of photosensitisers having been attached to the arginine–glycine–aspartic acid sequence (Figure 10.4) using solid-phase protocols (Boisbrun *et al*. 2008). In the same way, a small peptide known to be a nuclear localising sequence has been attached to purpurin 18, again via solid-phase synthesis (Walker, Vernon and Brown 2004).

10.14 Dendrimers

The use of polymers as drug delivery vehicles is well known, with the drug molecules either physically or chemically bound to the macromolecular chain. Dendrimers offer similar function but differ in shape, being approximately spherical or ovoid. In addition, positively charged dendrimers are known to interact with cell membranes, causing hole or pore formation, in a similar way to that of the cationic peptide antibiotics. Dendrimers are branched molecules with identical termini, e.g. amino groups, which can thus be attached in similar fashion to multiple molecules of the same drug/photosensitiser (Figure 10.5).

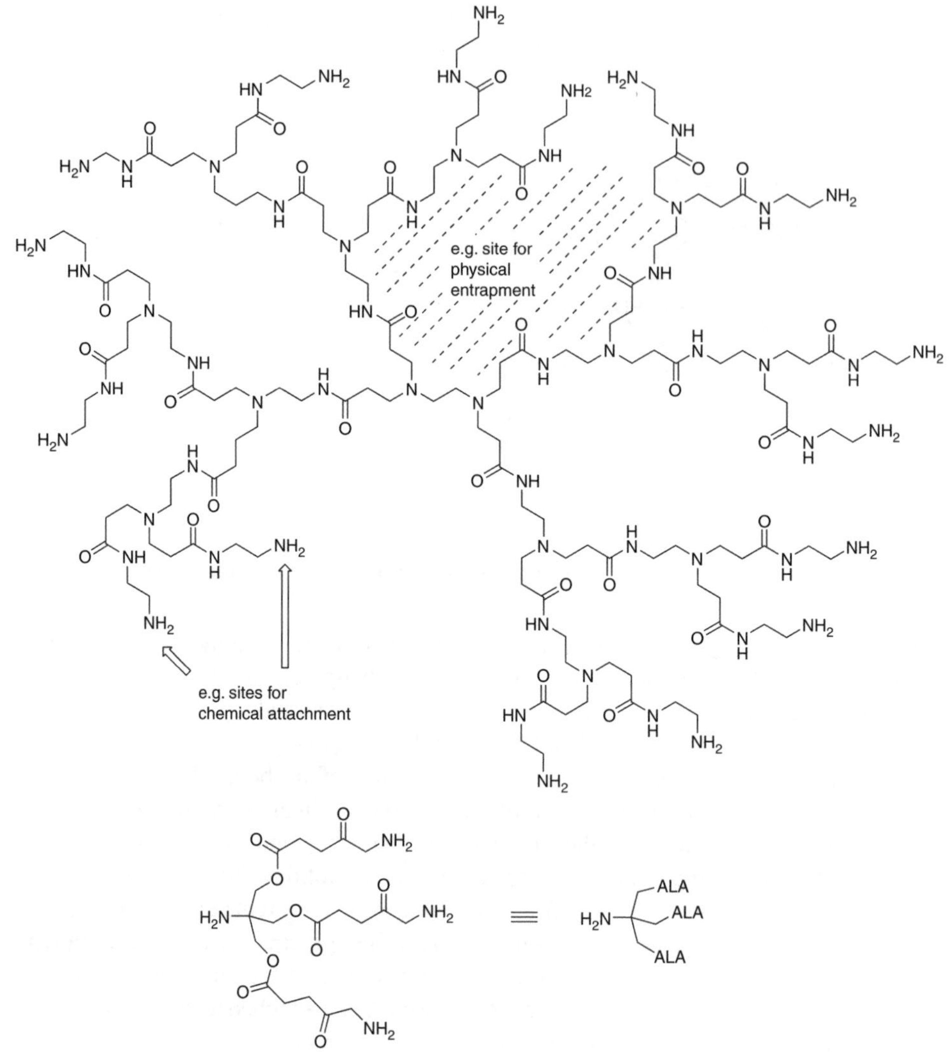

Figure 10.5 Dendrimer conjugate structure based on polyamidoamine, having 16 amino termini, and small ALA-based dendron.

Various small dendrimers (dendrons; Figure 10.5) have been investigated for the delivery of ALA to tumour cells *in vitro*, exhibiting, for example, increased uptake compared to the hexyl ester of ALA and increased PPIX production compared to ALA itself (Wolinsky and Grinstaff 2008).

A potential negative side to dendrimer-bound photosensitisers is that of rapid relaxation of the excited state due to the close proximity of neighbouring chromophores, thus leading to decreased singlet oxygen yields compared to the free photosensitiser

(Hackbarth *et al.* 2001). In addition, as with other delivery systems, singlet oxygen produced may react with the dendrimer structure, causing breakdown.

10.15 Nanoparticles

Much of the preceding topic deals with improved delivery of photosensitisers. However, each particular methodology has the disadvantage of size – often this makes the resulting therapeutic system obvious to the body's defences and/or too large to reach its target site without breaking down in some way.

Nanoparticles, usually with a diameter of less than 200 nm, have the ability to escape the attentions of circulating lipoproteins and opsonins and are normally small enough to be able to cross membranes such as the blood–brain barrier. The attachment of photosensitisers to nanoparticles thus offers considerable potential in terms of tumour targeting. Given the size limit, nanoparticles may be produced from a variety of materials – metals, salts, polymers, etc. This allows considerable scope for functionalisation of the particle, thus endowing further specificity or 'stealth' characteristics. A combination of this ability to 'customise' the particles and to attach photosensitiser molecules to them may thus be seen as an ideal photosensitiser delivery system.

A good example of the technological advance available here is given by the attachment of functionalised zinc phthalocyanine to a gold nanoparticle via thiol linkages. Organisation of the phthalocyanine molecules about the particle was such that molecules of a phase transfer agent could be interposed between the photosensitising moieties (Figure 10.6), in order to solubilise the highly hydrophobic phthalocyanine in the polar media around tumour cells, facilitating uptake. In addition, the situation of the photosensitiser molecules about the nanoparticle was reportedly monomeric rather than aggregated, leading to high yields of singlet oxygen on illumination (Wieder *et al.* 2006).

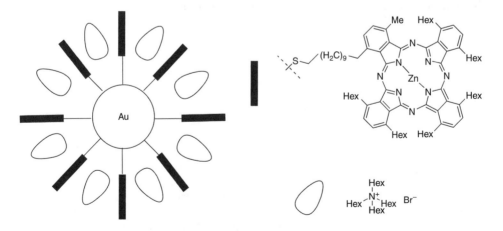

Figure 10.6   Solubilised zinc phthalocyanine attached to a gold nanoparticle.

Indeed, the PDT efficacy was twice that observed with the 'free' phthalocyanine derivative.

Nanoparticles may also be biodegradable, for example those produced from poly(lactic acid). Again, this type of nanoparticle loaded with photosensitiser, in this case hydrophobic fluorinated zinc phthalocyanine, exhibited increased tumour uptake in mice compared to the free active (Leroux *et al.* 1996). The use of related poly(lactide-co-glycolide) nanoparticles also led to improvements in the photodynamic activity of *meso*-tetra(4-hydroxyphenyl)porphyrin, particularly via prolonged vascular localisation (Vargas *et al.* 2004).

10.16 Magnetic targeting

The idea of using a magnetic field to concentrate a therapeutic agent in a particular region is attractive since the field itself can be transmitted through tissue, i.e. the approach is non-invasive (Figure 10.7). This, of course, relies on attaching the therapeutic to a particle that is susceptible to the field. Given the huge explosion in nanotechnology, covered in the previous section, it is unsurprising that magnetic nanoparticles have been investigated as targetable drug delivery systems for tumours. Increased uptake and retention in this way has been demonstrated in both *in vitro* and *in vivo* skin models for, for example, zinc phthalocyanine and Foscan (Primo *et al.* 2007, 2008).

In point of fact, several of the technologies covered in this chapter may be combined to produce – on paper at least – what amounts to, if not a magic bullet, then a smart, guided one! At the time of writing (2008), it is possible to produce a photosensitiser delivery system for PDT based on a magnetic nanoparticle having a PEG coating to confuse the defences and endow long serum half-life, and a molecular targeting agent, such as an antibody or hapten, specific for the aberrant cell type. In addition, the use of materials such as cadmium sulphide – *quantum dots* – can enhance the light delivery to the attached photosensitiser molecule via their inherent emission properties.

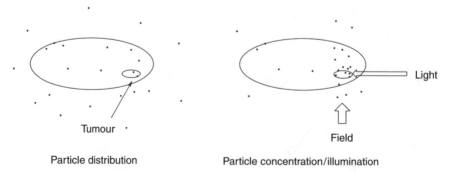

Particle distribution   Particle concentration/illumination

Figure 10.7   Magnetic nanoparticle use in PDT.

10.17   Light in photodynamic therapy (see Chapter 1)

One of the early observations that became a foundation stone of cancer PDT, and particularly for the porphyrin-based approach, was Policard's observation of porphyrin fluorescence during surgery. Plainly, the white lighting in the operating theatre contained wavelengths corresponding to absorption peaks in the porphyrins' action spectra. However, since the observation concerned the fluorescence of the porphyrins rather than tissue damage via photosensitisation, it is, in fact, more relevant to photodiagnosis than to photodynamic therapy. In addition, what Policard saw was due to the incidence of white light on an open surgical site, and this tells us little about light penetration through tissue, other than that porphyrins are endogenous absorbers.

As mentioned above, the delivery of light to the tumour site is of great importance, and much research has been carried out regarding both interstitial and superficial illumination. However, the choice is often straightforward, given proper understanding of the tumour target. A 'simple' SCC can usually be treated with topical photosensitiser, or ALA, and superficial illumination. A sarcoma may be thicker and require both systemic drug administration and interstitial illumination. Again, a spread of metastatic tumour nodules may be more easily illuminated superficially, while larger, internal tumours invariably require implanted optical-fibre light delivery. The duration of light treatment is also a consideration, the shortest optimal period being preferable form the patient's standpoint. The problems of tumour morphology and light transmission are covered in Chapter 3.

It is often stated in the literature that PDT offers *dual selectivity* in tumour destruction. This refers, of course, to the selectivity of the photosensitiser for tumour cells rather than healthy ones, and that of selective illumination of the tumour site. However, it should be realised that one is in fact dependent on the other. If the photosensitiser is administered systemically, there will be a period of distribution – in fact, of administration, distribution, metabolism and elimination – as is usual with drug therapy. Decisions must be made, then, on the drug–light interval, preferably when there is sufficient photosensitiser in the tumour to guarantee its eventual death, but there should also be a sufficient differential distribution between the tumour and its surrounding tissue at least, to allow selective damage. It is possible, obviously, to blank the peritumoural region with non-transmitting material, although this suggests a lack of proper selectivity for the tumour. Certainly with skin tumours, this would probably also indicate that the individual patient was photosensitive and would require strict management regarding light exposure. With the early porphyrins, the reality of post-treatment periods of weeks and even months where patients were required to stay out of bright light was one of the main drivers for the search for improved photosensitisers.

Clinically, the photodynamic therapy of cancer – although attempted on a sparse and scattered basis in the early 1900s – became an organised reality in the mid–late 1980s, employing haematoporphyrin preparations. The conservative nature of clinical medicine means that novel approaches to disease treatment in general must undergo a considerable period of testing before widespread acceptance. Although this can be somewhat shortened for cancer therapies, given the severity of symptoms and degree of associated mortality, the rate of acceptance for clinical PDT has been slow. The explanation of this phenomenon is most probably due to three factors: the fact that intensely coloured medication is used, the requirement for a combination of light and drug and the persistence of systemic photosensitivity associated with several photosensitisers used in the clinic.

The idea and arguments surrounding coloured therapeutics are given in greater detail elsewhere in this book (see Chapter 2). However, the fact remains that coloured drugs are not especially liked by clinicians, presumably being linked to old-fashioned (and thus ineffective?) 'dye therapy'. For some time, the front line treatment for methaemoglobinaemia has been 1% aqueous methylene blue, i.v. administered, and several vital stains (methylene blue again, or the triarylmethane dye, patent blue V) are used at similar levels for surgical demarcation. However, the fact remains that coloured drugs or aids to visualisation are seen to be 'low tech'.

The idea of combination therapy is not new – several therapeutics are, in fact, combinations of two drugs, and the use of 'drug cocktails' both in cancer and in serious microbial infection are relatively common, particularly where conventional drug resistance is a problem. PDT requires a combination of a photosensitising drug and light, but not synchronously – the variability of the drug-to-light interval is a problem, but not an unsurmountable one. Also, clinicians may see PDT as a combination of chemo- and radiotherapy. The experience of the side effects of both of these may well mitigate against acceptance, but the side effects of PDT are generally much milder than either. In addition, the costs of a suitable light source may be seen as prohibitive. However, lasers are no longer essential for PDT and, in any case, cost less than the particle accelerators required for radiotherapy. Overall, the experience of PUVA therapy in psoriasis is the closest in terms of clinical experience, and this presumably explains, at least in part, the more ready acceptance of PDT by dermatology departments.

It must be stated that early experiences with haematoporphyrin-based PDT generally featured resulting systemic photosensitivity, with patients having to stay indoors for extended periods. While this is problematic, and unpleasant for the patient, it is not as bad as the massive hair loss, bleeding gums and general nausea/morbidity associated with conventional cancer therapy. Foscan has also suffered from a poor side-effect record, and this may not have furthered PDT acceptance as expected, but experience with ALA has shown that treatment of a wide variety of presentations is possible, with excellent cosmesis and a lack of systemic photosensitivity.

So what is the future for PDT? In the author's opinion, the rate of clinical acceptance of this approach has increased and should continue to do so. The strides made with ALA have undoubtedly helped, as has the acceptance of non-oncological uses of PDT, such as verteporfin in the treatment of macular degeneration (see Chapter 12). The sister application of photoantimicrobial chemotherapy (Chapter 11), gaining more credence with the ever-increasing problem of clinical bacterial drug resistance, must also help – in fact, the strengthening of one discipline cannot but help the other.

As far as photosensitisers for use in PDT are concerned, this is less clear. There are often conference lectures and presentations entitled 'Why do we need more photosensitisers for PDT?' or some such, normally – and not surprisingly – from speakers not involved in photosensitising drug discovery. The point of these talks is, presumably, to point out the number of tested photosensitisers not yet in human trials compared to the time required to get the present handful to clinic. While it is true that there are limited resources to support photosensitiser research, we are nowhere near the perfect photo-therapeutic agent. It may be that such Nirvana might be attained by one of the existing clinical agents, suitably nanoformulated and with 'stealth' capability, but, certainly in the early 21st century, this does not seem likely in the short term.

The answer to the 'need' question is, of course, that we do need more and improved photosensitisers – for a variety of indications – but that we cannot at present afford to develop them all. Greater funding is required, and this is unlikely to be obtained until health services and large pharmaceutical companies become involved. Although several academic spin-outs have been launched, it is little easier to obtain funding from private equity, so new photosensitisers will continue to be developed in universities with the fervent hopes of the academic researchers involved that theirs will be 'the one'.

References

Alexiades-Armenakas MR. (2006) Laser-mediated photodynamic therapy. *Clinics in Dermatology* **24**: 16–25.

Allison RR, Cuenca RE, Downie GH, Camnitz P, Brodish B, Sibata CH. (2005) Clinical photodynamic therapy of head and neck cancers – a review of applications and outcomes. *Photodiagnosis and Photodynamic Therapy* **2**: 205–222.

Allison RR, Sibata C, Downie GH, Cuenca RE. (2006) Photodynamic therapy of the intact breast. *Photodiagnosis and Photodynamic Therapy* **3**: 139–146.

Boisbrun M, Vanderesse R, Engrand P, *et al.* (2008) Design and photophysical properties of new RGD targeted tetraphenylchlorins and porphyrins. *Tetrahedron* **64**: 3494–3504.

Booth S, Poole D, Moghissi K. (2006) Initial experience of the use of photodynamic therapy(PDT) in recurrent malignant and pre-malignant lesions of the vulva. *Photodiagnosis and Photodynamic Therapy* **3**: 156–161.

Braathen LR, Szeimies RM, Basset-Seguin N, *et al.* (2007) Guidelines on the use of photodynamic therapy for nonmelanoma skin cancer: an international consensus. *Journal of the American Academy of Dermatology* **56**: 125–143.

Brown SB, Brown EA, Walker I. (2004) The present and future role of photodynamic therapy in cancer treatment. *Lancet Oncology* **5**: 497–508.

Calzavara-Pinton PG, Szeimeis RM, Ortel B, Zane C. (1996) Photodynamic therapy with systemic administration of photosensitizers in dermatology. *Journal of Photochemistry and Photobiology B: Biology* **36**: 225–231.

Derycke ASL, de Witte PAM. (2004) Liposomes for photodynamic therapy. *Advanced Drug Delivery Reviews* **56**: 17–30.

Fantini S, Taroni P. (2008) Optical mammography. *Cancer Imaging* **1**: 445–453.

Hackbarth S, Horneffer V, Wiehe A, Hillenkamp F, Röder B. (2001) Photophysical properties of pheophorbide-a-substituted diaminobutane poly-propylene-imine dendrimer. *Chemical Physics* **269**: 339–346.

Huang Z, Chen Q, Dole KC, *et al.* (2007) The effect of Tookad-mediated photodynamic ablation of the prostate gland on adjacent tissues—in vivo study in a canine model. *Photochemical and Photobiological Sciences* **6**: 1318–1324.

Kamuhabwa A, Agostinis P, Ahmed B, *et al.* (2004) Hypericin as a potential phototherapeutic agent in superficial transitional cell carcinoma of the bladder. *Photochemical and Photobiological Sciences* **3**: 772–780.

Kloek J, Beijersbergen van Henegouwen GMJ. (1996) Prodrugs of 5-aminolevulinic acid for photodynamic therapy. *Photochemistry and Photobiology* **64**: 994–1000.

Khan SA, Dougherty TJ, Mang TS. (1993) An evaluation of photodynamic therapy in the management of cutaneous metastases of breast cancer. *European Journal of Cancer* **29**: 1686–1690.

Knobler E. (2004) Current management strategies for cutaneous T-cell lymphoma. *Clinics in Dermatology* **22**: 197–208.

Kolarova H, Nevrelova P, Bajgar R, Jirova D, Kejlova K, Strnad M. (2007) In vitro photodynamic therapy on melanoma cell lines with phthalocyanine. *Toxicology In Vitro* **21**: 249–253.

Kostron H, Fiegele T, Akatuna E. (2006) Combination of Foscan-mediated fluorescence guided resection and photodynamic treatment as new therapeutic concept for malignant brain tumors. *Medical Laser Application* **21**: 285–290.

Kübler A, Haase T, Rheinwald M, Barth T, Mühling J. (1998) Treatment of oral leukoplakia by topical application of 5-aminolevulinic acid. *International Journal of Oral & Maxillofacial Surgery* **27**: 466–469.

Lapeš M, Petera J, Jirsa M. (1996) Photodynamic therapy of cutaneous metastases of breast cancer after local application of meso-tetra-(para-sulphophenyl)-porphin (TPPS$_4$). *Journal of Photochemistry and Photobiology B: Biology* **36**: 205–207.

Leroux JC, Allémann E, DeJaeghere F, Doelker E, Gurny R. (1996) Biodegradable nanoparticles – from sustained release formulations to improved site specific drug delivery. *Journal of Controlled Release* **39**: 339–350.

Li LB, Luo RC, Liao WJ, Zhang MJ, Luo YL, Miao JX. (2006) Clinical study of Photofrin photodynamic therapy for the treatment of relapse nasopharyngeal carcinoma. *Photodiagnosis and Photodynamic Therapy* **3**: 266–271.

Link EM. (1999) Targeting melanoma with 211-At/131-I methylene blue: preclinical and clinical experience. *Hybridoma* **18**: 77–82.

Mang TS, Allison R, Hewson G, Snider W, Moskowitz R. (1998) A phase II/III clinical study of tin ethyl etiopurpurin (Purlytin)-induced photodynamic therapy for the treatment of recurrent cutaneous metastatic breast cancer. *Cancer Journal from Scientific American* **4**: 378–384.

McFadden K, Cruickshank M. (2005) New developments in the management of VIN. *Reviews in Gynaecological Practice* **5**: 102–108.

Meisel M, Kocher T. (2005) Photodynamic therapy for periodontal diseases: state of the art. *Journal of Photochemistry and Photobiology B: Biology* **79**: 159–170.

Moghissi K, Dixon K. (2005) Photodynamic therapy in the management of malignant pleural mesothelioma: a review. *Photodiagnosis and Photodynamic Therapy* **2**: 135–147.

Moghissi K, Dixon K. (2008) Update on the current indications, practice and results of photodynamic therapy (PDT) in early central lung cancer (ECLC). *Photodiagnosis and Photodynamic Therapy* **5**: 10–18.

Nyman ES, Hynninen PH. (2004) Research advances in the use of tetrapyrrolic photosensitizers for photodynamic therapy. *Journal of Photochemistry and Photobiology B: Biology* **73**: 1–28.

Oku N, Saito N, Namba Y, Tsukada H, Dolphin D, Okada S. (1997) Application of long-circulating liposomes to cancer photodynamic therapy. *Biological and Pharmaceutical Bulletin* **20**: 670–673.

Petersen BT, Chuttani R, Croffie J, *et al.* (2006) Photodynamic therapy for gastrointestinal disease. *Gastrointestinal Endoscopy* **63**: 927–932.

Primo FL, Michieleto L, Rodrigues MAM, *et al.* (2007) *Journal of Magnetism and Magnetic Materials* **311**: 354–357.

Primo FL, Rodrigues MMA, Simioni AR, Bentley MVLB, Morais PC, Tedesco AC. (2008) In vitro studies of cutaneous retention of magnetic nanoemulsion loaded with zinc phthalocyanine for synergic use in skin cancer treatment. *Journal of Magnetism and Magnetic Materials* **320**: e211–e214.

Rice L, Wainwright M, Phoenix DA. (2000) Phenothiazine photosensitisers III. Activity of methylene blue derivatives against pigmented melanoma cell lines. *Journal of Chemotherapy* **12**: 94–104.

Ris HB. (2005) Photodynamic therapy as an adjunct to surgery for malignant pleural mesothelioma. *Lung Cancer* **49**: S65–S68.

Schmitt IM, Chimenti S, Gasparro FP. (1995) Psoralen-protein photochemistry – a forgotten field. *Journal of Photochemistry and Photobiology B: Biology* **27**: 101–107.

Senior K. (2005) Photodynamic therapy for bladder cancer. *Lancet Oncology* **6**: 546.

Sessler JL, Miller RA. (2000) Texaphyrins: new drugs with diverse clinical applications in radiation and photodynamic therapy. *Biochemical Pharmacology* **59**: 733–739.

Sharman WM, van Lier JE, Allen CM. (2004) Targeted photodynamic therapy via receptor mediated delivery systems. *Advanced Drug Delivery Reviews* **56**: 53–76.

Sieber F. (1987) Merocyanine 540. *Photochemistry and Photobiology* **46**: 1035–1042.

Stylli SS, Howes M, MacGregor L, Rajendra P, Kaye AH. (2004) Photodynamic therapy of brain tumours: evaluation of porphyrin uptake versus clinical outcome. *Journal of Clinical Neuroscience* **11**: 584–596.

Stylli SS, Kaye AH. (2006) Photodynamic therapy of cerebral glioma – a review. Part II – clinical studies. *Journal of Clinical Neuroscience* **13**: 709–717.

Stylli SS, Kaye AH, MacGregor L, Howes M, Rajendra P. (2005) Photodynamic therapy of high grade glioma – long term survival. *Journal of Clinical Neuroscience* **2**: 389–398.

Sutedja TG, Postmus PE. (1996) Photodynamic therapy in lung cancer. A review. *Journal of Photochemistry and Photobiology B: Biology* **36**: 199–204.

Swamy N, James DA, Mohr SC, Hanson RN, Ray R. (2002) An estradiol-porphyrin conjugate selectively localizes into estrogen receptor-positive breast cancer cells. *Bioorganic & Medicinal Chemistry* **10**: 3237–3243.

Szygula M, Pietrusa A, Adamek A, *et al.* (2004) Combined treatment of urinary bladder cancer with the use of photodynamic therapy (PDT) and subsequent BCG-therapy: a pilot study. *Photodiagnosis and Photodynamic Therapy* **1**: 241–246.

Vargas A, Pegaz B, Debefve E, *et al.* (2004) Improved photodynamic activity of porphyrin loaded into nanoparticles: an evaluation using chick embryos. *International Journal of Pharmaceutics* **286**: 131–145.

Walker I, Vernon DI, Brown SB. (2004) The solid-phase conjugation of purpurin-18 with a synthetic targeting peptide. *Bioorganic & Medicinal Chemistry Letters* **14**: 441–443.

Wieder ME, Hone DC, Cook MJ, Handsley MM, Gavrilovic J, Russell DA. (2006) Intracellular photodynamic therapy with photosensitizer-nanoparticle conjugates: cancer therapy using a "Trojan horse". *Photochemical and Photobiological Sciences* **5**: 727–734.

Williams JL, Stamp J, Devonshire JR, Fowler GJS. (1989) Methylene blue and the photodynamic therapy of superficial bladder cancer. *Journal of Photochemistry and Photobiology B: Biology* **4**: 229–232.

Wolinsky JB, Grinstaff MW. (2008) Therapeutic and diagnostic applications of dendrimers for cancer treatment. *Advanced Drug Delivery Reviews* **60**: 1037–1055.

Yoshida T, Tokashiki R, Ito H, *et al.* (2008) Therapeutic effects of a new photosensitizer for photodynamic therapy of early head and neck cancer in relation to tissue concentration. *Auris Nasus Larynx* **35**: 545–551.

Zhu TC, Finlay JC. (2006) Prostate PDT dosimetry. *Photodiagnosis and Photodynamic Therapy* **3**: 234–246.

11

Antimicrobial application – photodynamic antimicrobial chemotherapy

[*Note*: the author's use of the acronym PACT for photodynamic antimicrobial chemotherapy rather than the alternative APDT (antimicrobial photodynamic therapy) or
PDD (photodynamic disinfection) arises from a wish to avoid confusion with anticancer
PDT and also to reflect the fact that normally the photosensitisers employed in PACT
have some activity against microbial species without illumination].

Unlike tumour cells, microbes have few similarities with a human host, and this
allowed the development of effective antimicrobial chemotherapy by the middle of
the twentieth century. Differences in cell exterior (cytology) and cellular biochemistry
provided the means of attack by chemicals inhibitory, for example, to the growing
bacterial cell wall (β-lactams) or to folic acid synthesis (sulphonamides). There are now
many structural types employed in antimicrobial chemotherapy, although the efficacy
of a steadily increasing number of drugs is being eroded due to resistant organisms,
typified by (but definitely *not* limited to) methicillin-resistant *Staphylococcus aureus*
(MRSA).

During much of the Victorian era of scientific breakthrough, the serious consequences
of microbial disease were an accepted part of everyday life, although it was not widely
appreciated that outbreaks of cholera or diphtheria, the high mortality rate in childbirth or
the ever-present spectre of tuberculosis were caused by bacteria. Although microscopy
was available to clinicians and scientists, this discipline did not come of age until late
in the nineteenth century, so many diseases were attributed to 'miasmas' or bad air
(sometimes this was not too far off the mark).

One of the major discoveries of the Victorian era in Britain was that of the synthetic
dye mauve by Perkin in 1856 (see Chapter 4). This was important in many ways for the
burgeoning industrial revolution due to the commercial possibilities of coloured materials, but also because it encouraged the discovery of new dyes through the emerging
science of organic chemistry.

Photosensitisers in Biomedicine   Mark Wainwright
© 2009 John Wiley & Sons, Ltd

The availability of the new 'aniline dyes' to scientists of the age interested in infectious disease allowed a quantum leap in disease aetiology. For example, Christian Gram used the differences in staining behaviour to develop a general classification for bacteria. The Gram stain – using crystal violet, iodine and safranin – remains an essential basic tool in microbiology (Horobin and Kiernan 2002). Robert Koch also used dyes to demonstrate the organism responsible for tuberculosis (*Mycobacterium tuberculosis*), which is not demonstrated by the Gram stain. However, it was the work of Paul Ehrlich that constituted the greatest breakthrough.

Ehrlich realised that the staining of microbial pathogens in human samples demonstrated a selectivity of the dyes used for the microbes, and that there must be a relationship between the chemical constitution of the dye molecules and their sites of concentration in the target cell. This is the basis of structure–activity, an essential tool in modern drug discovery, which, in Ehrlich's hands, also formed the basis of chemotherapy.

Salvarsan, or Ehrlich 606, is the best known of Ehrlich's contributions to infection control. The arsenical drug is based on an azoic dye structure with the two nitrogens of the azo group ($-N{=}N-$) replaced by another Group V element, arsenic. However, Ehrlich had demonstrated the antimicrobial capability of another dye, methylene blue, somewhat earlier in his career. Ehrlich with a co-worker, Guttmann, cured two people suffering from malaria (Guttmann and Ehrlich 1891). The idea that dyes could be so designed as to be specifically toxic for microbial species was revolutionary. It is mainly the non-specific colouration, i.e. normal pharmacological distribution, of systemically administered dyes that has discouraged their wider use in the clinic.

Selective staining led to selective toxicity, both thanks to Ehrlich. However, while dyes such as methylene blue have been widely used as lead compounds for drug development in the past, especially in antimalarial research (Wainwright and Crossley 2002), the pharmaceutical industry has little love for dyes or molecules derived from them.

As all good reviews of PACT recall, the first real demonstration of dyes and the photodynamic effect was carried out by Oskar Raab using examples such as acridine (which is not a dye) and eosin (which is a dye), as long ago as 1901 (Raab 1900), although his inactivation of the unicellular *Paramecium* flagellates was of little contemporary interest. However, the technique was demonstrated *in vitro*, again with methylene blue, against bacteria, bacteriophages and protozoa (Schultz and Kruger 1928; T'ung 1935, 1938) in the period 1928–1935, ironically covering both Fleming's discovery of the bactericidal action of penicillin and Domagk's discovery of Prontosil. Any significant development of this work was presumably quashed by the arrival of the aforesaid sulpha drugs and penicillins in the next few years, although an alternative use, that of whole blood disinfection with photoactivated methylene blue, was first suggested in 1955 in work carried out at the Walter Reed Institute.

11.1 Antimicrobial and photoantimicrobial action

Although, as has been described, photosensitisers pre-date conventional antimicrobial drugs, the action of the latter is much better understood. This is quite logical when one considers the depth of research entailed in antimicrobial development. Thus, modern antibacterial agents have well-established sites/modes of action before they are released into the clinic. For example, on its introduction in 1999 it was understood that the original oxazolidinone antibacterial, linezolid (Zyvox), would attack ribosomal protein synthesis in susceptible bacteria. Furthermore, this attack would take place at a site on the bacterial 23S ribosomal RNA fragment of the 50S subunit (Bozdogan and Appelbaum 2004). This binding inhibits formation of the bipartisan ribosomal 70S unit essential for bacterial protein synthesis. Similarly, precise descriptions are available for the remainder of the conventional antibacterial arsenal – β-lactams at the bacterial cell wall, quinolones at the enzyme DNA gyrase, sulphonamides at dihydrofolate synthase, etc. For photoantimicrobials, however, this is not the case.

From the description of early microbial staining above, it should be understood that the exhibition of selective staining represents gross dye uptake by the organism. Initially, this will be due to attraction between the cationic dye, e.g. crystal violet, methylene blue and anionic (carboxylate, $-CO_2^-$) groups in the cell exterior. Depending on the physicochemical characteristics of the dye, it may be fractionated into the cell and further compartmentalised. Such characteristics include molecular size and charge, hydrophilic/lipophilic balance and ease of reduction (neutral molecules are more easily taken up by cells than are charged ones). Given an irregular pattern of localisation, efficient photoactivation may cause damage at multiple and/or variable sites, as mentioned previously, usually via oxidation.

Bacteria are very simple, single-cell organisms. They are generally spherical, ovoid or rod-shaped with a rigid cell wall containing the cytoplasm, circular DNA and ribosomes. As mentioned, the Gram stain allows straightforward discrimination of the two main classes, according to their cell wall type, shown in Figures 11.1 and 11.2. As the cell exterior is, logically, the initial site of attack for an incoming drug, it is worth considering the structures involved.

Plainly, the Gram-negative cell wall (Figure 11.1) is more extensive and more complex than the Gram-positive version. This is, indeed, one of the reasons for the often greater intransigence of Gram negatives against conventional antibacterial agents. For example, penicillin derivatives (β-lactams) act on the growing peptidoglycan layer, which is situated beneath the outer membrane in Gram negatives. In other words, there is a barrier to action, and many β-lactams have poor activity against this class. There are ways through the outer membrane – porins and trans-membrane proteins, for example – but these are selective pathways, also used for ejecting unwanted or toxic molecules.

Comparatively, attacking the Gram-positive cell wall is a much more straightforward proposition, having no semi-permeable outer layer to protect the peptidoglycan structure (Figure 11.2).

Of course, not all antibacterial drugs attack the cell wall. There are also drug classes that act on the ribosome (tetracyclines, aminoglycosides, macrolides and oxazolidinones) inhibiting protein synthesis and others, such as the fluoroquinolones, that inhibit DNA replication. Such drugs, by definition, must get through the external barriers and into the cell proper, in order to act. The route involved for a drug such as the oxazolidine Zyvox, covered above, is thus both long and complex and provides a fine example of the drug design process, since this is an entirely synthetic therapeutic.

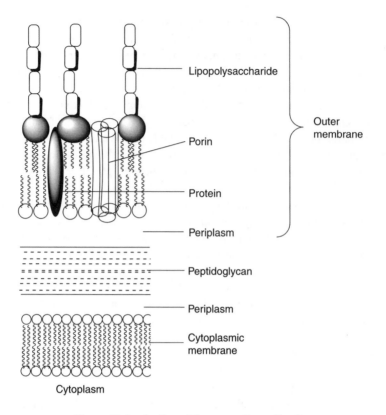

Figure 11.1   Section of Gram-negative cell wall.

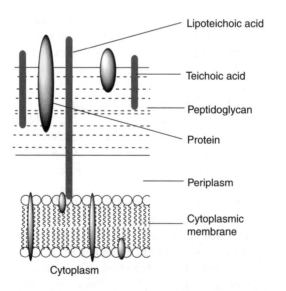

Figure 11.2   Section of Gram-positive cell wall.

Thus, while the aim of the industrial pharmaceutical scientist is to pinpoint the microbial target at the molecular level, those involved in antimicrobial photosensitiser research should pinpoint the microbe among host cells, but no more (Figure 11.3). As will be seen, this lack of targeting offers considerable advantages.

So, how does a photosensitiser get into a bacterial cell? As mentioned above, cationic photosensitisers such as methylene blue bind via ionic attraction to carboxylate or sulphonate residues in the various strata of the bacterial cell wall (Figures 11.1 and 11.2). Cations that remain unbound may enter the cell via tiny channels called porins, although there is a size limit – larger cations and aggregates would be prohibited. While the outer membrane excludes highly hydrophilic molecules, more lipophilic (i.e. amphiphilic and lipophilic) examples are likely to be able to cross into the cell via diffusion. In addition, the

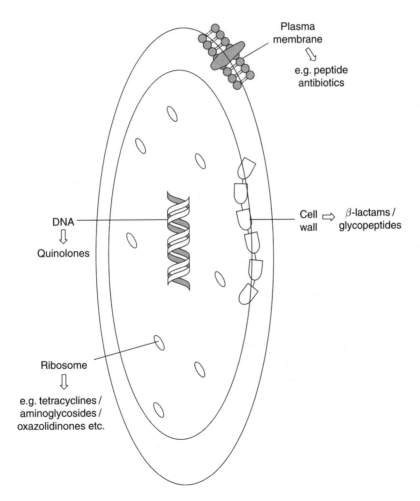

Figure 11.3 Specific target sites for different types of antibacterial agent. Any such site may be attacked by PACT.

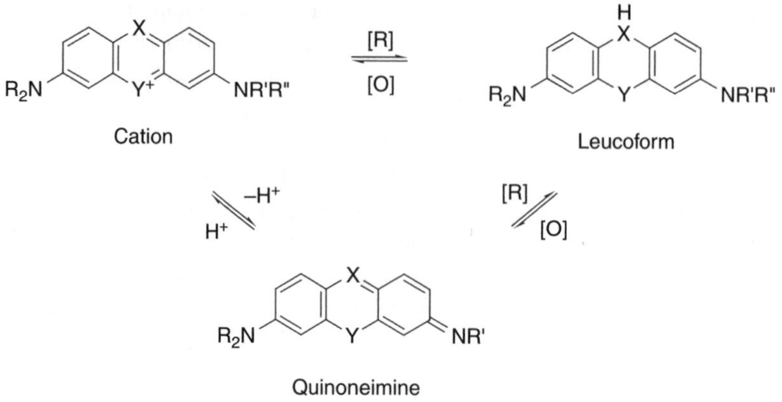

Figure 11.4 Speciation in azine photosensitisers. Key: X = CH and N; Y = NH, O, S and Se.

chemical alteration of a photosensitiser may furnish a resultant molecule that is more easily taken into the cell. This is probably the case with methylene blue, which is easily reduced by ubiquitous biochemical agents such as nicotinamide adenine dinucleotide (Digin *et al.* 2005). The reduced form of methylene blue (leucomethylene blue; Figure 11.4) is neutral and has much increased lipophilicity compared to the parent compound. Similarly, azines of this type, having primary or secondary amino groups as auxochromes, can form neutral quinoneimine species by deprotonation (Figure 11.4).

The uptake of leucomethylene blue should be much more straightforward than that for methylene blue itself and this might explain why there are sites of action for methylene blue in bacteria other than the cell wall alone (Figure 11.5) (Singh and Ewing, 1978; Webb, Hass and Kubitschek 1979). Such uptake phenomena for methylene blue have been suggested also in the malarial parasite *Plasmodium falciparum* (Wainwright and Amaral 2005).

It should also be mentioned at this point that fungi and viruses also have external/internal barriers, of varying complexity, as well as a range of internal organelle targets. Yeasts such as *Candida albicans* have a thick outer layer of protein and carbohydrate above an inner membrane that surrounds the cytoplasmic volume. However, yeasts and moulds contain far more mammal-like levels of cellular machinery than do bacteria. Viruses, on the other hand, are split into *enveloped* or *non-enveloped*, depending on the presence or absence of an outer, lipid-rich coating. An inner protein shell (capsid) then surrounds the viral nucleic acid and enzymes (e.g. reverse transcriptase) responsible for its reproduction after the parasitisation of a host cell.

Using the staining paradigm as a basis for photoantimicrobial action might be stated thus:

> If a microbe can be stained selectively with a dye that is also a photosensitiser, it should then be possible to destroy the microbe on illumination with light of the correct wavelength.

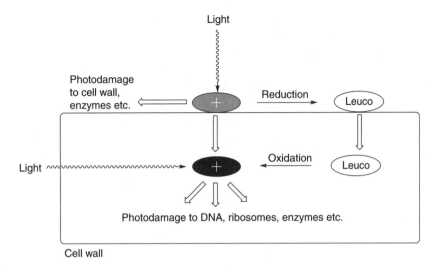

Figure 11.5 Routes to photodynamic damage in a microbial cell.

This automatically suggests a question: what about the Gram differentiation? Since both crystal violet and safranin are cationic dyes, surely the Gram classification is an indication that negative-staining bacteria will be more difficult to kill using cationic photosensitisers? This gives a fair approximation for hydrophilic types like crystal violet (and methylene blue) but not for more lipophilic examples or others that might exhibit alternative uptake pathways. For example, 1,9-dimethyl methylene blue (Chapter 4) is much more effective against Gram-negative organisms, such as *Yersinia enterocolitica*, than is methylene blue itself (Wainwright *et al.* 2001). Similarly, it has been shown that another Gram-negative pathogen, *Escherichia coli*, is more susceptible to the photobactericidal action of toluidine blue O (Chapter 4) than is the Gram-positive *Streptococcus sanguis* (O'Neill, Wilson and Wainwright 2003).

Dimethyl methylene blue, having two more methyl groups than methylene blue itself, is the more lipophilic cation (Chapter 4). In addition, it is harder to reduce than the parent compound (Wainwright *et al.* 1997). Consequently, it is possible that the increased activity of the dimethyl analogue is due to the greater lipophilicity facilitating entry to the bacterial cell. Toluidine blue O has a similar hydrophilic/lipophilic balance to methylene blue. The greater activity observed against *E. coli* may be due in part to the formation of a neutral quinoneimine species via auxochrome deprotonation (Figure 11.4) and increased uptake in the Gram-negative organism.

In the cases mentioned where molecular alteration modifies uptake, it should be recalled that neither the leucophenothiazine nor the quinoneimine is a photosensitiser (Chapter 4). Consequently, a mechanism for cation regeneration, i.e. oxidation or protonation, respectively, is required in the cell interior for effective photosensitisation (Figure 11.5). Such regeneration would, of course, lead to a concentration gradient and further uptake, as has been reported for the tumour cell uptake of certain benzo[*a*]phenothiazinium salts (Cincotta *et al.* 1994).

The photoantibacterial activity of cationic photosensitisers is maintained for a wide range of chromophoric types, certainly. Where neutral or anionic derivatives are inactive, the change of sign (via synthesis) to cationic normally endows the agent with improved photobactericidal performance. Thus, while haematoporphyrin and chlorin e_6 (see Chapter 6) are both anionic and mainly active against Gram-positive bacteria, porphyrin derivatives having pendant quaternary ammonium or pyridinium groups are highly active and broad spectrum (Lazzeri *et al.* 2004). The positive charge is plainly important in activity.

It is known, from exhaustive efforts in more traditional areas of antibacterial research, that drugs based on the tetracycline and quinolone chromophores are able to complex divalent metal ions (typically Mg^{2+}) in the bacterial cell membrane, thus interfering with the proper arrangement of lipid molecules (Vaara 1992). Cationic peptides such as daptomycin disrupt the external membrane (Figures 11.1 and 11.2), allowing its permeabilisation and the egress of ions from the cell, leading to membrane depolarisation and cell death.

Thus, it is well established that cationic interactions are important in some types of antibacterial action, and this is also one possible mode for cationic photosensitisers. This is supported by the fact that various workers have used the cationic peptide colistin (polymyxin) in conjunction with less inherently successful antibacterial photosensitisers, such as chlorin e_6, in order to allow the photosensitiser access to the cell interior (Nitzan *et al.* 1995). Another way to achieve such disruption is to attach a poly(lysine) chain to the photosensitiser, the high basicity of the amino acid residues guaranteeing cationic nature, and thus membrane disruption, at ambient pH (Demidova and Hamblin 2005). Reportedly, the conjugation of chlorin e_6 to poly(ethyleneimine) has a similar effect (Tegos *et al.* 2006). Similarly, targeting using specific antibodies is also possible and has been demonstrated for a number of photosensitisers, including chlorin e_6 and toluidine blue (Bhatti *et al.* 2000; Embleton *et al.* 2002).

11.2 Applications

How can photosensitisers be employed in everyday infection control? Photoantimicrobial action is, by now, well documented for a wide range of organisms *in vitro*, with an expanding portfolio of *in vivo* data, but this is for the main part a matter of academic endeavour and record alone. There remains quite a gap between this and a routine clinical reality, despite efforts of academics to make the research pertinent to daily application. At the time of writing, several groups have published research on the successful eradication of important bacterial pathogens such as MRSA and *Clostridium difficile* – two of the main causes of death from hospital-acquired infection. While there are still a small number of effective conventional antibacterial agents available to the infectious disease clinician, these drugs – linezolid, tigecycline, daptomycin, etc. – would be better used only for serious cases. Antibacterial conservation is no longer optional in the face of overwhelming drug resistance in both health care and community. The use of topical/local photoantimicrobials, alongside other approaches, can help to conserve this precious resource.

It is most likely that the problem of bacterial drug resistance will continue to increase and that it will be a case of attempting to manage the problem rather than eradicating it. It thus makes sense to conserve effective therapeutics for life-threatening infections, while lesser ones should be treated with alternatives. There are already effective topical antibacterial agents in use in hospitals – povidone-iodine, for instance, used in pre-operative skin disinfection – which might be more widely employed. As an example of local photosensitiser use, it is possible to apply PACT to clear the skin of organisms that, in conjunction with the rampant biochemical manufacture associated with adolescence, cause acne. The conventional treatment here involves extended periods of oral antibiotics (e.g. tetracyclines) or isoretinoids. 5-Aminolaevulinic acid (Chapter 6) may be employed to good effect as a number of implicated bacteria produce haem. The use of systemic antibacterial agents to treat topically or locally accessible sites makes very little sense if we are really trying to decrease antibiotic usage. Plainly, the expansion of local treatment or disinfection constitutes a sea change in terms of healthcare practice, but it is difficult to see how else the valuable systemics can be saved.

It should also be underlined that MRSA and *C. difficile* are merely the current *bêtes noires* of infectious disease specialists in the first decade of the 21st century. In subsequent decades, doubtless other pathogens will gain precedence. While it is unlikely that MRSA will be eradicated, concerted action against this particular problem may lessen its hold. However, it is equally unlikely that another Gram-positive organism would fail in filling the environmental niche so vacated. Methicillin-resistant *Staphylococcus epidermidis* (MRSE), a relative that – in its drug-susceptible form – covers much of the skin quite harmlessly, is increasingly implicated in septicaemia. For the Gram negatives, *Acinetobacter baumannii* and its relatives among the *Enterobacteriaceae* have considerably improved resistance profiles in recent times, while the virulence of *E. coli* (e.g. *E. coli* O104 in recent outbreaks of food poisoning) along with its possession of extended-spectrum beta-lactamase (ESBL) capability is a significant cause of concern. However, none of the microbial resistance mechanisms so far encountered has produced a deleterious effect on photosensitiser action.

It would be foolish, of course, to suggest that photoantimicrobials are the answer to all the problems of drug-resistant disease. However, the demonstrable activity of various cationic photosensitisers against pathogenic bacteria, regardless of conventional resistance status (Tang, Hamblin and Yow 2007), should be sufficient to make them at least an inclusive part of the antimicrobial armoury in future.

MRSA spread in hospitals usually occurs via hand contact, but the staphylococcal reservoir is generally accepted to be the nasal passage. Photodynamic disinfection is one possible approach, but again this represents a more involved protocol than the current practice, *viz.* application of antibacterial mupirocin cream to the nares. Photodynamic disinfection using methylene blue, for example, would require either a cream vehicle or another method of delivery, followed by illumination of the anterior nares using a light probe. Given the widespread requirement for disinfection in cases of MRSA colonisation, it is unlikely that the photodynamic approach would be routinely administered by healthcare staff, but presumably it would require initial instruction and oversight for compliance purposes.

The disinfective mode using photoantimicrobials is also applicable to surfaces. While this is somewhat less glamorous than patient treatment, the infective link to the patient environment is not in question and must therefore be an integrated part of disease control measures. Photosensitisers may be employed in direct application to surfaces or may be an integral part of the surface itself, i.e. as a constituent of the surface coating (Figure 11.6).

In either mode, photosensitisers would be finding a similar use to that of standard biocidal agents, so the requirement for improvement on the *status quo* would apply, as in the above-mentioned situation with conventional antibiotics. Since standard biocides are relatively inexpensive, and there is far less problem with resistance here than with antibacterial drugs (Maillard 2007), there is perhaps less of an argument for the inclusion of photoantimicrobials for general environmental disinfective use. However, their use in medical materials may have more potential, for example in photodynamic catheters (Figure 11.7) or in textiles and non-woven materials used in coverings and dressings.

An appealing combination of the disinfective approach lies in intra-operative prophylaxis. Invariably surgery involves brilliant illumination of the operation site. The instillation of an effective photosensitiser immediately following incision would logically provide an antiseptic environment and thus inhibit microbial colonisation. This approach would be particularly useful in joint replacement surgery, since infection of the new joint often entails further surgery and *de novo* replacement, as well as the usual sequelae associated with wound infection.

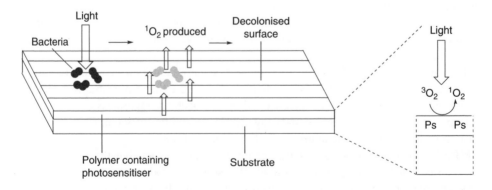

Figure 11.6 Photodynamic surface coating.

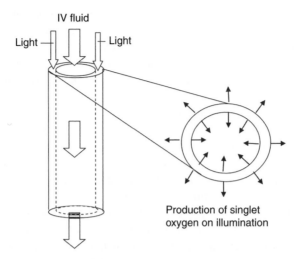

IV fluid

Light — — Light

Production of singlet
oxygen on illumination

Figure 11.7 Photodynamic catheter model, showing production of singlet oxygen at both the inner lumen and the outer surface.

11.3 Blood

Infection control is normally considered as referring to the treatment of infected individuals, although more recently the importance of environmental hygiene control has also been included.

However, on consideration of the various modes of infection transmission, it is clear that blood-borne pathogens constitute a significant threat in terms of disease aetiology. Consequently, the use of blood and blood products in transfusion and therapy offers a considerable potential pathogen reservoir. While this potential has always been understood by those intimately involved with transfusion, perhaps this was not the case in infection control at large until the advent of the HIV pandemic in the early–mid 1980s. In addition, the threat of transfusion–transmission infection (TTI) via blood and blood product usage is not limited to viruses such as HIV-1 and HIV-2. The range of potential infections is wide and varies with the complex mixture of substances constituting whole blood, and the treatment and storage conditions of the various fractions and derivatives.

As is generally accepted, whole blood collected at a typical public donation session may be used as whole blood in replacement, or as packed red cells, platelets or plasma, depending on clinical requirement. These fractions – typically plasma – may be further treated to supply, for example, clotting factors for the therapy of haemophiliacs. Indeed, the finding that AIDS was a consequence of TTI in clotting factor recipients as well as needle-sharing addicts was important in terms of blood product treatment.

Pathogen association with blood products is also influenced by the conditions in which materials are stored. For example, bacterial colonisation is more likely in platelet fractions (since these are usually stored at 20–24°C for up to 5 days) than in red blood cell concentrates (which are stored in red cell preservation solutions at 2–6°C for 42–49 days) although there are exceptions, such as *Pseudomonas fluorescens* and *Y. enterocolitica*, which flourish at lower temperatures.

The requirement for pathogen inactivation in blood products became urgent in the late twentieth century due to the AIDS pandemic, although hepatitis as a TTI was already established by this time. However, the task of clearing pathogens from donated blood and associated products is not a straightforward one.

As noted above, blood may contain a variety of pathogens – bacteria, viruses, yeasts, fungi and protozoa. Logically, the microbial load depends on both the health and the geographical location of the donor. Both pre-donation screening and post-donation blood analysis can remove unsuitable donors and donations, but occasionally the donor is asymptomatic or the microbial titre is too low to detect, or a pathogen may not be included in the testing protocol. There are also 'emerging pathogens', as HIV-1 was in the early 1980s, or West Nile virus was in the USA at the turn of this century.

Given the range of potentially infective agents, it is not logical to attempt their eradication via conventional antimicrobial chemotherapy. For example, very few antibacterial agents have a deleterious effect on viruses, etc. Consequently, a general antimicrobial agent is required, more along the lines of a biocidal agent. The search for such an agent is further complicated by the fact that treated blood fraction/product is required for systemic end use in human patients.

While several conventional approaches have been attempted here, e.g. solvent–detergent treatment or ultraviolet (UV) irradiation of plasma, these either have been ineffective across the range of products or have caused collateral damage.

The use of photosensitisers in blood product disinfection/pathogen eradication offers selectivity with more control over collateral damage. Long-wavelength visible light ($\geq$630 nm) can be used to illuminate all fractions of whole blood and – unlike UV – has no inherent effect on biomolecules.

At the time of writing there are three photodynamic procedures that either are in routine clinical use or have been cleared for clinical trials, using single-donor human plasma units for therapeutic use:

(1) Methylene blue/visible light (Lambrecht *et al.*, 1991).

(2) Amotosalen (psoralens – see Chapter 9) using UVA light (Snyder *et al.* 2005). This is a photochemical rather than a photodynamic procedure, based on the UV activation of a psoralen derivative in the presence of nucleic acid to form adducts where the psoralen nucleus is covalently bonded to one or two nucleotide bases.

(3) Riboflavin/UVB light (Ruane *et al.* 2004). Like methylene blue, illumination of riboflavin (vitamin B2) produces singlet oxygen (see *Natural product photosensitisers*, Chapter 9). Riboflavin, as an essential nutrient, has very low mammalian toxicity and is obviously well tolerated. However, like the psoralens, it is activated by long-wavelength UV and this light activation process suffers interference from endogenous light absorbers, for example those present in red blood cell cells.

As already mentioned, methylene blue can work in two ways: excitation with light can lead to transfer of the energy either via electron (Type I mechanism) or via energy transfer (Type II mechanism).

On intercalation with DNA or RNA, the Type I mechanism might progress via oxidised species such as hydroxyl radicals and the Type II mechanism via singlet oxygen. The results are nucleic acid breakages, mostly at the guanosine site. Therefore, in contrast to Amotosalen and other UV-absorbing psoralens, there is no covalent adduct formed but a Type I/II pathway similar to that for methylene blue is followed.

The methylene blue procedure was developed by Mohr and co-workers (Lambrecht *et al.* 1991) and improved in conjunction with MacoPharma GmbH (Williamson, Cardigan and Prowse 2003). The procedure is efficient enough to inactivate at least five logs of enveloped viruses relevant to transfusion medicine, e.g. HIV and HCV, but is less effective against non-lipid-enveloped viruses. In addition, the procedure has some influence on the efficacy of plasma coagulation proteins, e.g. fibrinogen and Factor VIII activity is reportedly decreased by around 25% as a result. Other factors including inhibitors are not reduced, as confirmed by proteomics analysis (Tissot *et al.* 1994). Considerable effort has been made to ensure that the pharmacology and toxicology of the MB-treated plasma is such that the use of treated plasma is suitable for specific clinical applications (Pohler *et al.* 2004).

The plasma decontamination system here uses European Pharmacopeia-quality methylene blue, and the procedure has found wide acceptance in most European countries. More than 4 million units of MB-treated plasma had been used clinically by 2005.

11.4 Thionin/light + low-dose UVB for the decontamination of platelet concentrates

Despite its documented efficacy as a photoantimicrobial, methylene blue is not suitable for photodynamic pathogen inactivation of platelet concentrates (PC) for two reasons:

1. It is not completely effective against bacteria and residual leucocytes.

2. Platelets are heavily damaged by illumination in the presence of MB.

It was, however, found that the fully demethylated derivative of MB, thionin (Th) was similarly effective in inactivating free viruses while leaving platelet functions almost fully intact (Mohr and Redecker-Klein 2002). A possible explanation is that Th is more hydrophilic than MB and may therefore exhibit less binding to platelet and other cell exteriors. Consequently, Th/light, even more than MB/light, is ineffective in the inactivation of leukocytes and bacteria in the presence of plasma. This is due to the fact that plasma is known to contain compounds that quench photodynamic Type II reactions involving singlet oxygen. Significant quenchers in plasma occur in both the lipoprotein fraction and the aqueous phase, e.g. urea, tocopherols, carotenoids, ascorbic acid and bilirubin (Karnofsky 1990). Nevertheless, the lipid-enveloped viruses tested are inactivated by Th/light treatment. In this respect, Th was at least as effective as MB. Obviously Th, similar to MB, has a high affinity for viral structures including the viral genome. It is also remarkable that in contrast to most non-enveloped viruses tested, namely the animal parvoviruses, the human parvovirus B19 was found to be highly sensitive to Th/light treatment (Mohr and Redecker-Klein 2002).

Conversely, in plasma-free suspensions, Th was effective in inactivating a number of Gram-positive and also some Gram-negative bacteria. This was not unexpected, given the long-established photobactericidal activity of other phenothiazinium dyes, particularly methylene blue and toluidine blue O (T'ung 1935; T'ung and Zia 1937), including their photodynamic inactivation of bacteria in blood components contaminated with *Y. enterocolitica* (Wainwright *et al.* 2001). However, reports suggesting that the photoactivity of TBO was unaffected by the presence of blood and serum (Wilson, Sarkar and Bulman 1993; Wilson and Pratten 1995) are not supported by the results of the investigations on the phototreatment of PC with Th. Note: Plasma is necessary to maintain the storage stability of the platelets in PC (Murphy 1999; Gulliksson *et al.* 2000).

The additional requirement for residual leucocyte inactivation in PC arises from the fact that they may contain cell-associated viruses and may cause alloimmunisation and refractoriness to further platelet transfusions (Eernisse and Brand 1981; Dzik 1995). However, like MB, Th is not effective in photoinactivating leucocytes, presumably due to the hydrophilicity of both compounds, making them unable to penetrate cellular membranes (see below). Earlier work by the author showed that the cytotoxicity of phenothiazine photosensitisers increases with increasing hydrophobicity, i.e. the more hydrophilic compounds are less cytotoxic (Wainwright *et al.* 1998; Wainwright 2000). This finding is supported by data published by Skripchenko and Wagner (2000), indicating that white blood cells were inactivated by illumination in the presence of dimethyl methylene blue, whereas the less hydrophobic MB did not cause significant changes in leukocyte viability.

11.5   The cellular problem

Clearly the historical use of vital stains such as methylene blue and the azures provides considerable foundation for the investigation of improved photosensitisers based on the phenothiazinium and derivative chromophores (see *Azines*, Chapter 4). Thorough knowledge of the chemistry entailed in phenothiazinium design and synthesis allows a more considered, logical approach to the problem of selectivity and, thus, to the minimisation of collateral damage. Given the amount of design and synthesis carried out on other chromophoric types involved in photodynamic therapy, e.g. phthalocyanines and porphyrins, similar strides should be possible here also. However, the fact remains that effective, safe red cell decontamination remains elusive. This is illustrated by the proposed use of either dimethyl methylene blue or of *meso*-phenyltris(4-(*N*-methylpyridinium))porphyrin (Chapter 6) as a red cell decontaminant. In both cases, the quantity of associated literature concerning the required addition of antioxidants was indicative of significant collateral damage – too significant, in fact, for either photosensitiser additive to reach clinical status.

Although both of the photosensitisers mentioned here were ultimately unsuccessful, it should be possible to gain the selectivity required. A related example is illustrative: In the chemotherapy of malaria, often the parasite (*Plasmodium* spp.) is attacked during the blood-borne phase of the disease, when plasmodial forms exist and multiply in the erythrocytes. It is possible to attack these forms via orally administered chloroquine (or other antimalarials where chloroquine resistance is present). This infers that chloroquine molecules can cross into the interior of the erythrocyte *and* target the parasite. However, it should be remembered that colonised red blood cells are altered by the parasite, which incorporates trans-membrane transporters, making the cell more permeable in order to allow the ingress of extra nutrients (Wainwright and Amaral 2005).

In red cell disinfection, the microbial target is far more likely to be a virus, although as has been mentioned, bacterial colonisation is also possible, while in tropical parts of the world protozoal colonisation (including *Plasmodium* and *Trypanosoma* spp.) should also be a consideration.

Two factors are particularly relevant in the disinfection of red cells. Given that the main target is viral, it is sensible to use nucleic acid-selective photosensitisers. Secondly, red cells do not contain useful nucleic acid. This should allow an unusual margin of safety, but thus far no satisfactory therapeutic has emerged (photodynamic or otherwise).

Part of the problem appears to be that, in molecular terms, photosensitisers effective in DNA intercalation are not efficient in crossing into the red cell without causing considerable damage to the membrane at some stage. For example, methylene blue itself is highly effective in the inactivation of HIV-1 in media. However, in red cell suspensions, internalised virus is not inactivated, and there is significant membrane damage and haemolysis. The 1,9-dimethylated derivative of methylene blue (DMMB) gives excellent levels of inactivation both inside and outside the cell, but again causes sufficient membrane/protein alteration to lead to storage lifetime shortening. In addition, DMMB is highly toxic to mammalian cells in culture, which would offer a potential risk to recipients of treated cells. Conversely, methylene violet (Chapter 4), which shares the methylene blue structure but with one of the dimethylamino auxochromes substituted by oxo ($=O$), exhibited improved intracellular activity compared to the parent compound, but lower activity in the presence of plasma proteins (Skripchenko, Robinette and Wagner 1997). In addition, methylene violet has very low aqueous solubility (Wainwright and Giddens 2003).

Each of the photosensitisers mentioned above has the correct molecular shape for DNA intercalation. For methylene blue and DMMB this is accompanied by the cationic charge required for effective binding to DNA via ionic linkage to phosphate groups in the nucleic acid backbone. The difference in antiviral performance in red cells (Figure 11.8) can be related to the greater lipophilicity endowed by the extra methyl groups in DMMB, which allows trans-membrane access. External but cell-associated methylene blue molecules stack at the membrane surface – the basis of methylene blue staining – and on illumination produce singlet oxygen *in situ*, thus causing haemolysis due to membrane damage.

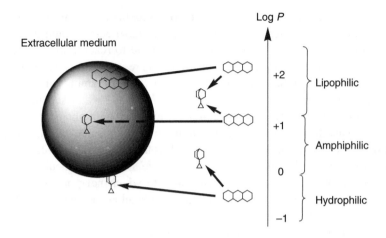

Figure 11.8 Antiviral capabilities of typical planar cations in red cell concentrate and variation with lipophilic character.

11.6   A role for photodynamic antimicrobial chemotherapy in tropical medicine?

There are, of course, many illnesses that are covered by the term 'tropical diseases'. Properly interpreted, this term relates to microbial infections usually limited geographically to tropical areas, infections not traditionally encountered in Western medicine. However, the globalisation caused by the ease of low-cost travel means that patients may present with almost any disease almost anywhere in the world. The danger is that an exotic viral disease, for example, may be transmissible in a new, crowded setting where there is, by definition, no herd resistance.

For many years, indeed since the end of rampant colonialism, tropical medicine has been a back number in terms of pharmaceutical research, military medical research groups aside. The fact is that there is little available profit associated with the supply of expensive therapeutics to the developing world. A massive political effort was required to start the supply of effective anti-AIDS medication to the poorer African states, and these were drugs used in Europe and the USA – AIDS is not a tropical disease. How much more difficult will it be, then, for effective modern medicines to be supplied by the pharmaceutical industry for diseases such as Hansen's disease (leprosy), trypanosomiasis, and leishmaniasis? The situation for malaria is somewhat different, being a threat to many millions of travellers from rich economies who thus require efficient prophylaxis. In addition, the Gates Foundation is supplying financial support for large-scale research in this area.

Although methylene blue is currently being re-evaluated as a clinical antimalarial (Meissner *et al.* 2006), this represents use as a traditional, not a photodynamic, drug. Noticeably, the acceptance of this coloured, systemic therapeutic has been excellent despite the obvious sequelae of coloured urine and stools, etc. Conversely, several tropical diseases have obvious, visible manifestations that might be amenable to local phototherapy. For example, cutaneous leishmaniasis and Hansen's disease have very evident skin lesions containing the causative organism – *Leishmania donovani* and *Mycobacterium leprae*, respectively. Local application of a photosensitiser to such presentations coupled with illumination should be able to clear the causative organisms. This has been demonstrated, at least *in vitro*, by both Boyle and Hassan for *Leishmania* spp. (Akilov *et al*. 2006; Bristow *et al*. 2006). As with drug-resistant bacteria, those leishmanial strains against which conventional therapies are ineffective are reportedly susceptible to the photodynamic approach (Sohl *et al*. 2007). Administered as a treatment by local health visitors, employing inexpensive portable light sources, this approach would be a simple and direct method to alleviate much suffering, so far seemingly overlooked by drug companies. Plainly with Hansen's disease, local therapy would be purely palliative.

Another important tropical disease is trypanosomiasis. In Africa, the most significant form (human African trypanosomiasis, HAT), caused by the bite of the tsetse fly, is fundamentally a disease of the nervous system: sleeping sickness. This is caused by the introduction of trypanosomes (usually *Trypanosoma brucei gambiense* or *T. brucei rhodesiense*). The American counterpart, Chagas' disease, which is normally caused by infection with *T. cruzi*, does not have a significant dermatological aspect (excepting the puncture wound caused by the triatomid bug, which is the disease vector), but infected persons may carry the protozoa in the bloodstream. *T. cruzi* must therefore be included in any photodisinfection protocols in affected regions, and the photokilling of the organism has been demonstrated with the silicon phthalocyanine Pc4 (Chapter 7) by Ben-Hur and co-workers (Gottlieb *et al*. 1995), and both *T. cruzi* and *L. donovani* have been photoinactivated using thiazole orange (Wagner *et al*. 2008). Interestingly, crystal violet is employed for the same purpose without photoactivation (Docampo, Moreno and Cruz 1988), which should be a potential lead for drug discovery against this organism.

What are we seeking to achieve with photoantimicrobial chemotherapy? To take a comparative example, consider the use of the beta-lactam antibiotic amoxycillin in the treatment of a tooth abscess. Here, the antibiotic is taken orally and is thus absorbed from the gastrointestinal tract into the bloodstream and then delivered to the site of infection via local capillaries. However, only a fraction of the administered antibiotic dose reaches the abscess. The remainder either is eliminated via normal metabolism or exerts its antibiotic effect against susceptible bacteria in the gastro-intestinal tract. Notably, there is no useful direct antibiotic effect at the site of administration. Conversely, photoantimicrobial application could be directed via the sub-gingival route, i.e. directly, thus being more efficient and not affecting the body's commensal flora other than in the small area around the tooth in question.

One of the main reasons for the widespread use of penicillins like amoxicillin is the high toleration level in humans – most patients have little problem in taking the tablets, and there are few incidences of side effects. However, in order to maintain an effective dose at our abscess site, a few days' worth of tablets is required – say 250 mg, three times daily for 7 days. In other words, if the course of tablets is completed (as directed), the system is exposed to 5.25 g of amoxycillin. In terms of the gut flora, this may be quite disastrous and might lead to what is euphemistically called a 'stomach upset' for the patient. In reality, this is probably the result of non-susceptible bacteria growing into spaces left by susceptible ones. Moreover, amoxicillin is well absorbed from the gut. Many other antibacterials are not well absorbed and may have considerable collateral effects.

A second reason for the use of antibiotics such as amoxicillin is that they are usually self-administered or at least that they do not require medical supervision beyond the accompanying instructions. Self-administration is, of course, a double-edged sword. How many millions of antibiotic prescriptions result in uncompleted courses of treatment because the patient forgets or simply feels better? Normally, this represents a sub-lethal dose to the targeted bacterial population as well as the toxic effects on commensals mentioned above.

Thus, both the low human toxicity and ease of administration of 'everyday' antibacterial agents expose the system to high levels of drugs that are highly damaging to our 'friendly flora'. This would be a small price to pay for very convenient therapy were it not for the contribution it makes to bacterial drug exposure and thus to resistance development.

Plainly there are infections that require systemic treatment, i.e. where the bacterial population is large and diffuse, as in septicaemia or pneumonia, or difficult to reach, as in meningitis. However, local infection treated locally would allow both the conservation of effective antibiotics and the slow resistance development. In addition, MRSA found on the skin surface or soft tissue is normally responsible for subsequent systemic infection. It would be advantageous to eradicate the pathogen at the earlier stage, doubly so if this were possible without recourse to antibiotics.

The obvious disadvantage of PACT is that it requires medical supervision. For example, the treatment of the abscess above would require accurate instillation of the photosensitiser, possibly via a cannula, followed by directed illumination of the infected site. Although it is now accepted that lasers are not required for such illumination, it is unlikely that effective, reliable light sources would be readily available in the patient's home. However, the requisite procedure might be undertaken routinely in dental or GP surgeries, or medical 'drop-in' centres. Indeed, since it is likely that oral photodynamic disinfection as described here would be the job of a dental hygienist, rather than a dental surgeon, it is also likely that routine photodynamic disinfection would be carried out by nursing staff. This would also be in line with the current devolution of a considerable amount of clinical practice.

References

Akilov OE, Kosaka S, O'Riordan K, *et al.* (2006) The role of photosensitizer molecular charge and structure on the efficacy of photodynamic therapy against *Leishmania* parasites. *Chemical Biology* **13**: 839–847.

Bhatti M, MacRobert A, Henderson B, Shepherd P, Cridland J, Wilson M. (2000) Antibody-targeted lethal photosensitization of *Porphyromonas gingivalis*. *Antimicrobial Agents and Chemotherapy* **44**: 2615–2618.

Bozdogan B, Appelbaum PC. (2004) Oxazolidinones: activity, mode of action, and mechanism of resistance. *International Journal of Antimicrobial Agents* **23**: 113–119.

Bristow CA, Hudson R, Paget TA, Boyle RW. (2006) Potential of cationic porphyrins for photodynamic treatment of cutaneous *Leishmaniasis. Photodiagnosis and Photodynamic Therapy* **3**: 162–167.

Cincotta L, Foley JW, MacEachern T, Lampros E, Cincotta AH. (1994) Novel photodynamic effects of a benzophenothiazine on two different murine sarcomas. *Cancer Research* **54**: 1249–1258.

Demidova TN, Hamblin MR. (2005) Effect of cell-photosensitizer binding and cell density on microbial photoinactivation. *Antimicrobial Agents and Chemotherapy* **49**: 2329–2335.

Digin Y, Dursun Z, Nisli G, Gorton L. (2005) Photoelectrochemical investigation of methylene blue immobilized on zirconium phosphate modified carbon paste electrode in flow injection system. *Analytica Chimica Acta* **542**: 162–168.

Docampo R, Moreno SNJ, Cruz FS. (1988) Enhancement of the cytotoxicity of crystal violet against *Trypanosoma cruzi* in the blood by ascorbate. *Molecular and Biochemical Parasitology* **27**: 241–247.

Dzik W. (1995) Use of leukodepletion filters for the removal of bacteria. *Immunological Investigations* **24**: 95–115.

Eernisse JG, Brand A. (1981) Prevention of platelet refractoriness due to HLA antibodies by administration of leukocyte-poor blood components. *Experimental Hematology* **9**: 77–83.

Embleton ML, Nair SP, Cookson BD, Wilson M. (2002) Selective lethal photosensitization of methicillin-resistant *Staphylococcus aureus* using an IgG-tin(IV) chlorin e_6 conjugate. *Journal of Antimicrobial Chemotherapy* **50**: 857–864.

Gottlieb P, Shen LG, Chimezie E, Bahng S, Kenney ME, Ben-Hur E. (1995) Inactivation of *Trypanosoma cruzi* trypomastigote forms in blood components by photodynamic treatment with phthalocyanines. *Photochemistry and Photobiology* **62**: 869–874.

Gulliksson H, Larsson S, Kumlien G, Shanwell A. (2000) Storage of platelets in additive solutions: effects of phosphate. *Vox Sanguinis* **78**: 176–184.

Guttmann P, Ehrlich P. (1891) Ueber die wirkung des methylenblau bei malaria. *Berliner Klinische Wochenschrift* **39**: 953–956.

Horobin R, Kiernan J. (2002) *Conn's Biological Stains: A Handbook of Dyes, Stains and Fluorochromes for Use in Biology and Medicine,* BIOS, Oxford, UK, 274.

Karnofsky JR. (1990) Quenching of singlet oxygen by human plasma. *Photochemistry and Photobiology* **51**: 299–303.

Lambrecht B, Mohr H, Knuever-Hopf J, Schmitt H. (1991) Photoinactivation of viruses in human fresh plasma by phenothiazine dyes in combination with visible light. *Vox Sanguinis* **60**: 207–213.

Lazzeri D, Rovera M, Liliana P, Durantini EN. (2004) Photodynamic studies and photoinactivation of Escherichia coli using meso-substituted cationic porphyrin derivatives with asymmetric charge distribution. *Photochemistry and Photobiology* **80**: 286–293.

Maillard J-Y. (2007) Bacterial resistance to biocides in the healthcare environment: should it be of genuine concern? *Journal of Hospital Infection* **65**: 60–72.

Meissner PE, Mandi G, Coulibaly B, *et al.* (2006) Methylene blue for malaria in Africa: results from a dose-finding study in combination with chloroquine. *Malaria Journal* **5**: 84.

Mohr H, Redecker-Klein A. (2002) Inactivation of pathogens in platelet concentrates by using a two-step procedure. *Vox Sanguinis* **84**: 96–104.

Murphy S. (1999) The efficacy of synthetic media in the storage of human platelets for transfusion. *Transfusion Medicine Reviews* **13**: 153–163.

Nitzan Y, Dror R, Ladan H, Malik Z, Kimel S, Gottfried V. (1995) Structure-activity relationship of porphines for photoinactivation of bacteria. *Photochemistry and Photobiology* **62**: 342–347.

O'Neill J, Wilson M, Wainwright M. (2003) Comparative antistreptococcal activity of a range of photobactericidal agents. *Journal of Chemotherapy* **15**: 329–334.

Pohler P, Walker WH, Reichenberg S, Mohr H, Gravemann U, Müller TH. (2004) Methylene blue treated plasma: pharmacokinetic and toxicological profile of MB and photoproducts. *Vox Sanguinis* **87**: 93.

Raab OZ. (1900) Uber die wirking fluoreszierender stoffe auf infusorien. *Zeitschrift Biologie* **39**: 524–546.

Ruane PH, Edrich R, Gampp D, Keil SD, Leonard RL, Goodrich RP. (2004) Photochemical inactivation of selected viruses and bacteria in platelet concentrates using riboflavin and light. *Transfusion* **44**: 877–885.

Schultz EW, Kruger AP. (1928) Inactivation of staphylococcus bacteriophage by methylene blue. *Proceedings of the Society of Experimental Biology and Medicine* **26**: 100–101.

Singh H, Ewing DD. (1978) Methylene blue sensitized photoinactivation of *E. coli* ribosomes: effect on the RNA and protein components. *Photochemistry and Photobiology* **28**: 547–552.

Skripchenko A, Robinette D, Wagner SJ. (1997) Comparison of methylene blue and methylene violet for photoinactivation of intracellular and extracellular virus in red cell suspensions. *Photochemistry and Photobiology* **65**: 451–455.

Skripchenko AA, Wagner SJ. (2000) Inactivation of WBCs in RBC suspensions by photoactive phenothiazine dyes: comparison of dimethylmethylene blue and MB. *Transfusion* **40**: 968–975.

Snyder E, McCullough J, Slichter SJ, *et al.* (2005) Clinical safety of platelets photochemically treated with amotosalen HCl and ultraviolet A light for pathogen inactivation: the SPRINT trial. *Transfusion* **45**: 1864–1875.

Sohl S, Kauer F, Pasch U, Simon JC. (2007) Photodynamic treatment of cutaneous leishmaniasis. *Journal of the German Society of Dermatology* **5**: 128–130.

Tang HM, Hamblin MR, Yow CMN. (2007) A comparative in vitro photoinactivation study of clinical isolates of multidrug-resistant pathogens. *Journal of Infection and Chemotherapy* **13**: 87–91.

Tegos GP, Anbe M, Yang C, *et al.* (2006) Protease-stable polycationic photosensitizer conjugates between polyethyleneimine and chlorin(e6) for broad-spectrum antimicrobial photoinactivation. *Antimicrobial Agents and Chemotherapy* **50**: 1402–1410.

Tissot JD, Hochstrasser DF, Schneider B, Morgenthaler JJ, Schneider P. (1994) No evidence for protein modification in fresh-frozen plasma after photochemical treatment: an analysis by high-resolution two-dimensional electropheresis. *British Journal of Haematology* **86**: 143–149.

T'ung T. (1935) Photodynamic action of methylene blue on bacteria. *Proceedings of the Society for Experimental Biology and Medicine* **33**: 328–330.

T'ung T. (1938) In vitro photodynamic action of methylene blue on *Trypanosoma brucei*. *Proceedings of the Society of Experimental Biology and Medicine* **38**: 29–31.

T'ung T, Zia SH. (1937) Photodynamic action of various dyes on bacteria. *Proceedings of the Society for Experimental Biology and Medicine* **36**: 326–330.

Vaara M. (1992) Agents that increase the permeability of the outer membrane. *Microbiological Reviews* **56**: 395–411.

Wagner SJ, Skripchenko A, Salata J, O'Sullivan AM, Cardo LJ. (2008) Inactivation of *Leishmania donovani infantum* and *Trypanosoma cruzi* in red cell suspensions with thiazole orange. *Transfusion* **48**: 1363–1367.

Wainwright M. (2000) Methylene blue derivatives – suitable photoantimicrobials for blood product disinfection? *International Journal of Antimicrobial Agents* **16**: 381–394.

Wainwright M, Amaral L. (2005) The phenothiazinium chromophore and the evolution of antimalarial drugs. *Tropical Medicine and International Health* **10**: 501–511.

Wainwright M, Crossley KB. (2002) Methylene Blue – a therapeutic dye for all seasons? *Journal of Chemotherapy* **14**: 431–443.

Wainwright M, Giddens RM. (2003) Phenothiazinium photosensitisers: choices in synthesis and application. *Dyes and Pigments* **57**: 245–257.

Wainwright M, Phoenix DA, Laycock SL, Wareing DRA, Wright PA. (1998) Photobactericidal activity of phenothiazinum dyes against methicillin-resistant strains of *Staphylococcus aureus*. *FEMS Microbiology Letters* **160**: 177–181.

Wainwright M, Phoenix DA, Rice L, Burrow SM, Waring JJ. (1997) Increased cytotoxicity and phototoxicity in the methylene blue series *via* chromophore methylation. *Journal of Photochemistry and Photobiology, B: Biology* **40**: 233–239.

Wainwright M, Phoenix DA, Wareing DRA, Smillie TE. (2001) Photobactericidal activity of phenothiaziniums against *Yersinia enterocolitica*. *Journal of Chemotherapy* **13**: 503–509.

Webb RB, Hass BS, Kubitschek HE. (1979) Photodynamic effects of dyes on bacteria. II. Genetic effects of broad-spectrum visible light in the presence of acridine dyes and methylene blue in chemostat cultures of *Escherichia coli*. *Mutation Research* **59**: 1–13.

Williamson LM, Cardigan RA, Prowse CV. (2003) Methylene blue-treated fresh-frozen plasma: what is its contribution to blood safety? *Transfusion* **43**: 1322–1329.

Wilson M, Pratten J. (1995) Lethal photosensitization of *Staphylococcus aureus* in vitro: effect of growth phase, serum and pre-irradiation time. *Lasers in Surgery and Medicine* **16**: 272–276.

Wilson M, Sarkar S, Bulman JS. (1993) Effect of blood on lethal photosensitization of bacteria in subgingival plaque from patients with chronic periodontitis. *Lasers in Medical Science* **8**: 297–303.

12

Non-oncological applications

While the oncological application of photosensitisers is dealt with exclusively under photodynamic therapy (PDT), non-oncological applications really grew out of anticancer PDT research and thus do not properly include the photoantimicrobial and related approaches (Chapter 11). Conversely, non-oncological applications do include photodiagnosis, which is mainly aimed at cancer detection, as well as applications discovered due to the clinical manifestation of neovascular or non-malignant epithelial tissue retention of lipophilic, anionic photosensitisers. Photocytotoxic agents are also covered in this chapter, mainly due to their proposed end use in oncology, the application again growing out of research in anticancer PDT.

As mentioned at various stages in this book, the early period of photodynamic therapy – the 1980s and 1990s – was dominated by haematoporphyrin derivative and its various refinements. As also mentioned, *ad nauseam*, there have been considerable problems for patients with post-treatment photosensitivity. This latter point has been made into a positive aspect via the recognition that the properties of the early photosensitisers, which led to deleterious side effects, could be used to target other non-economic cells, to the alleviation of various significant modern causes of morbidity.

The early clinical recognition that the target tissue for haematoporphyrin derivative (HpD) and its congeners is tumour vasculature and epithelium has since been seen as a consequence of the highly lipophilic nature of the preparations, in combination with the anionic charge provided by the porphyrin carboxylic acid residues. Such localisation is thus not limited to porphyrinoids and may be modelled by other chromophores having similar physicochemical properties.

12.1   Age-related macular degeneration

Vision relies on the incidence of light from the image on an area at the back of the eye containing the rods and cones necessary for the perception of colour and shading. This area is known as the macula (Figure 12.1).

Typically due to the decreased tissue maintenance and repair facility associated with increasing years, age-related macular degeneration (ARMD) is associated with a gradual loss of central, but not peripheral, vision (Figure 12.2).

Photosensitisers in Biomedicine   Mark Wainwright
© 2009 John Wiley & Sons, Ltd

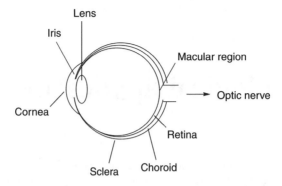

Figure 12.1 Simplified diagram of the eye.

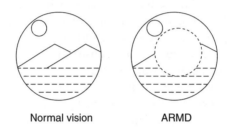

Normal vision   ARMD

Figure 12.2 Loss of central vision due to ARMD.

Photosensitisers can be used in the treatment of wet ARMD. In this condition, in addition to the retinal pigment cells degenerating, new blood vessels grow from tiny blood vessels in the choroid – choroidal *neovascularisation* – and break through into the macular region of the retina. The new vessels are fragile and leak blood and fluid, threatening the integrity of the rods and cones and leading to scarring of the macula. Loss of vision in wet ARMD occurs quite rapidly (months).

The use of lipophilic anionic photosensitisers such as verteporfin (Chapter 6) allows selective targeting and ultimate destruction of the neovasculature. Activity in this area has been fairly limited to the use of verteporfin, although novel benzoporphyrin derivatives, covered in Chapter 6, and texaphyrins have also been proposed.

12.2 Atherosclerosis

Cardiovascular disease is seen as a modern malady, the cause usually being cited as overindulgence and a diet too high in saturated fat. In practice, this is manifested as deposition of fatty plaques in the circulatory system, leading to impeded blood flow and eventually to blockage. The destruction of plaques within arteries may be carried out using a photodynamic approach (Figure 12.3). Again the properties of the photosensitiser can be so designed as to give preferential localisation within the atherosclerotic plaque, a good example of this being lutetium texaphyrin (Sessler and Miller 2000). After localisation, the plaque is illuminated intraluminally, using an optical fibre inserted into the requisite blood vessel. Given the unavoidable presence of blood at the site of action, the use of a long-wavelength absorber such as a texaphyrin (Lu derivative, $\lambda_{max} = 732\,nm$, see Chapter 6) is sensible, being outside the normal absorption spectrum of haem. This approach would be employed in place of conventional angioplasty – stenting via balloon insertion – and is thus known as *photoangioplasty*. Clearly such an approach is likely to be of great utility, given the widespread and increasing morbidity and mortality caused by blocked arteries (Rockson 1999).

Similarly, the damage caused by conventional angioplasty or stent insertion is often due to the accretion of smooth muscle tissue at the operation site, the build-up of which can lead to further occlusion of the blood vessel. The use of either 5-aminolaevulinic acid (ALA) or lutetium texaphyrin has been demonstrated to be beneficial in such cases, inhibiting the build-up, without significant effects on vessel integrity (Pai *et al.* 2005).

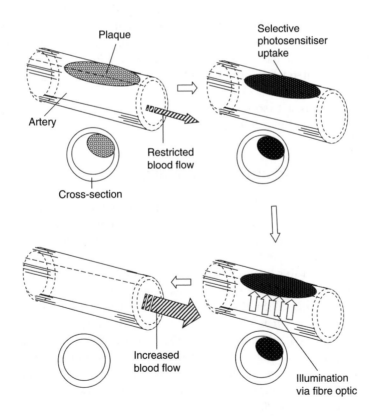

Figure 12.3 Destruction of arterial plaque.

12.3   Endometreosis

The growth of endometrial tissue – stroma and glands – outside the endometrial cavity is termed 'endometriosis' and is a cause of a range of symptoms, typically pain and excessive bleeding, normally in pre-menopausal women.

Again, due to the epithelial nature of the tissue involved, the aberrant, proliferative target may be attacked selectively using the photodynamic approach and lipophilic anionic photosensitisers. As with the other conditions mentioned in this chapter, porphyrins have been most widely used here, beginning with HpD. However, both ALA and verteporfin have been investigated as alternatives, in both cases being administered via intrauterine instillation (Fehr *et al.* 1996; Hornung *et al.* 1998).

12.4   Port wine stain

Skin features made up of grouped vascular lesions, port wine stains (PWS) usually appear at birth and may vary in colour (normally red or pink) and thickness. In an era both of increasing awareness of skin imperfection and elective cosmetic surgery, it is not surprising that the removal of PWS is also increasing significantly. The conventional approach here has been to use lasers or, more recently, pulsed dye lasers. However, given the vascular presentation, PWS may be treated in a similar way to ARMD (see above). Superficial illumination of the lesion a short time after i.v. administration ensures that the photosensitiser is excited while still in circulation in the targeted vascular compartment. This leads to thrombosis/vessel occlusion.

Such an approach is not (so far) widely employed outside China, although there is sporadic use in Europe (Yuan *et al.*, 2008).

12.5   Arthritis and autoimmune disorders

It has been shown by various groups that lymphocytes are inactivated by the photodynamic effect. Since activated lymphocytes are implicated in both arthritis and autoimmune disease, it may be possible to use a PDT approach for the alleviation of these conditions. Lymphocytes can be destroyed either in the bloodstream or in the inflamed joint, for example, using transcutaneous light delivery with verteporfin as the photosensitiser. It has been reported that non-diseased tissues are unaffected by the treatment (Nowis *et al.* 2005). This approach may be used for symptom reduction/ palliation in arthritis, but it is unclear whether cures are likely, especially given the variation in underlying pathologies here. The effects of psoralen–UVA therapy have been discussed in Chapter 9, along with the possibility of potentiated immune response with extracorporeal photochemotherapy. However, the testing of this latter protocol as a potential therapy for multiple sclerosis is not widely known and deserves further investigation (Besnier *et al.* 2002).

12.6   Photodynamic diagnosis

The use of synthetic dyes as stains for tissue/cell differentiation has been covered in Chapter 3, and there is little doubt that this technique is of enormous utility to medical science, both in terms of pathology and intraoperatively, for example in the demarcation of tumour margins.

As has been stated previously, given that a number of vital stains are also photosensitisers, it should be possible to stain malignant tissue and then to utilise the photodynamic effect to destroy it. A reasonable example of this would be the use of methylene blue to stain the supra-pyloric oesophagus for Barrett's tissue and then to eradicate the dysplastic cells using red light. There are various complications here, such as the amount of time that should be allowed for tissue staining, the depth of red light penetration and possible sub-epithelial tissue damage, among others, but the principle is reasonable. However, although methylene blue is used for the staining, *in situ*, of Barrett's oesophagus, analysis has shown that there are regular examples of aberrant staining or *false positives*. Clearly, the illumination of such presentations would result in the destruction of healthy tissue and is the result of breaking the cardinal rule of chemotherapy, that of cell selectivity.

In comparison to conventional (dye) tissue staining, fluorescent staining has the advantage that a much lower concentration of reagent is required. In addition, this approach is far more sensitive to changes in tissue caused by the presence of dysplastic or malignant cells and constitutes a more reliable method for earlier-stage disease than does visible staining (Stringer and Moghissi 2004).

Fluorescent tumour marking is thus a highly effective tool. However, the constantly stated problem of systemic photosensitisation with certain markers (typically porphyrins) means that there are complications concerning their administration. Much research has thus been dedicated to the examination of both endogenous, rather than exogenous, fluorescence and also to local application of fluorophores. The use of ALA in this technique, now named photodiagnosis, has provided considerable advances, utilising the greater conversion of the precursor to the fluorescent protoporphyrin IX (Chapter 6) by malignant tissue. For example, locally instilled into the bladder, ALA has been shown to lead to a transient but significant accumulation of protoporphyrin IX (PPIX) in both pre-malignant and malignant tissue. The excitation by blue light yields red PPIX fluorescence, thus distinguishing the target tissue from normal tissue, which appears blue (Jocham, Stepp and Waidelich 2008). The value of such a sensitive technique is obvious when considering surgical resection, particularly in malignancies having typically difficult presentations, such as those of the prostate or the brain. In addition, ALA has been used in identifying metastatic axillary lymph node involvement after breast cancer (Frei *et al.* 2004).

The development of photodynamic diagnosis must be considered as a significant advance in terms of the aid given to the surgeon. It is true that there are some drawbacks associated with the technique, mainly due to problems with tumour depth definition because of the attenuation of light by tissue, but overall the approach is a great step forward compared to conventional visible light demarcation of tissue.

12.7   Photocytotoxics and photochemical internalisation

The delivery of cytotoxic drugs to target cells has been much improved by the use of specific macromolecules as carriers, including antibodies and proteins as well as various stealth liposomal formulations of the type covered in Chapter 10.

In many cases, there are still membrane obstacles to be overcome by the cytotoxic molecule delivered to the relevant cell. Often molecules are taken into the cell via endocytosis, i.e. engulfed by a section of membrane and gathered into the cell's interior, but the release of the cytotoxic molecules from the formed endocytotic vesicle may be slow.

The photodynamic approach may be used to effect increased intracellular release rates in a number of ways, normally involving the opening of vesicles or otherwise liberation of drugs via the action of singlet oxygen (Berg *et al.* 2006).

As an example, the co-localisation of an endocytosed photosensitiser and a cytotoxic molecule may allow fusion of the vesicle containing the latter with the opened vesicle containing the photosensitiser post illumination, allowing the release of both into the cytosol (Figure 12.4).

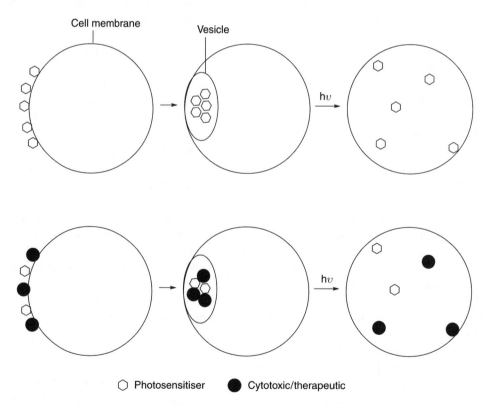

○ Photosensitiser   ● Cytotoxic/therapeutic

Figure 12.4 Internal cytotoxic release using photodynamic activation (= photochemical internalisation).

The use of photosensitisers in such a way also allows the expedited delivery of other therapeutic agents, with these linked to the photosensitiser itself via a dendritic carrier (see Chapter 10). In order for effective release of the therapeutic to occur, it is necessary to have a singlet oxygen-reactive moiety as the linker between the carrier and the agent. Either an amide ($-$CONH$-$) or an imino ($>$C$=$N$-$) bond is usually sufficient. Examples of such technology have been reported for gene transfection (Shieh *et al.* 2008) and cytotoxic (doxorubicin) delivery (Lai *et al.* 2007), both cases employing polyamidoamine dendrimers (Chapter 10).

A further idea involves the use of liposomes, which include bonds susceptible to singlet oxygen. Such bonds may obviously break in the presence of oxygen and an illuminated photosensitiser. Thus, liposomes may be constructed containing both the cytotoxic molecules and a photosensitiser, the release of both occurring post illumination, or construction may include an anchored photosensitiser as part of the liposomal membrane. In this case, only the free cytotoxic molecules would be released as a free agent after illumination (Figures 12.5 and 12.6).

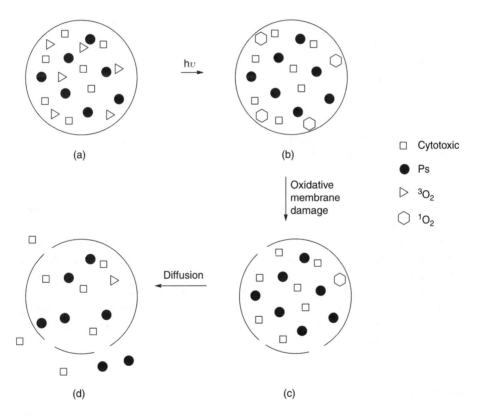

Figure 12.5 Cytotoxic delivery from liposome: (a) liposome containing mixture of photosensitiser and cytotoxic molecules, (b) production of singlet oxygen *in situ*, (c) damage to liposomal membrane caused by singlet oxygen and (d) diffusion of cytotoxic molecules and photosensitiser.

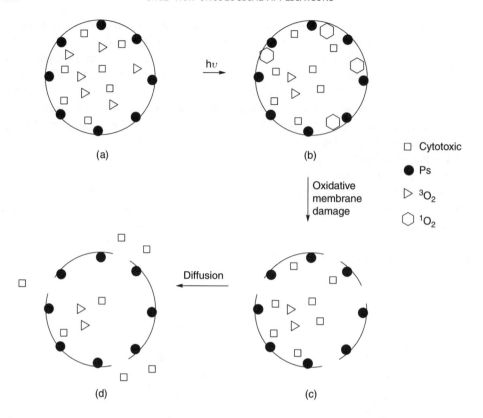

Figure 12.6 Cytotoxic delivery from functionalised liposome: (a) liposome with immobilised photosensitiser, containing cytotoxic molecules, (b) production of singlet oxygen *in situ*, (c) damage to liposomal membrane caused by singlet oxygen and (d) diffusion of cytotoxic molecules only.

Clearly, the biomedical uses of photosensitisers outside PDT and photodynamic antimicrobial chemotherapy represent a burgeoning area of research, since most applications represent spin-outs of activity in these fields. Non-oncological applications can increase the level of exposure of the 'biomedical population' to the photodynamic approach and should thus encourage its use across the board. It is important to build on experience gained, e.g. by ophthalmic surgeons using Visudyne for ARMD, to expose practitioners and colleagues to related photosensitiser use. In this way a critical-mass approach can be envisaged, which could increase the opportunities for photosensitiser use in the healthcare milieu.

References

Berg K, Hogset A, Prasmickaite L, *et al.* (2006) Photochemical internalization (PCI): a novel technology for activation of endocytosed therapeutic agents. *Medical Laser Application* **21**: 239–250.

Besnier DP, Chabannes D, Mussini JM, Dupas B, Esnault VL. (2002) Extracorporeal photochemotherapy for secondary chronic progressive multiple sclerosis: a pilot study. *Photodermatology Photoimmunology Photomedicine* **18**: 36–41.

Fehr MK, Wyss P, Tromberg BJ, *et al.* (1996) Selective photosensitizer localization in the human endometrium after intrauterine application of 5-aminolevulinic acid. *American Journal of Obstetrics and Gynecology* **175**: 1253–1259.

Frei KA, Bonel HM, Frick H, Walt H, Steiner RA. (2004) Photodynamic detection of diseased axillary sentinel lymph node after oral application of aminolevulinic acid in patients with breast cancer. *British Journal of Cancer* **90**: 805–809.

Hornung R, Fehr MK, Tromberg BJ, *et al.* (1998) Uptake of the photosensitizer benzoporphyrin derivative in human endometrium after topical application in vivo. *Journal of the American Association of Gynecologic Laparoscopists* **5**: 367–374.

Jocham D, Stepp H, Waidelich R. (2008) Photodynamic diagnosis in urology: state-of-the-art. *European Urology* **53**: 1138–1150.

Lai PS, Lou PJ, Peng CL, *et al.* (2007) Doxorubicin delivery by polyamidoamine dendrimer conjugation and photochemical internalization for cancer therapy. *Journal of Controlled Release* **122**: 39–46.

Nowis D, Stokłosa T, Legat M, Issat T, Jakóbisiak M, Gołąb J. (2005) The influence of photodynamic therapy on the immune response. *Photodiagnosis and Photodynamic Therapy* **2**: 283–298.

Pai M, Jamal W, Mosse A, Bishop C, Bown S, McEwan J. (2005) Inhibition of in-stent restenosis in rabbit iliac arteries with photodynamic therapy. *European Journal of Vascular and Endovascular Surgery* **30**: 573–581.

Rockson SG. (1999) Lutetium texaphyrin: a new therapeutic tool for human atherosclerosis. *Current Treatment Options in Cardiovascular Medicine* **1**: 199–202.

Sessler JL, Miller RA. (2000) Texaphyrins New drugs with diverse clinical applications in radiation and photodynamic therapy. *Biochemical Pharmacology* **59**: 733–739.

Shieh MJ, Peng CL, Lou PJ, *et al.* (2008) Non-toxic phototriggered gene transfection by PAMAM-porphyrin conjugates. *Journal of Controlled Release* **129**: 200–206.

Stringer M, Moghissi K. (2004) Photodiagnosis and fluorescence imaging in clinical practice. *Photodiagnosis and Photodynamic Therapy* **1**: 9–12.

Yuan KH, Li Q, Yu WL, Zeng D, Zhang C, Huang Z. (2008) Comparison of photodynamic therapy and pulsed dye laser in patients with port wine stain birthmarks: a retrospective analysis. *Photodiagnosis and Photodynamic Therapy* **5**: 50–57.

Index